Scientific and Technical Periodicals of the Seventeenth and Eighteenth Centuries: A Guide

compiled by

DAVID A. KRONICK

The Scarecrow Press, Inc.
Metuchen, N.J., & London
1991

British Library Cataloguing-in-Publication data available

Library of Congress Cataloging-in-Publication data

Kronick, David A. (David Abraham), 1917-
    Scientific and technical periodicals of the seventeenth and
eighteenth centuries : a guide / compiled by David A. Kronick.
        p. cm.
    Includes bibliographical references and indexes.
    ISBN 0-8108-2492-2
    1. Science--Periodicals--Early works to 1800--Bibliography--
Indexes. 2. Engineering--Periodicals--Early works to 1800--
Bibliography--Indexes.   I. Title.
Z7403.K76   1991
[Q1]
016.505--dc20   91-32012

Manufactured in the United States of America

Printed on acid free paper

TABLE OF CONTENTS

PREFACE

When we bury ourselves in the yellowed pages of old newspapers, it is like entering a ghost town, another Pompei, in which, as if we were turning back the wheel of time, we suddenly surprise a people long disappeared, in the midst of their daily existence, in their most intimate domestic tasks. (Ref.127, p.7)

1. Introduction

Scientific and technical periodicals, beginning with the *Philosophical Transactions* of the Royal Society of London in 1665 and constantly increasing in number to the end of the eighteenth century, represent a primary source for the study of one of the most interesting periods in our intellectual history. They are an important resource not only for investigating the diffusion of scientific and technical information in that period, but also for the study of the development of the methods and styles in which they were disseminated.

The "scientific revolution" was accompanied by, or perhaps better, gave rise to, scientific journals which made the new ideas in science and technology available to everyone who could read. In many ways they provide a more intimate view, as the quotation above suggests, than the books and monographs that were published during the same period. Though there are many lists of scientific and technical journals of this period (several of them will be cited later), these journals have been difficult to identify and to access. The titles are widely scattered in research libraries all over the world. A sample of 640 titles published between 1791 and 1800 derived from four source lists (Ref.128-131) was checked in three national union lists (Ref.133-135) with the following results: None of the lists included more than 43 percent of the titles and only 129 were found in all three. The titles unique to each list ranged from 29 to 64 although there was considerable overlap between any two of the three lists (Ref.132).

There are other reasons why these titles are difficult to find and to locate. The source list titles have been selected with different criteria for inclusion, as will be seen below. None of the lists seem to have been derived from an inspection of the actual volumes, but were taken for the most part from other lists (as is true in some measure of this list as well). In the seventeenth and eigthteenth centuries, it is frequently difficult to discriminate between titles which are periodicals and those which are not. The title catchwords such as *Beiträge, Bibliothèque, Journal,* and their cognates in other languages, are sometimes also used for monographic works, particularly if they appear in more than one volume. The distinction between books and periodicals was not made as firmly or as rigidly in the seventeenth and eighteenth centuries as it is today. Books and periodicals are listed and reviewed alongside each other in secondary publications and bibliographies. Some of the journal titles appeared in more than one edition and were reprinted when earlier editions were exhausted, sometimes with revised and augmented texts. This information is included in the records when the it was available to me. Some of the titles included cannot be considered periodical publications, but they are included in this list when they appear in one of the source or location lists, particularly when evidence to the contrary was not available. Another difficulty is that there are also wide variations in the bibliographic conventions used in the various source and location lists, compiled as they were at different times, with differences in scope and in the cataloging codes and rules followed.

The purpose of this guide is to try to bring together the periodicals and serial titles relevant to the history of science and technology which were published in the seventeenth and eighteenth centuries, and to indicate some of the places they may be located from available national union lists of periodicals and other sources. Access to the titles is also provided through indexes to names of editors, translators and the institutions which sponsored them. The periodicals of this period are much more specialized than one might expect. The subject(s) covered by each periodical is provided in an index and the preceding, succeeding and translated titles are listed in another index with references to the title in the main alphabetical list.

The information in the records in the *Guide* is not always derived from a description of books in hand (although I was lucky enough to have the opportunity to hold many of them briefly in various libraries in the U.S. and abroad). It is the result of a consolidation of information derived from various lists, examination of the volumes and secondary sources to produce as complete a record as I could for each title which is included.

2. Scope

The list includes titles related to science and technology published in the seventeenth and eighteenth century, cited in the source lists which are described below, and those which I have been able to locate elsewhere. One exception is Kirchner's list (Ref.130), which covers periodicals in all subjects published in German or in Germany. In this guide all the titles classified in Kirchner under the subjects science, natural history, mathematics, medicine and technology are included, as well as titles selected from the general subject field and other subject areas as they appeared to be relevant. Other titles were added to the guide as they were identified in other lists or in the location sources which were checked for this project or were discovered in various libraries I was able to visit. Periodical titles in all languages published between 1665 and 1800 which meet the selection criteria are included. Exceptions were made in the Garrison list where many of the titles listed under the headings: "Repositories of medical excerpts, reprints, reviews and translations" and "Serial collections of dissertations, cases and observations" could not appropriately be classed with periodicals. Many of them proved to be multi-authored works in one or more volumes or the collected papers of individual authors which were not issued as periodicals. Journals and society proceedings in the seventeenth and eighteenth century were frequently translated in complete or abridged form or in collections of excerpted editions. Although translations of individual titles are not included in most of the source or location lists, all translations of the selected titles which could be identified are listed as separate records with reference to the original publications.

Abridged editions in the original language as well as periodicals devoted to translated selections from several titles are also included. In this period when indexes and other means of retrospective literature searching were infrequently published, both abridgements and translations serve in a limited fashion for this purpose. Selected literary and general review journals published in this period are also included in the guide. Primarily they are those that appear in the source lists and those which represent the important review journals of a country or area. The review journals frequently reviewed journal issues in the same way they reviewed books and thus serve as another method of disseminating scientific and technical ideas and news. Occasionally they also provided an outlet for original scientific contributions as Duveen and Klickstein have shown in their bibliography of the works of Lavoisier (Ref. 90). Most of the almanacs and ephemerides in this period were excluded except for some titles which were found or appear to include information of a scientific or technical interest.

The subject scope of this compilation is discussed below in the section on the subject index. However, it must be admitted that the subject focus of some titles cannot always be determined from the title alone. Title words sometimes promise more than they deliver or else fail to disclose their actual subject coverage in the title words chosen. Errors of inclusion or omission of this kind therefore may occur. A good bibliographic principle to follow is that the sins of commission are less serious than those of omission. Titles which were published as continuations of titles which appear in the list but which were published after 1800 are cited in the record and appear in the title index but are not given a separate entry in the title list.

3. Methodology

Data drawn from the various source and location lists cited below were compiled on worksheets in the format outlined in the *Online systems serials format* produced by OCLC (Online Computer Library Center, Dublin, Ohio). The guide has attempted to use two international standards: MARC II (a machine-readable cataloging standard system developed by the Library of Congress which has been widely accepted) and AACR II (the *Anglo American Cataloging Rules*). The OCLC database was also searched for each title and used to revise the records when it provided useful additional information. Titles found in the OCLC database are designated in the Source field and provide an additional way to identify locations. (Two principal modifications were made to the MARC II format; the 040 field was used to indicate in which source list the title appeared, and the 049 field to designate the list in which the holdings of the title are provided. Some data were included in the "Notes" field (500) e.g., translations, new editions and reprints, instead of being assigned to other MARC II fields). Locations are designated even when the holdings of the title in the list is fragmentary, and in many cases all holding were consulted to construct the complete record.

Titles in the guide are listed alphabetically in word by word order both for periodicals sponsored by societies and other corporate bodies and independent periodicals. In the case of societies where the full title was not available the name of the Society is used as a qualifier after the title, e.g., *Mémoires. Société des arts de Géneve*. Changes of titles are entered under the new title with reference to preceding and succeeding titles and alternative titles which are all included in the title index. Most changes of title in the seventeenth and eighteenth centuries begin new series. Where dates of coverage and printing differ, as is often the case with the proceedings of societies, the dates of coverage are included in parentheses before the publication dates. Where publication is not continuous or regular, the date of the first and last issue is given, unless additional information is available relating to the publication record.

Filing order differs from list to list depending both on the rules existing at the time the list was complied, and the conventions adopted by the compiler. Variations in spelling are also treated differently by different compilers. An example is the interchangeability of the "i" and "y" (e.g., *Beiträge, Beyträge*) in German titles. In some lists this problem is dealt with by listing the terms as they are spelled in the sources used, while others list them uniformly under one form or the other, with a general reference from one to the other. Another common variation is the interchangeability of the "c" and "k" with words and names spelled differently in different lists. In this guide, although *i* and *c* are the preferred choices, especially when they affect the alphabetical filing order, some titles must be searched under both variants: e.g., *Arneikunst* and *Arneykunst* or *Oeconomische, Öconomische, Oekonomische*, and *Ökonomische*. Another convention sometimes adopted is that the name of the editor(s) which appears as the first part of a title is used to determine its place in the alphabetical listing. In this guide editor's names are dropped from the alphabetical list but can be accessed by consulting the Name index. Names of societies on the other hand are entered in the principal record as they are cited on the publication, whereas they are indexed under the names used for coordinating all the publications issued by a particular society (see section 7 below).

Where a particular title is not listed in one of the Source or Location lists used in the *Guide*, the library in which I consulted it is given as one location e.g., BN for the Bibliothèque Nationale in Paris, and BL for the British Library. Other locations of this nature are included in the List of Abbreviations. The fact that no holdings are shown in any of the countries covered by the union lists used does not, of course, mean that those titles are not to be found elsewhere, either in the country of origin or in other library catalogs. They may have been overlooked in the process of compiling a particular list or may have been considered out of scope for the list being compiled. Despite attempts to define and apply criteria to any group of publications, there is a wide range for disagreement on how the criteria are used. Thus a title may be listed in one source or location list and omitted from another because it is not seen as meeting the criteria adopted.

4. The Source Lists

Of the four basic source lists consulted for the *Guide*, the Garrison list (Ref 129) is the only one which is limited to titles published in 1800 and earlier. It includes 755 numbered items, but continuations are also entered under many of the numbered titles which raises the total to well over 1,000. (A list of "addenda et corrigenda" to the list was published in 1958 (Ref.142) and are included in the *Guide*). The subject scope is broader than that indicated in the title, because general titles like *Gentleman's magazine* are also included. It also includes "continuations, translations, abridgements" and two sections, called respectively "Repositories of medical excerpts, reprints, reviews and translations" and "Serial collections of dissertations, cases and observations" many of which do not qualify (as indicated earlier) for inclusion in a list of periodicals. The Garrison list is arranged by centuries (and in some instances by decades) and within centuries by subject divisions, so that although there is no subject index the subject groupings do supply some subject access e.g., Section X, "Brunonian doctrine". Locations for a few of the titles are added, primarily from an earlier edition of the *Union list of serials* (Ref.135). Many of the titles are derived from Callisen, a monumental nineteenth century medical bibliography in German in 33 volumes. It includes 28,061 authors and 68,404 titles, including both monographs and articles in journals. Volumes 23-24 are devoted to "Zeitschriften, Gesellschaftsschriften und gesammelte Schriften mehrer Verfasser seit 1780" (Ref.21). The series provides what is tantamount to an author index to the scientific literature from 1780 to 1830, although publications earlier than 1780 are also cited. Under many of the title entries for periodicals, lists of publications in which separate issues were reviewed are given, since literary and review periodicals in this period frequently reviewed journal issues in the same manner as monographs. Callisen promised a subject index to the entire series which unfortunately never materialized.

Each of the source lists, explicitly or implicitly, states the criteria for inclusion in the list, although they sometimes are ignored. Scudder, (Ref.131) for instance, states that his list includes only the serials of learned societies "in the natural, physical and mathematical sciences" published between 1663 and 1876, but many non-society serials as well as periodicals devoted to the applied sciences are included. This list is arranged by country of origin and under the country by the city in which they were printed. This means that translations are listed under a different city than the original publication, although reference numbers to the other places in the list are generously provided. Bolton's list (Ref.128), which is almost contemporary with that of Scudder, indicates that the scope includes scientific and technical periodicals published between 1665 and 1895. That does not prevent Bolton from including some society publications as well as some general periodicals. The Bolton list includes some 8600 titles for the designated period arranged in two alphabets, apparently representing an earlier and a later compilation.

An additional feature is a separate list of locations for some 3000 of the titles in the list in about 180 research libraries in the United States. The volume also includes a chronological table which indicates in what particular years a periodical was being published. It only covers about thirty of the principal scientific journals issued in 1800 or earlier. There is also a subject classification in which the title numbers are listed under the subject in the alphabetic order in which they appear in the list so that it may not be particularly useful for those interested in identifying those published in the seventeenth and eighteenth centuries.

Joachim Kirchner's list (Ref.130) includes all periodicals published in German or in Germany to 1900. The first volume covers the period to 1830. It is based primarily on an earlier list published in 1942 which covered the literature only to 1790 (Ref.111). It is arranged in 23 broad subject groups several of which are of direct interest to this guide: "Naturwissenschaftliche und mathematische Zeitschriften," "Medizinische Zeitschriften" and "Technische und gewerbetechnische Zeitschriften." Some titles also have been extracted for the *Guide* from other subject sections in Kirchner's list, particularly from "Ökonomische und forstwissenschafltiche Zeitschriften" and "Allgemeinwissenschaftliche Zeitschriften." Kirchner (who has published widely on the history of journalism in this period) indicated in the preface to this work that he had seen only those titles which are starred in his list, but had drawn many of the titles from bibliographies which he lists at the beginning of the book. This undoubtedly is true of most other lists of this kind as it is of this *Guide*. He also consulted special regional and subject catalogs. The locations listed for most of the titles were retained from an earlier bibliography which encompassed a united Germany. Writing in the divided Germany of 1969, he comments that many of the titles he had held in hand at the *Staatsbibliothek* in Berlin before the war were divided between East and West Berlin and some were destroyed in the war. There are some 200 German libraries included in his list of locations, as well as some Danish, Austrian Swiss and English libraries where he could not designate a German collection. Locations are provided for many titles which do not yet appear in the *Zeitschriften Datenbank* (Ref.146). Some of the entries have no location probably because they may have been derived from other bibliographies. There is a title index (pp.371-489) compiled in the German manner with a single entry under a selected key word that appears in the title at one point or another. Thus all titles which have the word *Abhandlungen* imbedded in it appear under that word in the index, with the other title words listed below.

Because it has been used to provide descriptive data for the records, OCLC (Online Computer Library Center, Dublin, Ohio), a centralized computer database which includes the holdings which have been added by many of the research and academic libraries of the United States, has also been cited as a "source." Many of the titles located in the *Union List of Serials* (Ref.135), and therefore presumed to be held by some library in the U.S. or Canada, were not found in OCLC. Two reasons are possible: OCLC is a database which is always in process of construction and libraries (particularly the large research libraires) may not have given priority to adding the older material which is our interest in the *Guide*. Secondly publications of scientific societies are especially difficult to search in this system. Nevertheless a fairly large number of the titles have been identified with OCLC as a source. Whenever OCLC appears as a source it also provides locations which sometimes may not appear in the *Union List of Serials*. Sources, with the exception of OCLC are indicated by the initial letter (see Appendix II) of the author of the source title i.e., B234 indicates that the title is listed as number 234 in Bolton's list.

5. Locations

There are a large number of periodical holdings and location lists which have appeared over the years, including holdings of individual libraries, regional union lists of libraries in various geographic areas, from city lists, to state and national lists, as well as lists by subject coverage in any of the above categories. I have chosen to search only those national union lists which were available to me, and which are also available in major research libraries. I have also chosen to indicate only which union lists reports the title rather than trying to provide the holdings of individual libraries. While this would have been desirable, it was not feasible to do so in a guide of this kind. Instead the records reflect a composite of the most complete records cited in all of the lists consulted. Each of the union lists has its own conventions concerning the order in which the titles are listed, i.e., whether the title is listed under its original form or changed form, word by word as it appears in the title, or by omitting the unimportant words. In another variation which occurs all the publications of a society or other corporate body are placed under a selected name for that organization with references from all other forms of the actual title. These variations occur because the lists were compiled at different times under differing cataloging rules in different countries. It is advisable therefore, before using a list, to read the preface carefully to determine the rules and conventions under which the particular list has been compiled.

I have used the most current edition of the national union lists I could acquire, although older listings are frequently useful because early journals tend to be rather stable in their physical location. Thus locations indicated in such lists as Bolton's and the BUC can be relied upon for designated locations for the most part unless, for example, your call slip is returned with the unhappy notification "destroyed in the war," or, more ambiguously, "not on shelf." In some cases where I was not able to locate a title in any of the lists searched, I indicated from which library or reference source I had derived it. Some of the entries may indeed be "ghosts," etherial documents which have or had no existence except in an error of transcription or in the fatigued mind of some bibliographer.

a. *British Union Catalogue of Periodicals* (BUC) (REF.133)

The BUC contains more than 140,000 titles held in about 440 British libraries. Titles are listed alphabetically by the first word which is not an article. Periodicals associated with corporate bodies as well as all periodical titles are generally entered under the earliest known name of the periodical or the society with references from other forms. Only the significant words (which are printed in heavy type) are considered in the alphabetization, i.e., particles are printed but ignored in the filing order. Variant spellings e.g., "bulletino" and "bollettino" are listed under one form with a general reference from the other. Designations such as "royal," "imperial" and their cognates in languages other English are omitted in the alphabetical listing as well as the insignificant words, thus The *Académie royal des belles-lettres de Caen* is filed as *Académie belles lettres Caen*.

Translations are often not cited under the translated title but may be noted under the original language title. Diacriticals which may affect the filing order are ignored if they appear on the first character of a title, e.g., "Ökonomische" or "öconomische" is filed as if spelled "okonomische." Titles which begin with the variant spelling "teusche" and "deutsche" are all filed under "deutsche." Titles which are prefaced by editor's names are filed under the name as it appears in the publication, e.g., *D. Christoph Wilhelm Jacob Gatterer's Technologische Magazin* is listed under "D." in the alphabet.

b. *Catalogue Collectif des Périodiques...Dans les Bibliotheques de Paris* (CCF) (Ref.134)

This title is coded CCF to distinguish it from the Union List published in the Netherlands (coded CCN Ref. 144) which has similar beginning initials. The CCF is the result of a project started by the Bibliothèque Nationale in Paris in 1939 (which is the terminal starting date of the titles included). It was not published until after World War II, beginning with volume 4 in 1967. The fifth volume (published 1980) consists of addenda and a very useful list of the societies included in the compilation, each followed by a list of the titles by which the society is represented. It serves in lieu of cross references from variant names of the society. The list includes 75,000 French and foreign periodicals in 75 of the major academic and research libraries in France. Among the exclusions are newspapers published after 1849, almanacs, other kinds of non-substantive publications and periodicals of limited interest such as annual directories and church bulletins. Titles in cyrillic characters are included in separate volume (Ref.145). Titles are listed in alphabetical order including all the particles (unless they are the first word, although in a few cases this convention is ignored). Titles are listed (for both independent periodicals and publications of societies) according to the title of the first issue, thus titles beginning with common words such as *Acta, Bulletin,* etc. frequently must be identified by consulting the fifth volume.

Among some of the other conventions: Diacriticals are not taken into consideration in the filing order e.g., German titles beginning with "O" and "Oe" or "Ö" are filed together, and German words beginning with "C" are filed according to their modern form e.g., "K," "Z," etc. Changes in title are listed in chronological order along with indexes, when they have been identified, in a combined note which precedes the title entry and is marked with an asterisk. German titles which begin either with *Beyträge* or *Beiträge* are listed under the later form. When the date(s) of coverage differ from the publication date(s) the dates of coverage are used with reference to the dates of publication. Terms of royal and governmental association such as *Kungliga*, *Königliche*, *Imperial*, etc. are used in the filing order. An unusual (and very useful) feature of the list is that it supplies the shelf number for the title supplied by the participating library.

c. *Centrale Catalogus van Periodieken en Serienwerken in Nederlandse Bibliotheken* (Ref.144)

The Dutch national union catalog includes some 24,000 titles representing the holdings of about 250 libraries in that country. It contains titles of periodicals as well as newspapers and serials which are sometimes omitted from other lists. In general, however, it is not very strong on titles which were published outside of the Netherlands. The list follows current conventions in filing titles under the first word in the title whether it includes the name of a society or not, unless the society's name appears as the first part of the title. Like the British list it is alphabetized by the key words in the title omitting all particles, e.g., "of," "von," etc. Titles are entered under *Beyträge* or *Beiträge* depending on the form used in the publication. This convention is followed also with titles that contain the words *medizinische* and *medicinische*. This catalog does not attempt to provide a complete description of a title, but only supplies the holdings of the libraries included.

d. *Union List of Serials in Libraries of the United States and Canada* (Ref. 135)

Equivocal as it may seem, the U.S. list appears to contain a larger percentage of the periodicals of this period than any of the other lists consulted (Ref.132). The first edition was published in 1927 and the third and last edition (used in this guide), much expanded, appeared in 1965. It includes 156,449 serial titles at 956 libraries in the U.S. and Canada. All types of serial publications are included except government publications, administrative reports of institutions and newspapers published after 1820. Titles are listed under its last form with a description of the changes which occurred with cross references from the earlier titles.

Except for serials with a distinctive title, publications of learned societies and academies in Europe are entered under the first word after the adjective which denotes royal or other privilege, e.g., *Kaiserlich*, *Reale*, *Imperiale*.

e. *Zeitschriften-Datenbank* (ZDB) (Ref.146)

The *Zeitschriften-Datenbank* is a machine readable online database which is still in progress of compilation. It is produced in what was West Germany and is also made available in quarterly updates on microfiche. The version used in the guide was published in October 1988, but since the updates include primarily new titles, the October listing probably contains most of the titles within the scope of this guide that are included. The Berlin collections apparently were divided East and West, it appears by a method resembling the judgment of Solomon, so that Kirchner (Ref.130) must be depended upon for those titles which were held in East Germany. Since it was designed as an online system it is probably easier to consult in that form since the alphabetical order of titles is then not as much a consideration. Its cataloging rules are prescribed by the *Regeln für alphabetische Katalogisierung* (RAK) which requires that titles which contain corporate names be entered under the name of the society or institution. They seem to be listed under the name which the society or institution had at the time the publication appeared, although cross references are sometimes supplied from the variants.

Translations of the proceedings of scientific societies are sometimes listed under the original name of the society, and sometimes under the translated title as is the case with *Abhandlungen zur Naturgeschichte Physik und Oekonomie aus den philosophische Transaktionen*. This kind of ambivalence is demonstrated by the other lists, some of which do not include translations. Among the other conventions: title words including the variant forms "z" or "c" (e.g., *medizinische* or *medicinische*) are filed as they appear in the title as is the case with other variant spellings ("y," "i"). This is true for other variants such as *Westfälische* and *Westphälische*. The spelling conventions concerning the use of the apostrophe in such titles as *Journal de l'histoire...*" and the hyphen as in titles beginning *Kaiserlich-Königlich* should also be determined in this list as in other lists.

f. Microforms and Reprints

Microforms and reprint editions provide another form of access to the early periodicals included in the *Guide*, and where this information has been available to me it has been included in the records. There is currently underway a concerted effort to preserve the books and records of the past, which is being supported by the National Endowment for the Humanities and other institutions. A *Register of Microform Masters* has been published by the U.S. Library of Congress from 1965 to 1983 and is continuing with the assistance of the Online Computer Library Center in Dublin, Ohio (OCLC). There are many titles in this form that are also available from commercial microform publishers. Titles listed in the *Guide* which are included in the publication *Guide to microforms in print; incorporating international microforms in print* are included in the title records. Microforms from other sources are also included. Sources are indicated with the same code used by the *Guide to microforms*. Microforms are issued in various formats which are also indicated by number after the code for publisher in the microform guide, e.g., the code RM(3) indicates that the title is available from Readex Microprint Corp. on micro-opaque cards 6 inches by 9 inches at a reduction of 74 times. The pre-1801 holdings of the Goldsmiths' Library of Economic Literature and the Kress Library of Business and Economics which have been microfilmed by Research Publications Inc. are identified in the title record. Microforms are produced as both negative and positive film and may require special types of readers some of which are also equipped to make full size prints from the images. Microform publishers cited and the list of microprint variants are listed in Appendix 1.

A number of titles in the *Guide* are also available in the form of reprints. Where this information was available it is noted in the title record. There undoubtedly are others listed in the *Guide to Reprints* (Ref.151) but frequently they are titles which are fairly widely available in original printings in research libraries.

6. Subject Index

The subject scope of the *Guide* is briefly discussed in section 2 above, but a few comments must be made about the terms used in the subject index. The primary subdivision is broadly between "periodicals," those titles published independently, and "societies," to distinguish them from those published under the auspices of an organization or society, although in some cases a title may not strictly be considered a periodical or the proceedings of a society. Periodicals are subdivided by the country in which they were published e.g., "Botany-Periodicals-Germany." Society publications are subdivided by the place in which society had its seat, e.g., "Surgery-Societies-Paris," although there are some exceptions e.g., the *Academia naturae curiosa* which was based in the town in which the then current president lived, and did not have a permanent home until the nineteenth century. Geographic names are those of their usages currently, and are generally derived from the imprints of the publications, although imprints are not always to be trusted in this period. Where Frankfurt appears in the imprint, unless otherwise indicated the reference is to Frankfurt am Main.

Germany and the cities that are now part of Germany appear to dominate the subject index, although Germany cannot be said to have achieved a status in the development of science and technology comparable to that of some of the western European countries. This may perhaps be attributed to the use of the Kirchner list (Ref.130) as a primary source list for the *Guide*. Another explanation may lie in the degree of political fragmentation which existed in this area. In the eighteenth century, Germany consisted of numerous kingdoms, principalities and other kinds of governmental units. Many of them sought to establish their own periodicals and learned societies. Many of the German periodicals of this period are largely derivative and devoted to reports and translations of work published elsewhere. The preponderance of German periodicals can perhaps also be attributed to what the Germans themselves call *Schreiblustigkeit*, a kind of compulsive joy in writing.

Where possible U.S. Library of Congress subject terms have been used but with the meanings which appear to have been assigned to them in the seventeenth and eighteenth centuries. Most of the subject terms assigned are straightforward and clearly identified by their titles, but there a few which are used in the index which may appear ambiguous. The terms "arts," "Kunst" and their cognates in other languages in the seventeenth and eighteenth century often denoted the practical arts and crafts as well as the fine arts and in some cases titles bearing these terms especially when other evidence is available, have been assigned to the subject "technology," a term which was coined by Johann Beckmann (1739-1811) who wrote a text book on the subject in 1777 (Ref.147,p.36). One example is *Transactions of the Society instituted at London for the encouragement of the arts, manufactures and commerce*, later the Royal Society of Arts. This title is also to be found in the subject index under "agriculture," which was clearly a major interest of the society in its early days, and also under "economics," which is a catch-all term which requires some explanation in the usage applied to the term in the *Guide*. Although "economics" as a theoretical and practical field of inquiry existed as a discipline in the eighteenth century in the sense it has today, in this guide is serves to characterize those periodicals and societies which concerned themselves with the broader economic interests of the community. The societies which proliferated everywhere in Europe and the British Isles were called variously, societies for the improvement of arts (in the sense of applied arts) and/or agriculture, *oeconomische Gesellschaften, patriotische Gesellschaften, sociétés d'emulation*, and many similar titles. (Refs.148,149). In some ways the activities of these societies resemble somewhat those of the chambers of commerce today, and represent an interesting field of inquiry (Ref.6). In fact one of the publications is the guide is clearly identified as that of a *chambre de commerce* (*Recueil des procès verbaux des séances de la Chambre de commerce du Dunkerque*). Further study is necessary to find a better way to characterize and to classify the titles which have been grouped under "economics" and which here encompass various aspects of economic life including fiscal policy, commerce and manufacturing.

When the society or periodical had narrower interests in those areas with which the "economical societies" concerned themselves, such as manufactures, mineral industries (used for titles concerned with mining or mineralogy), beekeeping, technology, trade or commerce, they are assigned to those subjects. Titles with the word "Wissenschaft" or "Physics" or, if they concern themselves with several disciplines in science or with natural science in general, are assigned the subject tag "science". Some titles have been tagged "military art and science" or "naval art and science" particularly when an examination has indicated that they are also concerned with scientific and technological issues. Titles with the German term *Seelenkunde*, have been assigned to "psychology" although it was then essentially a branch of philosophy (as it is in some respects today), because some of the reports appeared to employ scientific observational methods. There were many more specialized scientific and technical journals in the seventeenth and eighteenth centuries than one might expect before studying them carefully. Garrison identified seven titles with "Brunonianism," a medical theory which was advanced in the eighteenth century, and six with "animal magnetism" which also had its vogue in that period (Ref.129). Use of such terms as "biological chemistry" may seem to be anachronistic in the period in which the title was published, but it seems the best available to describe the focus of one the periodicals.

As indicated above many periodicals which are tagged "general" have been included, particularly if they were included in the source lists, because they played an important role in the seventeenth and eighteenth centuries in disseminating news of scientific and technical developments. General periodicals have been included selectively depending on their duration and representation of a particular geographic region. When used with the societies the term "general" indicates that the society concerned itself with arts and letters as well as with science as a great many of them did in this period.

Many of the titles have been classified according to their function as well as subject. Abridgements of journals, particularly those of societies, were produced frequently in this period especially when the publication from which they were produced were available only in a foreign language. Abridgements served a dual function, as a replacement for titles no longer available, and to some extent as indexes to the contents of the titles from which they were abridged. Almanacs are among the earliest form of serial publication, but only those which seemed to contain substantive material which varied from year to year and were relevant to the subject scope of this list are included, i.e., *Almanach oder Taschenbuch für Scheidekünstler und Apotheker*, an important pharmacy journal of the period. Dissertations (and academic disputations) played a much larger role in academic education and information dissemination in this period than they do currently and were widely collected and reported on in periodicals, some of which were devoted solely to that purpose. Prize essays are a form of publication for which it is difficult to find counterparts today. The winning essay submitted for prizes offered by many of the societies of this period were sometimes published as separates, but they also made up a class of serial literature when societies sometimes published them as collection of their *prix*. Translations of serial publications occurred frequently during this period. Their existence are noted as part of the record to which they refer, although frequently published much later than the original. The subcategory "Popular" is used only when the periodical is clearly designated for a broad audience, something which is not always easy to determine in a period when science had not yet become so specialized or inaccessible to the general public.

7. Institutional, Personal Name, and Title Indexes

A comprehensive guide to the learned and scientific societies of the seventeenth and eighteenth centuries which will provide access to all the variants of the names of societies in this period does not seem to exist, although there are some publications which are useful in this regard. Societies, especially those of long duration, underwent numerous changes of name in response to political changes, changes of scope, or sometimes, it seems, in response to a whim on the part of their members. The different source and location lists used in the *Guide* treat this problem in different ways when they wish to relate different titles to the same organization. This issue has been briefly discussed in the sections relating to the lists which are represented in this compilation. The method followed in the alphabetical list of titles in the *Guide* is to use the actual names of the organization as it is given in the title of the publications, but to adopt some of the most frequently used names with some modifications for listing them in the index. For the most part all terms denoting official sponsorship, which represent the most frequent reason of changes in names, (*Königliche, Imperial, Royal, Churfürstliche*, etc.) have been omitted in the index, with the exception of those in which the term is required for identifying the society, e.g., Royal Society of London, Royal Society of Edinburgh. The same index entries are used for all titles associated with the societies that appear in the list. Thus all the publications related to the *Akademie der Wissenschaften* in Berlin are entered in the index under: *Académie des sciences et belles-lettres, Berlin*; *Academia regia scientiarum Berolinensis*; *Deutsche Akademie der Wissenschaften zu Berlin*; *Akademie der Wissenschaften, Berlin*; and *Societas regia scientiarum Berolinensis*.

All variant forms of a title appears both in the record and in the title index with the number of the record in the alphabetical list. This is also true for titles which precede or follow the title record. Titles which follow but which were not published until after 1800 are included in the record and in the title index, but do not appear as separate records. Where titles include the editor's name at the beginning e.g., *Rudolph Augustin Vogel's medicinische Bibliothek*, they are entered under the first word of the title, here *Medicinische Bibliothek*, and can be accessed by the author's name through the name index.

All names associated with the editorial role in producing a title, with the exception of printers or publishers (frequently the same individual in seventeenth and eighteenth century publications) are included both in the record and in the name index. Editor's names in this period are particularly important because "editors" frequently functioned also as authors. Translators, when they have been identified, are also included. Birth and death dates for individuals are added in the index when they were identified.

8. Conclusion and Acknowledgments

This *Guide* is the result of an interest which I have followed sporadically for almost 40 years, since my dissertation subject was approved at the University of Chicago in 1952. Since then, during my career as a medical librarian, I have spent many pleasant hours in research libraries both in the United States and abroad pursuing this interest. I owe a large debt of gratitude for all the help I have received from the staffs of these libraries, principally the British Library where I spent five months (and later a whole year) and the *Bibliothèque Nationale* in Paris where I spend one month in 1972, thanks to a grant from the Council for Library Resources. The National Library of Medicine, which was still called the Armed Forces Medical Library when I began working there in 1953, provided me with an invaluable opportunity to get acquainted with many of early periodicals in their collections. A second extended opportunity to use the rich resources of this library came in 1989 when I was invited to serve as a visiting scholar in the National Library of Medicine's History of Medicine Division.

I was fortunate to have the MARCIVE Corporation in San Antonio where much of the technical work on the Guide was carried out. MARCIVE, a company which has long had experience working with Marc II data, sorted all the random records with which I provided them and produced the indexes. The University of Texas Health Science Center in San Antonio, where I had the privilege of serving for 25 years as Director of the library, and Virginia Bowden, its current Director, who graciously provided space and staff assistance also deserve much gratitude. I am much indebted to several people in completing this project, including William Terry of MARCIVE Inc. and James McKinney, on the University of Texas Health Science Center library staff, who makes solving difficult technical computer problems seem easy. Among those who provided assistance in the project are Helen Irwin of the Library staff and Rebecca Holloway, of the University's Computing Resources department, who was responsible for producing the final format. I am also fortunate in having a wife like Marilyn, who without complaint busied herself in foreign cities while I buried myself among the great libraries of London, Paris, Strasbourg, Göttingen, etc. She claims to have walked more streets in Europe than others who are occupationally associated with such activities. Above all I am grateful to the National Endowment for the Humanities which provided the means which made it possible for me to produce this *Guide* by their generous support in Grant RC 21509-88) and to the patience and helpfulness of the members of its staff.

The list of periodicals in the *Guide* can not, of course, be regarded as complete or definitive. Some titles of scientific or technological interest in this period have undoubtedly eluded me. It is also incomplete in the respect that I have not been able to indicate location for all of the titles. Some of them indeed may exist as "ghosts" in printed lists only. I hope, nevertheless, that the Guide will be useful to those who are interested in and work in this fascinating period of history, when new ideas and new ideas were proliferating along with the periodicals which provide one of their most accessible records. Although there is an extensive list of references appended to this guide, a great deal of work remains to be done. Some of the records undoubtedly require further clarification and research. This *Guide* can only be regarded as one rung of the ladder toward bringing the periodicals and society proceedings of the seventeenth and eighteenth century under control and to attempting to define the territory. I think it may even help to elucidate and understand some of the problems we are having today in scientific journalism. I hope there will be others who are tempted to climb higher.

David A. Kronick
The University of Texas
Health Science Center
Briscoe Library
San Antonio, Texas

1   Abbildungen und Beschreibungen der Insekten in der Grafschaft Hanau-Münzenberg. - Jg. 1 bis 3, 1777-1779. Hanau: Waisenhaus-Buchhandlung, 1777-1779. **OTHER TITLE:** Naturgeschichte der europäischen Schmetterling. **EDITOR:** Johann Andreas Bergstrasser. **ALSO ISSUED AS:** Naturgeschichte der europäischen Schmetterling. Hanau, 1779-1780. **SOURCES:** K3263 **LOCATIONS:** LC **SUBJECT:** Entomology - Periodicals - Germany.

2   Abbildungen und Beschreibungen naturhistorischer Gegenstände, mit farbigen und schwarzen Kupfern. Vol.1-4, 1795-1802. Berlin: 1795-1802. Supplement No.1-17, 1796-1802. **SOURCES:** B3, S2261 **LOCATIONS:** ZDB **SUBJECT:** Natural history - Periodicals - Germany.

3   Abhandlungen aus der Naturgeschichte, praktischer Arzneikunst und Chirurgie aus den Schriften der Haarlemer und anderer holländischen Gesellschaften. - Bd.1-2, 1775-1776. Leipzig: Johann Friedrich Junius, 1775-1776. **CONTINUATION OF:** Abhandlungen der Holländischen Gesellschaft der Wissenschaften zu Haarlem. Translated by Abraham Gotthelf Kästner who planned to cover the proceedings of other Dutch scientific societies as well. **TRANSLATION OF:** Verhandelingen uitgegeven door de Hollandsche maatschappij der wetenschappen te Haarlem. **SOURCES:** B13, G474a, K3570, S2878 **LOCATIONS:** BUC, CCF, ULS, ZDB **SUBJECT:** Science - Societies - Harlem - Translations.

4   Abhandlungen aus der Naturlehre, Haushaltungskunst und Mechanik auf die Jahre 1739-1779 / Kgl. Schwedischen Akademie der Wissenschaften. Bd.1 bis 43, (1739-1784) 1749-1784. Hamburg: Georg Christian Grund; Leipzig: Adam Heinrich Holle, 1749-1784. v.1-4 was translated by Abraham Gotthelf Kästner, 5-9 by Joachim Dietrich Brandis, 10-12 by Heinrich Friedrich Link. Bd.1-3 is also in a second edition. **TRANSLATION OF:** Handlingar Kungliga Svenska Vetenskaps-Akademiens. **CONTINUED BY:** Neue Abhandlungen aus der Naturlehre, Haushaltungskunst und Mechanik. **INDEXES:** Zweifaches Universalregister über die ersten xxv Bande (1739-1763) von der Abhandlungen. Leipizig, Hollens Witwe, 1771. There is also an index to vols. 26-41. **MICROFORMS:** RM(3). **SOURCES:** B14, G477a, K3203, S2768 **LOCATIONS:** BUC, CCF, CCN, ULS, ZDB **SUBJECT:** Science - Societies - Stockholm - Translations. **SUBJECT:** Natural history - Societies - Stockholm - Translations. **SUBJECT:** Home economics - Societies - Stockholm - Translations.

5   Abhandlungen der baierischen Akademie über Gegenstände der schönen Wissenschaften. - Bd.1, 1781. Munich: J.B. Strobel, 1781. **CONTINUATION OF:** Abhandlungen der Chürfurstlich-baierischen Akademie der Wissenschaften. **CONTINUED BY:** Neue philosophische Abhandlungen der baierischen Akademie der Wissenschaften. **INDEXES:** Indexed in Vezenyi (Ref.1, p.xiv). **SOURCES:** K349, OCLC **LOCATIONS:** BUC, CCF, CCN, ULS, ZDB **SUBJECT:** General - Societies - Munich.

6   Abhandlungen der Churfürstlich-baierischen Akademie der Wissenschaften. -
      Bd.1-10, 1763-1776. Munich; Nürnberg: Stein, 1763-1776. The first 4 vols.
      (1763-1767) included both 'philosophische' and 'historische' classes and were
      paged separately from 1768. Publisher varies. CONTINUED BY: Neue
      philosophischen Abhandlungen der baierischen Akademie der Wissenschaften.
      CONTINUED BY: Neue historische Abhandlungen der baierischen Akademie
      der Wissenschaften. INDEXES: Index to v.1-10, 1763-1776 (Ref.1,p.xiv).
      MICROFORMS: RM(3). SOURCES: G459, K235, OCLC, S3114
      LOCATIONS: BUC, CCF, CCN, ULS, ZDB SUBJECT: Science - Societies
      - Munich.

7   Abhandlungen der freyen ökonomischen Gesellschaft in St. Petersburg, zur
      Aufmunterung des Ackerbaues und der Hauswirtschaft in Russland. - Teil 1-11,
      1767-1777. St. Petersburg: 1767-1777. TRANSLATION OF: Trudy vol'nago
      ekonomicheskoi obchestvo, St. Petersburg. SOURCES: K2887
      LOCATIONS: CCF, ULS, ZDB SUBJECT: Economics - Societies - St.
      Petersburg - Translations. SUBJECT: Agriculture - Societies - St. Petersburg
      - Translations.

8   Abhandlungen der Gesellschaft der Künste und Wissenschaften in Batavia. - Bd.1,
      1782. Leipzig: Weygand, 1782. Translated by Johann Erich Biester.
      TRANSLATION OF: Verhandelingen, Bataviaasch genootschap der konsten
      en wetenschapen. SOURCES: G460a, K1122, S2934 LOCATIONS: CCF,
      ZDB SUBJECT: Science - Societies - Amsterdam - Translations.

9   Abhandlungen der Halleschen Naturforschenden Gesellschaft. - Bd.1, 1783.
      Dessau; Leipzig: Barth, 1783. CONTINUED BY: Neue Schriften der
      Naturforschenden Gesellschaft zu Halle. MICROFORMS: RM(3).
      SOURCES: G483, K3291, OCLC, S2493 LOCATIONS: BUC, CCF, CCN,
      ULS, ZDB SUBJECT: Natural history - Societies - Germany.

10   Abhandlungen der Holländischen Gesellschaft der Wissenschaften zu Harlem. -
      Bd.1, 1758. Altenburg: Richter, 1758. Translated by Abraham Gotthelf
      Kästner. TRANSLATION OF: Verhandelingen uitgegeven door de Hollandsche
      maatschappij der wetenschappen te Haarlem. CONTINUED BY:
      Abhandlungen aus der Naturgeschichte, praktischer Arzneikunst und Chirurgie
      aus den Schriften der Haarlemer und anderer holländischen Gesellschaften.
      SOURCES: G474a, K214, S2595 LOCATIONS: BUC, ZDB SUBJECT:
      Science - Societies - Haarlem - Translations.

11   Abhandlungen der Königlich-Preussischen Akademie der Wissenschaften in Berlin.
      - Bd.1-6, (1788-1803) 1788-1804. Berlin: Reimer, 1788-1804. A translation of
      the Histoire et mémoires? INDEXES: Kohnke, Otto, ed. Gesamtregister über
      die in den Schriften der Akademie von 1700-1899 erscheinen wissenschaftlichen
      Abhandlungen und Festreden. Berlin, 1900 (Ref.126). SOURCES:
      GXIII:476e LOCATIONS: CCF SUBJECT: General - Societies - Berlin.

12   Abhandlungen der königlichen böhmischen Gesellschaft der Wissenschaften zu
      Prag. - Jahrgang 1-4, 1785-1788. Prague; Dresden: Walther, 1785-1788.

CONTINUATION OF: Abhandlungen einer Privatgesellschaft in Böhmen. **CONTINUED BY:** Neuere Abhandlungen der Königlichen böhmischen Gesellschaft der Wissenschaften zu Prag. **INDEXES:** Repertorium sämtlicher Schriften der königlichen böhmischen Gesellschaft der Wissenschaften vom Jahre 1769 bis 1868, zusammengestellt von Wilhelm Rudolph Weitenwever, Prague, 1860. **INDEXES:** Generalregister zu den Schriften den Königlichen böhmischen Gesellschaft der Wissenschaften, 1784-1884, zusammengestellt von Georg Wegner, Prague, 1884. **SOURCES:** G4563a, K380, OCLC, S3492 **LOCATIONS:** CCF, ULS, ZDB **SUBJECT:** Science - Societies - Prague.

13   Abhandlungen der königlichen medicinisches Gesellschaft in Copenhagen. - Bd.1, 1787. Offenbach: Ulrich Weiss und C. Ludwig Brede, 1787. Translated from the Latin. **TRANSLATION OF:** Acta regiae societatis medicae Havniensis. **SOURCES:** GI3b, K3631 **SUBJECT:** Medicine - Societies - Copenhagen - Translations.

14   Abhandlungen der königlichen Parisischen Academie der Chirurgie. - Bd.1-5, (1743-1774) 1754-1776. Altenburg: Akademische Buchhandlung, 1754-1776. A translation of the Mémoires de l'Académie royale de chirurgie by Johann Ernest Zeyher (v.1-3), Johann Ernst Greding (v.4) and Georg Heinrich Königsdorfer. **SOURCES:** G9a, K3535 **LOCATIONS:** ZDB **SUBJECT:** Surgery - Societies - Paris - Translations.

15   Abhandlungen der Londoner Gesellschaft zur Rettung Verunglückter und Scheintodter von das Jahr 1774 bis 1784. - Bd.1, 1798. Breslau: Hirschberg & Lissa, 1798. Translated by Christian August Struve. **TRANSLATION OF:** Transactions of the Royal Humane Society from 1774 to 1784. **SOURCES:** G499b **SUBJECT:** Resuscitation - Societies - London - Translations.

16   Abhandlungen der Londonschen Gesellschaft zur Vermehrung des medicinischen und chirurgischen Wissenschaften. - Bd.1, 1793, Bd.2, 1800, Bd.3, 1812. Braunschweig: Christian Friedrich Thomas; Leipzig: Schmidt und Nauk, 1793-1812. Translated by Theodor Georg August Roose. **SOURCES:** GI18a **LOCATIONS:** ZDB **SUBJECT:** Medicine - Societies - London - Translations. **SUBJECT:** Surgery - Societies - London - Translations.

17   Abhandlungen der Naturforschenden Gesellschaft in Zürich. - Bd.1-3, 1761-1766. Zürich: Heidegger, 1761-1766. Material for a fourth volume was prepared but never printed. **MICROFORMS:** RM(3). **SOURCES:** G484, K3232, OCLC, S2206a **LOCATIONS:** CCP, BUC, ULS, ZDB **SUBJECT:** Agriculture - Societies - Zürich. **SUBJECT:** Natural history - Societies - Zürich. **SUBJECT:** Science - Societies - Zürich.

18   Abhandlungen der ökonomischen Gesellschaft in Basel. - St. 1-3, 1796-1798. Basel: Schweighofer, 1796-1798. **SOURCES:** K3007 **LOCATIONS:** ZDB **SUBJECT:** Economics - Societies - Basel.

19   Abhandlungen der Römisch-kaiserlich-königliche-josephinische medicinisch-chirurgische Akademie, Vienna. - v.1, 1787, v.2, 1802. Vienna: R. Gräffer,

1787-1802. **TRANSLATED AS:** Actes de l'Académie C.R. Josephina medico-chirurgicale de Vienne. **TRANSLATED AS:** Dissertationi medico-chirurgico-practiche. v.1 also published in Latin as Acta Academiae. **OTHER TITLE:** Acta Academiae Caesareae regia Josephinae medico-chirurgicae Vindobonensis. **SOURCES:** G75, K3630, OCLC **LOCATIONS:** BUC, CCF, ULS, ZDB **SUBJECT:** Medicine - Societies - Vienna. **SUBJECT:** Surgery - Societies - Vienna.

20   Abhandlungen der schwedischen Aerzte, oder, Sammlung seltener Beobachtungen und Fälle aus allen Theilen der Medicin. - T.1, 1785. St. Gallen: Huber, 1785. Selectively translated by Johann Jacob Römer. **TRANSLATION OF:** Acta medicorum Suecicorum. **SOURCES:** G53a. K3612 **LOCATIONS:** ZDB **SUBJECT:** Medicine - Periodicals - Sweden - Translations.

21   Abhandlungen der Seeländlischen Gesellschaft der Wissenschaften zu Vlissingen. - Bd.1, 1775. Giessen: Krieger, 1775. Contains only medical and surgical papers. Translated by Andreas Böhm. **TRANSLATION OF:** Verhandelingen uitgeven door het zeeuwsch genootschap der wetenschappen te Vlissingen. **SOURCES:** G522a, K3569, S3083 **LOCATIONS:** ZDB **SUBJECT:** Medicine - Societies - Vlissingen - Translations.

22   Abhandlungen die Verbesserung der Landwirthschaft bretreffend / Kaiserlich Königlich ökonomischen Gesellschaft in Böhmen. - Bd.1-8, 1797-1808. Prague: J. Herrl, 1797-1808. **CONTINUED BY:** Verhandlungen der k.k. patriotisch-ökonomischen Gesellschaft im Königreich Böhmen. **SOURCES:** K3012 **LOCATIONS:** CCF **SUBJECT:** Agriculture - Societies - Prague.

23   Abhandlungen die von der königliche Dänischen Gesellschaft den Preis erhalten haben. - Sammlung 1, 1781. Copenhagen: Gyldendal, 1781. **TRANSLATION OF:** Skrifter, som udi det kjöbenhavnske selskab af laerdoms og videnskabs elstere ere fremlagte og oplaeste. **SOURCES:** G475b, K350, S613b **LOCATIONS:** BUC **SUBJECT:** Science - Societies - Copenhagen - Translations.

24   Abhandlungen einer Privatgesellschaft in Böhmen, zur Aufnahme der Mathematik, der vaterländischen Geschichte und der Naturgeschichte. - Bd.1-6, 1775-2784. Prague: Gerle, 1775-1784. **EDITOR:** Ignaz Edler von Born. **CONTINUED BY:** Abhandlungen der königlichen böhmischen Gesellschaft der Wissenschaften zu Prag. **CONTINUED BY:** Neuere Abhandlungen der Königlichen böhmischen Gesellschaft der Wissenschaften zu Prag. **INDEXES:** Repertorium sämtlicher Schriften der königlichen böhmischen Gesellschaft der Wissenschaften von Jahre 1769 bis 1868 zusammengestellt von Wilhelm Rudolph Weitenwever. Prague, 1860. **INDEXES:** Generalregister zu den Schriften den Königlichen böhmischen Gesellschaft der Wissenschaften, 1784-1884 zusammengestellt von Georg Wegner. Prague, 1884. **SOURCES:** G463, K3257, S3482 **LOCATIONS:** BUC, CCF, CCN, ULS, ZDB **SUBJECT:** Science - Societies - Prague.

25   Abhandlungen einer Privatgesellschaft von Naturforschern und Oeconomen in
      Oberdeutschland. - Bd.1, 1792. Munich: Joseph Lindauer, 1792.
      CONTINUATION OF: Oberdeutsche Beiträge zur Naturlehre und Oeconomie.
      Intelligenzblatt nos. 1-7 issued as a supplement is cited. EDITOR: Franz von
      Paula Schrank. SOURCES: G438, K3346 LOCATIONS: BUC, ZDB
      SUBJECT: Natural history - Societies - Munich.

26   Abhandlungen ökonomischen, technologischen, naturwissenschaftlichen und
      vermischten Inhalts. - Heft. 1-52, 1798. Erfurt: Hennings, 1798. EDITOR:
      Johann Jonathan Gellermann. SOURCES: K3016 LOCATIONS: ZDB
      SUBJECT: Science - Periodicals - Germany. SUBJECT: Technology -
      Periodicals - Germany.

27   Abhandlungen über die Geschichte und Alterthümer, die Künste, Wissenschaften
      und Literatur Asiens. - Bd.1-4, 1795-1797. Riga: Johann Friedrich Hartknoch,
      1795-1797. The first volume is based on Johann Georg Christian Fick's
      translation published by Vossische Buchhandlung in Berlin; v.4 contains a
      treatise by Johann Friedrich Kleuker "Das Brahmische Relgionsystem."
      TRANSLATION OF: Asiatic researches or transactions of the Society instituted
      in Bengal for inquiring into the history and antiquities, the arts, sciences and
      literature of Asia. SOURCES: B15, G458a, S3698 LOCATIONS: BUC
      SUBJECT: General - Societies - Calcutta.

28   Abhandlungen über Gegenstände der Natur und Kunst. - Jahrg.1, nos. 1-10,
      Jan.-Nov. 1777. Osnabrück: Christian Ludolf Reinhold, 1777. Monthly.
      OTHER TITLE: Minerva oder Abhandlungen über Gegenstände der Natur und
      Kunst. EDITOR: Christian Ludolf Reinhold. SOURCES: K4083
      LOCATIONS: BUC SUBJECT: General - Periodicals - Germany.

29   Abhandlungen und Beobachtungen aus der Arzneigelahrtheit von einer Gesellschaft
      von Aerzten in Hamburg. - Bd.1, 1775. Hamburg: Carl Wilhelm Meyn, 1775.
      Other editions in Hamburg and in 1776 in Berlin are cited. EDITOR: Paul
      Dieterich Gisseke. SOURCES: GIII:37, K3568 LOCATIONS: ULS
      SUBJECT: Medicine - Periodicals - Germany.

30   Abhandlungen und Beobachtungen durch die ökonomische Gesellschaft zu Bern
      gesammelt. - Jahrg.3-7 (each in 4 issues), Jahrg.8-14 (each in two issues)
      1762-1773. Bern: Oekonomische Gesellschaft, 1762-1773. CONTINUATION
      OF: Sammlungen von landwirthschaftlichen Dingen der Schweizerischen
      Gesellschaft in Bern. Published simultaneously in French and German.
      OTHER TITLE: Mémoires et observations recueillies par la Société
      oeconomique de Bern. EDITOR: Elie Bertrand (Ref.29,p.152).
      CONTINUED BY: Neue Sammlung physisch-ökonomischer Schriften
      herausgegeben von der ökonomischen Gesellschaft des Kantons Bern.
      CONTINUED BY: Neueste Sammlung von Abhandlungen und Beobachtungen
      von der ökonomischen Gesellschaft in Bern. SOURCES: K2874
      LOCATIONS: BUC, CCF, CCN, ULS, ZDB SUBJECT: Agriculture -
      Societies - Bern. SUBJECT: Economics - Societies - Bern.

31   Abhandlungen und Erfahrungen der fränkisch-physicalisch-ökonomischen
      Bienengesellschaft. - Theil 1-4, 1770-1774. Nürnberg: Johann Eberhard Zeh,
      1770-1774. EDITOR: Johann Leonhard Eyrich. SOURCES: K2900, S3158
      LOCATIONS: CCF, ZDB SUBJECT: Bee culture - Societies - Nürnberg.

32   Abhandlungen und Erfahrungen der physikalisch-oeconomischen
      Bienengesellschaft in Oberlausitz zur Aufnahme der Bienenzucht in Sachsen. -
      Theil 1-4, 1766-1771. Dresden: Walther, 1766-1771. Place and publisher vary.
      OTHER TITLE: Abhandlungen der Ökonomischen Bienengesellschaft in der
      Oberlausitz. EDITOR: Adam Gottlob Schirach. CONTINUED BY:
      Gemeinnützige Arbeiten der Churfürstliche Bienengesellschaft in der
      Oberlausitz. INDEXES: Index in v.4. SOURCES: K2884, S2519
      LOCATIONS: CCF, ZDB SUBJECT: Bee culture - Societies - Oberlausitz.

33   Abhandlungen von Verbesserung der Niederösterreichischen Landwirthschaft. -
      Bd.1, 1767. Vienna: Johann Thomas Edlen, 1767. EDITOR: Martin Nicolaus
      Baumann. LOCATIONS: GB SUBJECT: Agriculture - Periodicals -
      Austria.

34   Abhandlungen zur Naturgeschichte, Chemie, Anatomie, Medizin und Physik aus
      den Schriften des Instituts der Künste und Wissenschaftenzu Bologna. - Bd.1-2,
      1781-1782. Brandenburg: J.W. Halle und J.C. Halle, 1781-1782. Translated by
      Nathanael Gottfried Leske. TRANSLATION OF: De Bononiensi scientiarum
      et artium instituto atque academia commentarii, Reale Accademia delle Scienze
      dell' Istituo di Bologna. SOURCES: G493, K3273, S2417 LOCATIONS:
      BUC, CCF, ZDB SUBJECT: Science - Societies - Bologna - Translations.

35   Abhandlungen zur Naturgeschichte der Thiere und Pflanzen. - Bd.1-2, 1757-1758.
      Leipzig: 1757-1758. A translation of Perrault's Mémoires pour servir à
      l'histoire naturelle des animaux et des plantes. SOURCES: G523, S2928
      LOCATIONS: ULS SUBJECT: Natural history - Societies - Paris -
      Translations.

36   Abhandlungen zur Naturgeschichte, Physik und Oekonomie, aus den
      Philosophischen transactionen. - Bd.1, in two parts, 1779-1780. Leipzig:
      Weygand, 1779-1780. Translated by Nathanael Gottfried Leske. It covers the
      first 14 volumes of the Philosophical transactions, i.e., to 1693, omitting those
      articles which were abstracted and translated from foreign journals. The preface
      to pt.2 promised a further volume which does not appear to have been
      published. TRANSLATION OF: Philosophical transactions. SOURCES:
      B748, G585, K3269 LOCATIONS: ULS, ZDB SUBJECT: Science -
      Societies - London - Translations.

37   Abrégé des Transactions philosophiques de la Société de Londres. - T. 1-15,
      1787-1791. Paris: Buisson, 1787-1791. An abridged translation of the
      Philosophical transactions. Under the general direction of Jacques Gibelin, it
      was projected to be published in 12 parts each with its own subject(s) and
      editor(s). Some parts appeared in more than one volume. Part 9 does not
      appear to have been published (Ref.21,v.25 p.484). SOURCES: GII:4k, S1597

LOCATIONS: BUC, CCF   SUBJECT: Science - Societies - London - Translations.

38   An Account of the benevolent institution for the sole purpose of delivering poor married women at their own habitation. - London: 1783-1806. Issues have been recorded in 1783, 1791 (under the title: List of the governors), 1801 and 1806. OTHER TITLE: List of the governors of the benevolent institution. LOCATIONS: NUC   SUBJECT: Obstetrics - Societies - England.

39   Acta Academia Electoralis Moguntinae Scientiarum utilium quae Erfordiae est. - T.1, 1757-T.2, 1761. Erfurt; Gotha: Mevius, Dietrich, Weber, 1757-1761. OTHER TITLE: Akten der Akademie gemeinützigeN Wissenschaften zu Erfurt. CONTINUED BY: Acta Academia Electoralis Moguntinae 1777ff. CONTINUED BY: Nova Acta Academia Electoralis Moguntinae 1797ff. CONTINUED BY: Abhandlungen der Churfürstlichen Mainzischen Akademie nützlicher Wissenschaften zu Erfurt. INDEXES: Graesel, Arnim. Repertorium zu den Acta und Nova Acta der Akademie, Halle, 1894. MICROFORMS: RM(3). SOURCES: G439, K210, OCLC, S2546a LOCATIONS: BUC, CCF, CCN, ULS SUBJECT: Science - Societies - Erfurt.

40   Acta Academia Philo-exoticorum naturae et artis, Brescia. - Brescia: Mariam Ricciardum, 1686-1687. Appeared monthly from March 1686 to February 1687 (Ref.20). French translation in Collection académique v.7. OTHER TITLE: Acta nova Academiae Philo-exoticorum naturae et artis. Edited by Francesco Lana. SOURCES: GII:13 LOCATIONS: BUC SUBJECT: Science - Periodicals - Italy.

41   Acta Academiae Caesareae regia Josephinae medico-chirurgicae Vindobonensis. - v.1, 1788. Vienna: R. Gräffer, 1788. TRANSLATED AS: Actes de l'Académie C.R. Josephina medico-chirurgicale de Vienne. TRANSLATED AS: Dissertazioni medico-chirurgico-practiche. Also published in German as Abhandlungen. OTHER TITLE: Abhandlungen der Römisch-kaiserlich-königliche-josephinische medicinisch-chirurgische Akademie, Vienna. SOURCES: GIII75a, K3642 LOCATIONS: BUC, CCF, ULS, ZDB SUBJECT: Medicine - Societies - Vienna. SUBJECT: Surgery - Societies - Vienna.

42   Acta Academiae Electoralis Moguntinae Scientiarum utilium quae Erfurti est. - T.1(1776-77) 1777, t.2(1778-79) 1780, t.3(1780-81) 1782, t.4(1782-3) 1784, t.5-7(1784-87), t.8(1788-89) 1790, t.9(1790-91) 1791, t.10(1792) 1792, t.11(1793) 1794, t.12(1794-5) 1796. Erfurt: Wittekind, Keyser, 1777-1796. CONTINUATION OF: Acta Academia Electoralis Moguntinae 1757. OTHER TITLE: Akten der Akademie gemeinnützigen Wissenschaften zu Erfurt. CONTINUED BY: Nova Acta Academia Electoralis Moguntinae 1797. CONTINUED BY: Abhandlungen der Churfürstlichen Mainzischen Akademie nützlicher Wissenschaften zu Erfurt. INDEXES: There is a cumulative index for the years 1776-1791. MICROFORMS: RM(3). SOURCES: G439, K320, OCLC, S2546a LOCATIONS: BUC, CCF, CCN, ULS, ZDB SUBJECT: Science - Societies - Erfurt.

43   Acta Academiae scientiarum imperialis Petropolitana. - T.1 -6 (1777-82) 1778-86 in 12 vol. St. Petersburg: Academia Scientiarum; Leipzig: Hartknoch, Cnobloch, 1778-1786. **TRANSLATED AS:** Physikalische und medicinische Abhandlungen. **CONTINUATION OF:** Commentarii Academia scientiarum imperialis Petropolitana. **CONTINUATION OF:** Novi commentarii Academiae scientiarum imperialis Petropolitanae. Some of the volumes appear in other editions. Issued in two volumes for the year. The articles are chiefly in Latin, but the proceedings, called "Histoire de l'académie imperiale des sciences" are in French. **CONTINUED BY:** Nova acta Academiae scientiarum imperialis Petropolitanae. **MICROFORMS:** IDC(8). **SOURCES:** G442c, OCLC, S3706 **LOCATIONS:** BUC, CCF, CCN, ULS, ZDB **SUBJECT:** Science - Societies - St. Petersburg.

44   Acta academica praesentem academiarum, societarum, litterarium, gymnasiorum et scholarum statum illustrantia. - v.1-6, 1733-1738. Leipzig: 1733-1738. **SOURCES:** G586, K94 **LOCATIONS:** BUC, CCF, ULS, ZDB **SUBJECT:** General - Periodicals - Germany.

45   Acta der Ostpreussische Mohrungsche physikalisch- ökonomische Gesellschaft. - Bd.1-3, 1792-1800. Königsberg: G.L.Hartung, 1792-1800. Publisher varies. **OTHER TITLE:** Schriften und Verhandlungen der Ostpreussichen phyikalisch-ökonomischen Gesellschaft; Beiträge zur Kultur der Ökonomie in Preussen. **CONTINUED BY:** Kleine Schriften der königliche Ostpreussischen physikalisch-ökonomischen Gesellschaft. **SOURCES:** S2868 **SUBJECT:** Economics - Societies - Königsberg.

46   Acta eruditorum. - Vol. 1-50, 1682-1731. Leipzig; Grosse: Gleditsch, 1682-1731. It began as a monthly (Ref.2). An Italian abridgement was published in Venice in 7 vols. (1740-1746) under the title: Actorum eruditorum synopsis (S2084) which may be the same as the Opuscula omnia actis eruditorum, also published in Venice in 7 vols. during the same years covering 1684-1740 of the Acta. There was a new edition of the first 12 vols. in Leipzig 1683-1692 and a French translation in Collection académique v.4, 7. Reprints: Hildesheim, Olms and New York: Johnson. **OTHER TITLE:** Actorum eruditorum synopsis; Opuscula omnia actis eruditorum; Neue Zeitungen von gelehrten Sachen; Acta eruditorum supplementa. **EDITORS:** Otto Mencke (1641-1707), Johann Burchard Mencke (1707- 1732). **CONTINUED BY:** Nova acta eruditorum. A weekly supplement appeared in German under the title: Neue Zeitungen von gelehrten Sachen. Supplements in Latin appeared under the title: Acta eruditorum supplementa. **INDEXES:** Indices generales appeared in 1693, 1704, 1714, 1723, 1733, and 1745 for the preceeding years in Leipzig, Gleditsch (publisher varies). **MICROFORMS:** SMS(1), GO(1), RP(1). **SOURCES:** B24, G12, K1, OCLC, S2880 **LOCATIONS:** BUC, CCP, CCN, ULS, ZDB **SUBJECT:** General - Periodicals - Germany - Reviews.

47   Acta Helvetica, physico-mathematico-anatomico-botanico-medica, Societas physico-medica basiliensis. - Bd.1(1751), Bd.2(1755), Bd.3(1758), Bd.4(1760), Bd.5(1762), Bd.6(1766), Bd.7(1768), Bd.8(1777). Basle: Johann Rudolph Imhof, 1751-1777. Title varies slightly. **EDITORS:** Daniel Bernoulli, Johann

Rudolph Zuingerus (Ref.23). **CONTINUED BY:** Nova Acta Helvetica, physico-mathematico-botanico-medica. **MICROFORMS:** RM(3). **SOURCES:** B26, G508, K3208,S2128 **LOCATIONS:** BUC, CCF, CCN, ULS, ZDB **SUBJECT:** Science - Societies - Basle.

48   Acta literaria et scientiarum Suecicae / Konglige Vetenskaps-Societeten i Upsala. - v.3-4, 1730-1739. Upsala: Kiesewetter, 1730-1739. **TRANSLATED AS:** Recueil des mémoires les plus interessans de chemie et d'histoire naturelle. **CONTINUATION OF:** Acta literaria et scientiarum Suecicae. Publisher varies. **CONTINUED BY:** Acta regiae societatis scientiarum Upsaliensis. **CONTINUED BY:** Acta nova regiae societatis scientiarum Upsaliensis. **MICROFORMS:** RM(3). **SOURCES:** B27, G478, OCLC, S704 **LOCATIONS:** BUC, CCF, ULS, ZDB **SUBJECT:** Science - Societies - Upsala.

49   Acta literaria societatis rheno-trajectinae. - T.1-4, 1793-1803. Leyden: Luchtmans, 1793-1803. Also published at Utrecht. **CONTINUED BY:** Nova acta literaria societatis rheno-trajectinae. **SOURCES:** S925 **LOCATIONS:** BUC, CCF, CCN, ULS, ZDB **SUBJECT:** General - Societies - Utrecht.

50   Acta literaria Sueciae Upsaliae publicata / Konglige Vetenskaps-Societeteten i Upsala. - v.1-2, 1720-1729. Upsala; Stockholm: Johannes Henricus Russwarm, 1720-1729. **TRANSLATED AS:** Recueil des mémoires les plus interessans de chemie et d'histoire naturelle. **CONTINUED BY:** Acta literaria et scientiarum Suecicae. **CONTINUED BY:** Acta regiae societatis scientiarum Upsaliensis. **CONTINUED BY:** Acta nova regiae societatis scientiarum Upsaliensis. **MICROFORMS:** RM(3). **SOURCES:** B27, G478, S704 **LOCATIONS:** BUC, CCF, CCN, ULS **SUBJECT:** Science - Societies - Upsala.

51   Acta medica et philosophica Hafniensia. - v.1(1671/72)-v.5(1677/79), 1673-1680. Copenhagen: Petri Haubold, 1673-1680. Articles in Latin or Danish. French translation in Collection académique, 4(1757) 185-376, 5(1776) 362-440, 11: 140-386. **OTHER TITLE:** Thomae Bartholini acta medica et philosophica. **CONTINUED BY:** Collectanea societatis medicae Havniensis. **CONTINUED BY:** Acta societatis medicae Havniensis. **CONTINUED BY:** Acta regiae societatis medicae Havniensis. **CONTINUED BY:** Acta nova regiae societatis medicae Hafniensis. **MICROFORMS:** RM(3). RP(1). **SOURCES:** B28, G3a, OCLC, 609 **LOCATIONS:** BUC, CCF, CCN, ULS, ZDB **SUBJECT:** Medicine - Societies - Copenhagen.

52   Acta medicorum Berolinensium in incrementum artis et scientiarum collecta et digesta. - Dec. 1: v.1-10, 1717-22; Dec. 2: v.1-10, 1723-30; Dec. 3: v.1, 1731. Berlin: Gottfried Gedicke, 1717-1731. New editions of some of the volumes have been recorded. Edited by Johann Daniel Gohl. **INDEXES:** Index to Dec. 1 in v.Dec.1:10. **SOURCES:** GIII:3, K3516 **LOCATIONS:** BUC, CCF, ULS, ZDB **SUBJECT:** Medicine - Periodicals - Germany.

53   Acta medicorum Suecicorum, seu sylloge observationum et casuum rariorum in variis medicinae partibus praesertim in historia naturali, praxi medica et

chirurgia. - T.1, 1783. Upsala: Schwederi, 1783. **TRANSLATED AS:** Abhandlungen der schwedischen Aerzte. There are two variants; one published in Upsala, Stockholm and Abo in 1783 and the other in Stockholm at the same time in an incomplete version (Ref.24). **SOURCES:** G53,OCLC **LOCATIONS:** BUC, CCF, CCN, ULS, ZDB **SUBJECT:** Medicine - Periodicals - Sweden.

54   Acta nova regiae societatis scientiarum Upsaliensis. - v.1-14, 1773-1850. Upsala: Edmann, 1773-1850. **TRANSLATED AS:** Recueil des mémoires les plus interessans de chemie et d'histoire naturelle. **CONTINUATION OF:** Acta literaria et scientiarum Suecicae. **CONTINUATION OF:** Acta literaria Sueciae Upsaliae publicata. **CONTINUATION OF:** Acta regiae societatis scientiarum Upsaliensis. **OTHER TITLE:** Nova acta regiae societatis scientiarum Upsaliensis. **INDEXES:** Kungliga vetenskaps-societeten i Upsala. Catalogue methodique des Acta et Nova Acta.1744-1889. Redigé par Aksel G.S. Josephson. Upsala, 1889. **MICROFORMS:** RM(3). **SOURCES:** B27, G478, OCLC, S704 **LOCATIONS:** BUC, CCF, ULS, ZDB **SUBJECT:** Science - Societies - Upsala.

55   Acta philosophica societatis regia in Anglia. - v.1-7, 1665-1781. Amsterdam: Henry and Theodor Boom, 1665-1681. Variants are noted which indicate there were several printings and editions. A Frankfurt edition in octavo in 1671 (4th ed. 1674) and a Leipzig edition in quarto in 1675. Christopher Sand translated the volumes for 1666-1668, and Johann Sterpin the volume for 1665 (Ref.27). Title varies. **TRANSLATION OF:** Philosophical transactions. **SOURCES:** GII4h, OCLC, S3017, S766 **LOCATIONS:** BUC, CCF, CCN, ZDB **SUBJECT:** Science - Societies - London - Translations.

56   Acta philosophicao-medica / Hessische Akademie der Wissenschaften. - v.1, 1771. Frankfurt; Leipzig: Fleischer, 1771. **SOURCES:** K3556 **LOCATIONS:** BUC, CCF, CCN, ULS, ZDB **SUBJECT:** Medicine - Societies - Giessen. **SUBJECT:** Science - Societies - Giessen.

57   Acta physico-medica / Academia Caesarea Leopoldina-Carolina naturae curiosorum. - T.1(1727), t.2(1730), t.3(1733), t.4(1737), t.5(1740), t.6(1742), t.7(1746), t.8(1748), t.9(1752), t.10(1754). Nürnberg: W.N. Endter, 1727-1754. **TRANSLATED AS:** Auszüge medicinisch-chirurgischer Beobachtungen. **CONTINUATION OF:** Ephemerides sive observationum medico-physicarum, Academiae Caesareae Leopoldinae naturae curiosoum. **CONTINUATION OF:** Miscellanea curiosa, Academia naturae curiosorum. Publisher varies. French translation in Collection académique. **CONTINUED BY:** Nova acta physico-medica academiae Caesareae Leopoldinae- Carolinae naturae curiosorum. **INDEXES:** Cumulative index for v.1-10, 1727-1754. **MICROFORMS:** IDC(8),RP(1). **SOURCES:** B29, GII:7c, K3185, S3146a **LOCATIONS:** BUC, CCF, CCN, ULS **SUBJECT:** Science - Societies - Halle.

58   Acta regia societatis medicae Havniensis. - v.1 (1783), v.2(1791), v.3(1792), v.4(1803) Copenhagen; Schubothe: Brummer, 1783-1803. **TRANSLATED AS:**

Abhandlungen der königlichen medicinischen Gesellschaft in Copenhagen. **CONTINUATION OF:** Collectanea societatis medicae Havniensis. **TRANSLATED AS:** Merkwürdige Krankengeschichte und seltene practische Beobachtungen berühmter Aerzte. **CONTINUATION OF:** Acta medica et philosophica. **CONTINUATION OF:** Acta societatis medicae Havniensis. **CONTINUED BY:** Acta nova regiae societatis medicae Hafniensis. **SOURCES:** GI:3 **LOCATIONS:** BUC, CCF, CCN, ULS, ZDB **SUBJECT:** Medicine - Societies - Copenhagen.

59   Acta regiae societatis scientiarum Upsaliensis. - v.1-5 (1740-1750), 1744-1751. Stockholm: Saluci, 1744-1751. **TRANSLATED AS:** Recueil des mémoires les plus interessans de chemie et d'histoire naturelle. **CONTINUATION OF:** Acta literaria et scientiarum Suecicae. **CONTINUATION OF:** Acta literaria Sueciae Upsaliae publicata. **CONTINUED BY:** Acta nova regiae societatis scientiarum Upsaliensis. **INDEXES:** Kungliga vetenskaps-societeten i Upsala. Catalogue methodique des Acta et Nova Acta 1744-1889. Redigé par Aksel G.S. Josephson. Upsala, 1889. **MICROFORMS:** RM(3),IDC(8). **SOURCES:** B27, G478, OCLC, S704 **LOCATIONS:** BUC, CCF, ULS, ZDB **SUBJECT:** Science - Societies - Upsala.

60   Acta societatis Jablonoviannae / Fürstlich-Jablonowskische Gesellschaft der Wissenschaften. - T.1-5, (1770-1774)1772-1780. Leipzig: Schwickert, 1772-1780. **CONTINUED BY:** Acta nova Societatis Jablonoviannae. **SOURCES:** G466, K287 **LOCATIONS:** BUC, CCF, ULS, ZDB **SUBJECT:** General - Societies - Leipzig.

61   Acta societatis medicae Havniensis. - v.1-v.2, 1777-1779. Copenhagen: C.A.Proft, 1777-1779. **TRANSLATED AS:** Merkwürdige Krankengeschichte und seletene practische Beobachtungen. **CONTINUATION OF:** Acta medica et philosophica Hafniensia. **CONTINUATION OF:** Collectanea societatis medicae Havniensis. **CONTINUED BY:** Acta regiae societatis medicae Havniensis. **CONTINUED BY:** Acta nova regiae societatis medicae Hafniensis. **SOURCES:** GI3 **LOCATIONS:** BUC, CCF, CCN, ULS, ZDB **SUBJECT:** Medicine - Societies - Copenhagen.

62   Actas de la Sociedad médica de nuestra Señora de la Esperanza. - v.1, 1754. Madrid: 1754. Articles are in Latin and Spanish. **EDITOR:** Antonio Fernandez de Lozoya. **SOURCES:** OCLC. **SUBJECT:** Medicine - Societies - Madrid - Prize essays.

63   Actas y memorias / Sociedad económica de los amigos del país de la provincia de Segovia. - v.1-3, 1785-1793. Segovia: Espinosa, 1785-1793. **OTHER TITLE:** Continuación de las memorias, Sociedad de los amigos del país de la provincia de Segovia. **LOCATIONS:** BUC, CCF, ULS **SUBJECT:** Economics - Societies - Segovia.

64   Actes de la société de médecine, chirurgie et pharmacie établie à Bruxelles. - Parts 1-2, 1797-1799. Paris: Croullerus, 1797-1799. **SOURCES:** GIII:139,OCLC **LOCATIONS:** BUC, CCF, ZDB **SUBJECT:** Medicine - Societies - Brussels.

SUBJECT: Surgery - Societies - Brussels.   **SUBJECT:** Pharmacy - Societies - Brussels.

65   Actes de la société d'histoire naturelle de Paris. - T.1, 1792. Paris: Reynier, 1792. **CONTINUED BY:** Mémoires de la Société d'histoire naturelle de Paris. **MICROFORMS:** RM(3), IDC(8). **SOURCES:** G549, OCLC, S1583a **LOCATIONS:** BUC, CCF, CCN, ULS, ZDB **SUBJECT:** Natural history - Societies - Paris.

66   Actes de l'Académie Caesarea regia Josephine médico- chirurgicale de Vienne. - v-.1, 1792. Montpellier: Jean Martel, 1792. A translation of the society's Acta Academiae Casearea by J. Escudé. **TRANSLATION OF:** Acta Academiae Caesareae regia Josephinae medico-chirurgicae Vindobonensis. **LOCATIONS:** CCF, ULS   **SUBJECT:** Medicine Societies - Vienna - Translations. **SUBJECT:** Surgery - Societies - Vienna - Translations.

67   Actorum eruditorum quae Lipsiae publicantur supplementa. - v.1-10, 1692-1734. Leipzig : Johann G. Haeredes, J.F. Gleditsch, 1692-1734. **OTHER TITLE:** Acta eruditorum, supplementum. **INDEXES:** Indexed in Indices generales auctorum et rerum, Acta eruditorum. **MICROFORMS:** SM(1),RP(1). **LOCATIONS:** BUC, CCF, CCN, ULS **SUBJECT:** General - Periodicals - Germany.

68   Actorum eruditorum synopsis, id est, opuscula actis eruditorum inserta, quae ad mathesin, physican, medicinam, anatomiam, chirurgiam et philologiam pertinent. - v.1-7 (1682-1742) 1693-1745. Venice: J. Baptista Pasquali, 1693-1742. **CONTINUATION OF:** Acta eruditorum. An abridgement of the Acta eruditorum. There may have been two separate editions published in Venice. **OTHER TITLE:** Opuscula omnia actis eruditorum lipsiensis inserta. **SOURCES:** B30, G12b, S2084 **LOCATIONS:** ULS **SUBJECT:** General - Periodicals - Italy - Abridgements.

69   Addresses to the medical society of students at Edinburgh. - Edinburgh: 1771-1775. **SOURCES:** (Ref.17:14). **SUBJECT:** Medicine - Societies - Edinburgh.

70   Adversaria medico-practica. - Bd.1-3 (each in 4 parts), 1769-1772. Leipizg: Weidmann & Reich, 1769-1772. **EDITOR:** Ludwig, Christian Gottlieb. **SOURCES:** K3552, OCLC **LOCATIONS:** NUC, ZDB **SUBJECT:** Medicine - Periodicals - Germany.

71   Die Aerzte, ein medicinisches Wochenblatt. - T.1-2 (52 issues) 1785. Hamburg: Hermann, 1785. **EDITOR:** Christian G. Donatius. **CONTINUED BY:** Deutsche Gesundheits-zeitung. **SOURCES:** GIII:63, K3613 **LOCATIONS:** ZDB **SUBJECT:** Medicine - Periodicals - Germany - Popular.

72   Aesculap, eine medicinisch-chirurgische Zeitschrift. - Theil 1, 1790. Leipzig: Weygand, 1790. **EDITORS:** Friedrich August Weber and Markus Philipp

Ruhland. SOURCES: GIII85, K3658 SUBJECT: Medicine - Periodicals - Germany.

73 Akademicheskïia izviestïia / Academia scientiarum imperialis Petropolitanae. - v.1-8, 1779-1781. St. Petersburg: 1779-1781. Included some medical articles (Ref.31). LOCATIONS: ULS SUBJECT: Science - Societies - St. Petersburg.

74 Algemeen magazijn van wetenschap, konst en smaak, Behelzende: I. Wysbegeerte en zedekunde. II. Natuurkunde en natuurlyke historie. III. Historiekunde. IV. Beschaade letteren, fraaije kunste en mengelwerk. - Deel 1-5, 1785-1791. Amsterdam: D'erven P. Meyer en G. Warnars, 1785-1791. CONTINUED BY: Nieuwe algemeen magazijn. CONTINUED BY: Vaderlandsch magazijn van Wetenschap, konst en smaak. INDEXES: Index to v.1-5 in 5. MICROFORMS: RM(3). SOURCES: B80, S721 LOCATIONS: CCF, CCN, ULS SUBJECT: General - Periodicals - Netherlands.

75 Algemeen verslag der Nederlandsche maatschapij tot nut van't algemeen / Amsterdam. - Deel 1-12, 1786-1839. Amsterdam: 1786-1839. A general welfare society which included popular education in science among its activities (Ref.30). CONTINUED BY: Bijdragen tot bevordering van de kennis en den bloie. LOCATIONS: BUC, CCF, CCN, SUBJECT: General - Societies - Amsterdam.

76 Algemeene genees- natuur- en huishoudkundige jaarboeken / Genootschap van genees-en natuurkundigen, Dordrecht. - Deel 1-4, in 6 vols., 1785-1788. Dordrecht en Amsterdam: A. Blusse en W. Holtrop, 1785-1788. CONTINUATION OF: Genees- natuur- en huishoudkundige jaarboeken. CONTINUATION OF: Nieuwe genees- natuur- en huishoudkundige jaarboeken. SOURCES: B1872a, G259, S808 LOCATIONS: CCN, ULS SUBJECT: Medicine - Societies - Dordrecht. SUBJECT: Natural history - Societies - Dordrecht. SUBJECT: Home economics - Societies - Dordrecht.

77 Algemeene konst-en letter-bode, voor meer-en min- geoeffenden behelzende berigte int de geleerde waereld van alle landen. - Haarlem: A. Loosjes, 1788-1793. Weekly. Imprint varies. CONTINUED BY: Nieuwe algemeene konst-en letter-bode. INDEXES: Algemeene register op de 25 deelen (1788-1899). Haarlem,1803. MICROFORMS: IDC(8). SOURCES: B81, OCLC LOCATIONS: CCF, CCN, ULS, ZDB SUBJECT: General - Periodicals - Netherlands.

78 Algemeene Oefenschoole van konsten en wetenschappen. - Theil 1-30, 1763-1783. Amsterdam: Pieter Meijer, 1763-1783. Includes index. SOURCES: OCLC LOCATIONS: CCN SUBJECT: Science - Periodicals - Netherlands.

79 Allerneueste Mannigfaltigkeiten, eine gemeinnützige Wochenschrift. - v.1-4, 1781-1784. Berlin: Johann Carl Franz Einfeld, 1781-1784. Weekly. CONTINUATION OF: Mannigfaltigkeiten. CONTINUATION OF: Neues Mannigfaltigkeiten. CONTINUATION OF: Neueste Mannigfaltigkeiten.

EDITOR: Johann Friedrich Wilhelm Otto. **SOURCES:** B2831c, K5605, S2347.4 **LOCATIONS:** BUC, CCF, ULS, ZDB **SUBJECT:** General - Periodicals - Germany.

80   Allgemeine Beiträge zur Beförderung des Ackerbaues, der Künste und Gewerbe. - Pt.1-2, 1800. Leipzig: 1800. **EDITOR:** Johann Gottlieb Geissler. **SOURCES:** B85 **SUBJECT:** Agriculture - Periodicals - Germany. **SUBJECT:** Manufactures - Periodicals - Germany.

81   Allgemeine deutsche Bibliothek. - Bd.1-118, 1765-1796. Berlin: Christoph Friedrich Nicolai, 1765-1796. Imprint varies. **EDITOR:** Christoph Friedrich Nicolai. **CONTINUED BY:** Neue allgemeine deutsche Bibliothek. Supplements published at 5 or 6 year intervals under the title: Anhang which include a list of books and an index to the years covered. **INDEXES:** Anhang Bd.1-12 (1764-1768) 1771; Anhang Bd.13-24 (1769-1773) 1777, 3v.; Anhang Bd.25-36 (1774-1777) 1780, 6v.; Anhang Bd.37-52 (1778-1781) 1785, 4v.; Anhang Bd.53-86 (1782-1878) 1791 6v.; Anhang Bd.87-117 (1788- 1791) 1796. **MICROFORMS:** GO(9). **SOURCES:** G588, K248, OCLC **LOCATIONS:** BUC, CCF, CCN. ULS, ZDB **SUBJECT:** General - Periodicals - Germany - Reviews.

82   Allgemeine geographische Ephemeriden. - Jahrg.1-19 (Bd.1-51) 1798-1816. Weimar: Verlag des Industrie Comptoir, 1798-1816. **EDITORS:** Franz Xaver von Zach (Jahrg.1-2), Adam Christian Gaspari and Friedrich Justin Bertuch (Jahrg.3-19). **CONTINUED BY:** Neue allgemeine geographische Ephemeriden. From 1812 supplemented by Allgemeiner typographischer Monatsbericht für Teutschland. **SOURCES:** B100, K1308, OCLC, S3295.1 **LOCATIONS:** BUC, CCF, CCN, ULS, ZDB **SUBJECT:** Geography - Periodicals - Germany.

83   Allgemeine Geschichte der Natur in alphabetischen Ordnung. - Bd.1-11, 1774-1793. Berlin: Schlesinger, 1774-1793. Based on Valmont de Bomare's Dictionnaire raisonné universel d'histoire naturelle. Continued only as far as "Cas-Coq" in the alphabet. Includes citations to periodicals and proceedings of scientific societies. **EDITOR:** Friedrich Heinrich Wilhelm Martini. **TRANSLATION OF:** Dictionnaire raisonné universel d'histoire naturelle. **SOURCES:** G532a **LOCATIONS:** NUC **SUBJECT:** Natural history - Encyclopedias - Germany.

84   Allgemeine Gesundheitsregeln: eine Wochenschrift. - St.1-52, 1790. Lohenstein und Leipzig: 1790. A reprint appeared in Leipizg in 1793. **EDITOR:** Jakob Ludwig Gebhard. **SOURCES:** GIV:189, K3663 **SUBJECT:** Medicine - Periodicals - Germany - Popular.

85   Allgemeine Handlungs-Zeitung und Anzeigen, nebst monatlichen Beiträge für das Neueste und Nutzlichtste der Chemie, Fabrikswissenschaft, Apothekerkunst, Oekonomie und Warrenkenntniss. - Jahrg.1-44, 1794-1837. Nürnberg: Comptoir der Handlungzeitung, 1794-1837. Weekly. **OTHER TITLE:** Kaiserlich-privilegirte allgemeine Handlungs-Zeitung und Anzeigen. **EDITOR:**

Johann Michael Leuchs.   **SOURCES:** K2767   **LOCATIONS:** NUC
SUBJECT: Economics - Periodicals - Germany.

86   Allgemeine Historie der Reisen zu Wasser und zu Lande. - Bd.1-21, 1747-1774.
Leipizig: Arkstee und Merkus, 1747-1774. Translated by Johann Joachim
Schwabe from collections on voyages in various languages including a new
general collection of voyages (London, T.Astley, 1745-1747) and Histoire
génerale des voyages by Prevost d'Exiles and others. **EDITOR:** Johann
Joachim Schwabe. **MICROFORMS:** UM(1), Lost Cause Press.
LOCATIONS: NUC   **SUBJECT:** Geography - Periodicals - Germany -
Translations.

87   Allgemeine Literatur-zeitung. - Jena: Expedition; Leipizg: Churfürstliche.
Sächsische Zeitungsexpedition, 1785-1803.   Imprint varies. **EDITORS:**
Christian Gottfried Schütz, Friedrich Justin Bertuch and Gottlieb Hufeland.
CONTINUED BY: Jenaische allgemeine Literatur-zeitung. Supplements were
issued under the title: Intelligenzblätter (1785ff) and Ergänzungsblätter (or
Revision der Literatur für die Jahre) (1785ff). **INDEXES:** Index for 1785-1790
in Intellegenzblatt no.4, April, 1790, and later for the years 1796-1800.
MICROFORMS: GO(9). **SOURCES:** K388 **LOCATIONS:** BUC, CCF,
CCN, ULS, ZDB **SUBJECT:** General - Periodicals - Germany - Reviews.

88   Allgemeine medicinische Annalen. - v.1, 1800. Altenberg: Brockhaus, 1800.
Monthly.   **CONTINUATION OF:** Medicinische Nationalzeitung für
Deutschland. **EDITORS:** J.F. Sierer and Johann F. Pierer. **CONTINUED
BY:** Allgemeine medecinische Annalen des neunzehnten Jahrhunderts.
SOURCES: GIII:177, K3713 **LOCATIONS:** BUC, CCF, ULS, ZDB
SUBJECT: Medicine - Periodicals - Germany.

89   Allgemeine ökonomische und mit philosophischen Abhandlungen vermischte
Nachrichten. - St. 1-37, 1754-1757. Nürnberg: 1754-1757. **EDITOR:** Johann
Wolfgang Brenk. **SOURCES:** K2865 **SUBJECT:** Economics - Periodicals -
Germany.

90   Allgemeine theoretische Stadt-und Landwirthschaftskunde. - St. 1-3, 1789-1790.
Leipizg: Haug, 1789-1790.   **EDITOR:** Friedrich Gottlieb Leonhardi.
SOURCES: K2964 **SUBJECT:** Agriculture - Periodicals - Germany.

91   Allgemeiner ökonomischer oder Landwirthschafts-kalender auf das Jahr, das ist,
nützlicher und getreuer Unterricht für den Land- und Bauersmann, besonders
in Schwaben. - Jahrg.1-2, 1770-1771. Stuttgart: Johann Benedict Mezler,
1770-1771. **OTHER TITLE:** Allgemeines landwirthschaft Calender.
EDITOR: Balthasar Sprenger. **CONTINUED BY:** Nützlicher und getreuer
Unterricht für den Land-und Bauersmann. **CONTINUED BY:** Ökonomische
Beiträge und Bemerkungen zur Landwirthschaft. **MICROFORMS:** RP(1)
(Goldsmiths' Kress Library). **SOURCES:** K2902 **LOCATIONS:** CCF, ZDB
SUBJECT: Agriculture - Periodicals - Germany - Popular. **SUBJECT:**
Economics - Periodicals - Germany - Popular. **SUBJECT:** Medicine -
Periodicals - Germany - Popular.

92   Allgemeines Archiv für die Länder und Völkerkunde. - Bd.1-2, 1790-1793.
      Leipizg: Christian Gottlob Hillscher, 1790-1793. EDITOR: Friedrich Carl
      Gottlob Hirsching. SOURCES: B5124, G245, K1225 LOCATIONS: BUC,
      ZDB SUBJECT: Geography - Periodicals - Germany.

93   Allgemeines Helvetisches Magazin zur Beförderung der inländischer Naturkunde.
      - v.1,1799. Winterthur: Steiner, 1799. CONTINUATION OF: Magazin für
      die Naturkunde Helvetiens. EDITOR: Johann Georg Albrecht Höpfner.
      CONTINUED BY: Helvetisches Monatschrift. SOURCES: B125, G524,
      S2184 LOCATIONS: ULS SUBJECT: Natural history - Periodicals -
      Switzerland.

94   Allgemeines historisches Magazin zur Beförderung der Erdbeschreibung, und der
      Natur-, Staats- und Kirchengeschichte. - St.1,1762. Gotha: 1762. SOURCES:
      B126, K1018, S2657 SUBJECT: Geography - Periodicals - Germany.
      SUBJECT: General - Periodicals - Germany.

95   Allgemeines Jahrbuch der Geographie und Statistik. - Weimar: Verlag des
      Industrie-Comptoir, 1800. EDITOR: Adam Christian Gaspari. SOURCES:
      B127, G590, K2788 LOCATIONS: BUC, ZDB SUBJECT: Geography -
      Periodicals - Germany.

96   Allgemeines Journal der Chemie. - Jahrg.1-5, in 60 Hefte in 10 Bde i.e. 6 issues
      to a vol. Leipzig: Breitkopf und Härtel, 1793-1803; Berlin: Heinrich Erdlach,
      1798-1803. Monthly. CONTINUATION OF: Allgemeines Journal der Chemie
      Intelligenzblatt.Vols. 7-10 published: Berlin, Heinrich Erdlach. EDITOR:
      Alexander Nicolaus Scherer. CONTINUED BY: Neues allgemeines Journal
      der Chemie. CONTINUED BY: Journal für die Chemie und Physik.
      Supplements published as: Intelligenzblatt, Nr. 1-6, 1798-1799.
      MICROFORMS: RM(3). SOURCES: B128, G591, K3772, OCLC, S2882
      LOCATIONS: BUC, CCF, CCN, ULS, ZDB SUBJECT: Chemistry -
      Periodicals - Germany.

97   Allgemeines Journal für die Handlung, oder gemeinnützige Aufsätze, Versuche
      und Nachrichten für die Kaufleute. - v.1-3 (6 issues each) 1785-1788. Scherin:
      Bödner, 1785-1788. CONTINUATION OF: Ephemeriden der Handlung.
      EDITOR: Johann Christian Schedel. SOURCES: K2743 LOCATIONS: ZDB
      SUBJECT: Commerce - Periodicals - Germany.

98   Allgemeines Journal für Handlung, Schiffahrt, Manufaktur und die darauf
      beziehung habenden Gewerbe überhaupt.- St.1-12, 1799-1800. Leipzig:
      Breitkopf und Härtel, 1799-1800. SOURCES: K2789 LOCATIONS: BUC,
      ZDB SUBJECT: Economics - Periodicals - Germany.

99   Allgemeines Magazin der Natur, Kunst und Wissenschaften. - T.1-12, 1753-1767.
      Leipzig: Gleditsch, 1753-1767. Consists largely of translations from other
      languages covering a broad spectrum of agriculture, domestic economy,
      mathematics and science. A popular journal on the model of the Universal
      Magazine of Knowledge and Pleasure (Ref.3:46). EDITOR: Johann Daniel

Titius.   **INDEXES:** Index v.1-12 in 12.   **SOURCES:** B131, K191,
OCLC,S2883 **LOCATIONS:** CCF, CCN, ULS, ZDB **SUBJECT:** Science -
Periodicals - Germany - Popular.

100   Allgemeines Magazin für bürgerliche Baukunst. - Bd.1-2, 1789-1796. Weimar:
1789-1796. **EDITOR:** Gottfried Huth. **SOURCES:** K4092 **LOCATIONS:**
CCN, ZDB **SUBJECT:** Building - Periodicals - Germany.

101   Allgemeines Magazin für Jäger und Jagdfreunde. - Grätz: Kienreich, 1794.
SOURCES: K2997 **LOCATIONS:** ZDB **SUBJECT:** Hunting - Periodicals
- Austria.

102   Allgemeines ökonomisches Forstmagazin, in welchem allerhand nützliche
Beobachtungen, Vorschläge und Versuche über die wirthschaftliche Policey -
und Cameral - Gegenstände des samtlichen Wald -, Forst- und Holzwesens
enthalten sind. - Bd.1-12, 1763-1769. Frankfurt; Leipzig: J.B. Mezler,
1763-1769. Bd.1-2 also issued in an augmented and revised edition, published
by Mezler, Frankfurt and Leipizg, 1783. **EDITOR:** Johann Friedrich Stahl.
CONTINUED BY: Forst-und Jagdbibliothek. **INDEXES:** Indexes in v.2,4,6,8
and 10 to that volume and to the preceeding volume. **SOURCES:** K2876
LOCATIONS: CCF, CCN, ULS, ZDB **SUBJECT:** Forestry management -
Periodicals - Germany.

103   Allgemeines ökonomisches Magazin. - Bd.1-2, 1782-1783. Hamburg: 1782-1783.
Monthly. **EDITOR:** Johann Heinrich Pratje. **LOCATIONS:** BUC,ZDB
SUBJECT: Economics - Periodicals - Germany.

104   Allgemeines Repertorium der Literatur. - Ser.1, Bd.1-3 (1785-1790) 1793-1794;
ser.2, Bd.1-3 (1791-1795) 1799-1800, ser.3, Bd.1-2 (1796-1800) 1806-1809.
Jena: Academische Buchhandlung; Weimar: Industrie Comptoir, 1793-1809.
A classified list of the scholarly literature including that of science and
technology. **EDITOR:** Johann Samuel Ersch. **LOCATIONS:** BUC, CCF,
CCN, ULS, ZDB **SUBJECT:** General - Periodicals - Germany - Indexes.

105   Allgemeines Repertorium für empirische Psychologie und verwandte
Wissenschaften. - Bd.1-6, 1792-1801. Nürnberg: Felsecker, 1792-1801.
EDITOR: Immanuel David Mauchart. **CONTINUED BY:** Neue allgemeines
Repertorium für empirische Psychologie. **SOURCES:** B132, G247, K569,
S3150 **LOCATIONS:** CCN, ZDB **SUBJECT:** Psychology - Periodicals -
Germany.

106   Allgemeines Repertorium zur praktischen Beförderung der Künste und
Manufakturen. - Bd.1-2, 1797-1798. Zittau: Schöps, 1797-1798. **EDITOR:**
Johann Gottlieb Geissler. **CONTINUED BY:** Neues Repertorium zur
praktischen Beförderung der Künste und Manufacturen. **SOURCES:** K3918
SUBJECT: Manufactures - Periodicals - Germany.

107   Allgemeines Sachregister über die wichtigsten deutschen Zeitschriften und
Wochen-Schriften. - v.1-2, 1790. Leipzig: Weygand, 1790. A classified list of

German periodicals for the period 1700- 1789 which includes a subject index to the following 8 periodicals: Ephemeriden der Menscheit, Schlözers Staatsanzeigen, Deutsche Museum, Göttingsche Magazin der Wissenschaften, Deutsche Merkur, Schlözers Briefwechsel, Hannöverische Magazin and Berliner Monatschrift. **EDITORS:** Johann Heinrich Christian Beutler and Johann Christian Friedrich Gutsmuth. **SOURCES:** G539   **LOCATIONS:** NUC **SUBJECT:** General - Periodicals - Germany - Indexes.

108   Allgemeines Schwedisches Gelehrsamkeits Archiv unter Gutafs des Dritten Regierung. - Bd.1-7, 1781-1796. Leipizg: Johann Friedrich Junius, 1781-1796. **EDITOR:** Christoph Wilhelm Lüdeke. **SOURCES:** G594, K354, OCLC **LOCATIONS:** BUC, CCF, ULS, ZDB **SUBJECT:** General - Periodicals - Germany - Reviews.

109   Allgemeines Wochenblatt zur Erhaltung der unschätzbaren Gesundheit. - Qu. 1-4, 1787-1788. Frankfurt: Weiss, 1787-1788. **EDITOR:** Johann Michael Hofmann. **SOURCES:** K3641a **SUBJECT:** Medicine - Periodicals - Germany - Popular.

110   Almanacco d'economia pel Grandveato di Toscano. - Florence: 1791. **SOURCES:** (Ref.76) **SUBJECT:** Economics - Periodicals - Italy.

111   Almanach contenent le précis de l'agriculture du jardinier- fleuriste. - Paris: Desnos, 1776. **SOURCES:** (Ref.35:558). **SUBJECT:** Agriculture - Periodicals - France.

112   Almanach d'agriculture necessaire à tout laboureur. - Paris: 1774-1775. **SOURCES:** (Ref.35:491). **LOCATIONS:** CCF **SUBJECT:** Agriculture - Periodicals - France.

113   Almanach de Lisboa / Academia real das sciencias de Lisboa. - v.1-29, 1782-1826, Lisbon: Typografia da Academia, 1782-1826. Title varies slightly. **SOURCES:** OCLC **LOCATIONS:** NUC **SUBJECT:** Science - Societies - Lisbon.

114   Almanach de medécine. - Paris: Grange et Dufour, 1763. **SOURCES:** (Ref.35:352). **SUBJECT:** Medicine - Periodicals - France.

115   Almanach de santé. - Paris: 1774. **EDITOR:** Jean Jacques Gardane (Ref.36, p.65). **CONTINUED BY:** Étrennes d'un médecin. **SOURCES:** (Ref.35:503) **SUBJECT:** Medicine - Periodicals - France.

116   Almanach de santé, ou, annuaire. - Paris: 1796-1797. **EDITOR:** Jean Baptiste Lecouteux de Cantileu. **SOURCES:** G200 **SUBJECT:** Medicine - Periodicals - France.

117   Almanach der Fortschritte in Wissenschaften, Manufakturen, Handwerken und Künsten. - Jahrg.1-15, 1795-1810.Erfurt: Keyser, 1795-1810. The first volume was in alphabetical order by subject, which was abandoned for a classified order in the second volume. A reissue of the first volume in the same order was

promised in the preface to vol. 4 (1800). **OTHER TITLE:** Übersicht der Fortschritte in Wissenschaften; Almanach der Fortschritte, neuesten Erfindungen und Entdeckungen in Wissenschaften; Annalen der Fortschritte in Wissenschaften. **EDITORS:** Jahrg.1-12, Gabriel Christoph Benjamin Busch, Jahrg.13-15, Johann Bartholma Trommsdorff. **CONTINUED BY:** Annalen der Fortschritte in Wissenschaften, Manufakturen, Handwerken und Künste. **INDEXES:** Indexes to v.1-6 in 6, and 7-12 in 12. **SOURCES:** G595, K2771, S2550 **LOCATIONS:** BUC, CCF, CCN, ZDB **SUBJECT:** Manufactures - Periodicals - Germany.

118 Almanach der Landwirthschaft und Industrie für die K.K. Erbstaaten. - Prague: 1798. **SOURCES:** (Ref.37) **SUBJECT:** Agriculture - Periodicals - Czechoslovakia. **SUBJECT:** Manufactures - Periodicals - Czechoslovakia.

119 Almanach des ballons, ou globe aerostatique, étrenne du jour physico-historique et chantante. - Annonay: Langlois, 1784. **SOURCES:** (Ref.35:759) **LOCATIONS:** BN **SUBJECT:** Balloon ascension - Periodicals - France.

120 Almanach des bastiments pour l'annëe contenant les noms & demeures de Messieur les Architectes-Experts-Bourgeois. - Paris: Stoupe, 1774-1792. **SOURCES:** (Ref.35:504) **LOCATIONS:** BN **SUBJECT:** Building - Periodicals - France.

121 Almanach des Ernstes und des Schertzes für Aerzte, Chirurgen und Geburtshelfer. - Bd.1-2, 1800-1801. Erfurt: Georg Adam Keyser, 1800-1801. **OTHER TITLE:** Alte Zeit und neue Zeit, oder, Herzenserleichterung über medizinische Neuigkeiten. **EDITOR:** Johann Ludwig Andreas Vogel. **SOURCES:** GIII:171, OCLC **LOCATIONS:** BUC, NUC, ZDB **SUBJECT:** Medicine - Periodicals - Germany - Popular.

122 Almanach des gourmands, ou, Calendrier nutrif etc. - v.1-8, 1799-1806. Paris, 1799-1806. At least 3 other editions were issued. **EDITORS:** A.B.L. Grimod de la Reynìre and Coste. **LOCATIONS:** BUC, ZDB **SUBJECT:** Nutrition - Periodicals - France.

123 Almanach du commerce, des arts et métiers. - Lille: Jacques et Ranackere, 1786-1788. **SOURCES:** (Ref.38,p.377) **SUBJECT:** Economics - Periodicals - France.

124 Almanach für Ärzte und Nichtärzte. - Bd.1-15, 1782-1796. Jena: Christian Heinrich Cuno, 1782-1796. **EDITOR:** Christian Gottfried Gruner. **CONTINUED BY:** Neues Taschenbuch für Ärzte und Nichtärzte. **SOURCES:** G52, K3595 **LOCATIONS:** BL, CCN, ULS, ZDB **SUBJECT:** Medicine - Periodicals - Germany - Popular.

125 Almanach für Bienenfreunde, oder, Anweisung zur praktischen Bienenzucht auf jeden Monath in Jahr. - Halberstadt: Gossent, 1792-1973. Imprint varies. **SOURCES:** (Ref.37) **SUBJECT:** Bee culture - Periodicals - Germany.

126   Almanach für deutsche Landwirthe. - Franfurt am Oder: 1782. **EDITOR:** Georg Heinrich Borowski. **SOURCES:** Cited in Allgemeines ökonomisches Magazin, 1782,1:1063. **SUBJECT:** Agriculture - Periodicals - Germany.

127   Almanach für die medicinische Polizei, gerichtliche Arzneiwissenschaft und Volksarzneikunde, mit besonderer Rücksicht auf die Medicinalbedürfnisse Mecklenburgs. - Schwerin: Bärensprung, 1797. **EDITOR:** Georg Heinrich Masius. **SOURCES:** G136 **LOCATIONS:** NUC **SUBJECT:** Medical jurisprudence - Periodicals - Germany.

128   Almanach historique, geographique, genealogique des voyages, des arts et des sciences. - Amsterdam: 1760-1781. **LOCATIONS:** CCN **SUBJECT:** Geography - Periodicals - Netherlands.

129   Almanach oder Taschenbuch für Scheidekünstler und Apotheker. - Bd.1-23, 1780-1802. Weimar: Karl Ludolf Hoffmann, 1780-1802. **OTHER TITLE:** Taschenbuch für Scheidekünstler und Apotheker. **EDITOR:** Johann Friedrich August Göttling. **CONTINUED BY:** Chemisches Taschenbuch für Ärzte, Chemiker und Pharmaceuten. **CONTINUED BY:** Trommsdorff's Taschenbuch für Ärzte, Chemiker und Pharmaceuten. **INDEXES:** Cumulative index for 1780-1803 in Trommsdorff's Taschenbuch für Chemiker, Jena, 1820-1829. **SOURCES:** B153, G209, OCLC, S3296 **LOCATIONS:** BUC, CCF, CCN, ZDB **SUBJECT:** Pharmacy - Periodicals - Germany.

130   Almanach van het chirurgyns gilde der stadt Amsterdam. - Amsterdam: 1707-1784. **OTHER TITLE:** Chirurgyns almanach. **LOCATIONS:** CCN **SUBJECT:** Surgery - Societies - Amsterdam.

131   Almanach vétérinaire, contenant l'histoire abrégé des progrès de la medécine des animaux depuis l'établissement des écoles vétérinaires en France. - v.1, 1782, 1v.1792. Paris: Vallat-la-Chapelle, 1782-1792. The first vol. was published in Paris in 1782; a revised and augmented edition covering the years 1782-1790 was published in Paris in 1792. There were at least 2 other editions and a reprint in 1809 with a change of title (Ref.35:673). **EDITORS:** Philibert Chabert, Pierre Flandrin and Jean Baptiste Huzard. **SOURCES:** G223 **LOCATIONS:** ZDB **SUBJECT:** Veterinary medicine - Periodicals - France.

132   Almanach voor beminnars van wettenschappen geleerheid en gooden smaak. - Amsterdam: 1779-1785. **LOCATIONS:** CCN **SUBJECT:** General - Periodicals - Netherlands.

133   Almanaco per gli medici, chirurgi. - Bergamo: Locatelli, 1789. **SOURCES:** (Ref.136) **SUBJECT:** Medicine - Periodicals - Italy. **SUBJECT:** Surgery - Periodicals - Italy.

134   Almanaque náutico y efemérides astronómicas / Observatorio astronómico de marina, San Fernando. - v.1-39(1792-1838)1781-1837. Madrid: Imprenta Real, 1791-1837. Subtitle varies. **EDITOR:** Ciprano Vimercati. **SOURCES:** B166,

S1730 LOCATIONS: BUC, CCF, NUC, ZDB SUBJECT: Astronomy - Societies - San Fernando.

135 Alt-Märkisches ökonomische-physicalisches Magazin. - Berlin: Arnold Wever, 1747. SOURCES: K2912 SUBJECT: Economics - Periodicals - Germany.

136 Altes und neues aus dem Erzgebirge, nebst ausführlichen Nachrichten von den in Erzgebirge herauskommenden neuen Schriften. - Bd.1-3, 1747-1749. Freiburg: Theodor Gottlieb Reinhold, 1747-1749. Issued twice a month. OTHER TITLE: Erzgeburgische und insonderheit Freybergische Merkwürdigkeiten. SOURCES: B174. K4399, S2615 LOCATIONS: BUC SUBJECT: Mineral industries - Periodicals - Germany.

137 American museum, or, Universal magazine. - v.1-13, 1787-1798. Philadelphia: Mathew Carey, 1787-1798. Monthly. Subtitle varies. Suspended 1793-1797. EDITOR: Mathew Carey. MICROFORMS: UM(1). SOURCES: B234, OCLC LOCATIONS: CCF, CCN, ULS SUBJECT: General - Periodicals - United States.

138 L'Ami de l'Art défensif, ou,observations sur le Journal Polytechnique de l'École central des travaux publics - No.1-6, 1793-1798. Paris: Louvet, 1793-1798. Publisher and subtitle vary. OTHER TITLE: Journal de l'École polytechnique; Art défensif. EDITOR: Marc René Montalembert. LOCATIONS: NUC SUBJECT: Military art and science - Periodicals - France.

139 Amoenitates academicae seu dissertationes variae physicae, medicae, botanicae. - v.1-10, 1749-1790. Stockholm: Godfried Kieswetter, 1749-1790. Linnaeus collected and published the first 7 v.1749-1769, Schreber added the last three volumes (1785-1790). The entire collection contains 199 dissertations, disputations and programs of Linnaeus and his students. Imprints vary and other editions appeared of some of the volumes. EDITORS: Carl von Linnaeus and Johann Christian Daniel Schreber. MICROFORMS: IDC(8). SOURCES: B272, G335, S677 LOCATIONS: NUC SUBJECT: Science - Periodicals - Sweden - Dissertations.

140 Amoenitates medicae. - T.1-5, 1745-1747. Leipzig: 1745-1747. EDITOR: Heinrich Friedrich von Delius. SOURCES: K3526 SUBJECT: Medicine - Periodicals - Germany - Dissertations.

141 Analecta transalpina seu epitome commentariorum / Regia scientiarum academia suecica. T.1-t.2, 1762. Venice: 1762. TRANSLATION OF: Handlingar, Kungliga Svenska Vetenskapsakademien. SOURCES: S2102 LOCATIONS: BUC SUBJECT: Science - Societies - Stockholm - Translations.

142 Analekten zur Naturkunde und Ökonomie für Naturforscher, Aerzte und Ökonomen. - Heft 1, 1789. Zittau: Schöps, 1789. Includes dissertations and other works translated primarily from the Latin. EDITOR: Johann Hermann Pfingsten. SOURCES: B274, G597, K3329, S3338 SUBJECT: Natural history - Periodicals - Germany.

143   Anales de ciencias naturales. - v.1-7, 1799-1804. Madrid: Imprenta Real, 1799-1894. v.1-2 as Anales de historia natural. **OTHER TITLE:** Anales de historia natural. **SOURCES:** B280a, G598, OCLC, S1734   **LOCATIONS:** BUC, CCF, CCN, ULS, ZDB   **SUBJECT:** Natural history - Periodicals - Spain.

144   Anales, ó colección de memorias sobre las artes, la artillería, la historia natural de España y Amèricas / Real laboratorio de quimica, Segovia. - v.1-2, 1794-1795. Segovia: 1794-1795. **SOURCES:** B285, S1774   **SUBJECT:** Science - Societies - Segovia.

145   Analyse des travaux, première partie, sciences, mathématiques et physiques / Société royale des arts, LeMans. - v.1(1794-1819)1820. Le Mans: 1820. **EDITOR:** André Pierre Ledru. **SOURCES:** S1205a   **LOCATIONS:** NUC **SUBJECT:** Science - Societies - LeMans.

146   Analytical essays towards promoting the chemical knowledge of mineral substances. - v.1-v.2, 1801-1804. London: T.Cadell and W. Davies, 1801-1804. **EDITOR:** Martin Heinrich Klaproth. **TRANSLATION OF:** Beiträge zur chemischen Kenntniss der Mineralkörper. **MICROFORMS:** RM(3). **SOURCES:** OCLC   **LOCATIONS:** NUC **SUBJECT:** Chemistry - Periodicals - Germany - Translations. **SUBJECT:** Mineralogy - Periodicals - Germany - Translations.

147   The analytical review, or, History of literature domestic and foreign. - v.1 (May 1788)-v.28 (Dec. 1798), n.s. v.1 (June 1799). London: Joseph Johnson, 1787-1799. Monthly. Subtitle and imprint varies. **EDITOR:** Thomas Christie. **MICROFORMS:** UM(1), RP(1) (Goldsmiths'-Kress Library). **SOURCES:** B291, OCLC, S182   **LOCATIONS:** BUC, CCF, ULS, ZDB   **SUBJECT:** General - Periodicals - England - Reviews.

148   Anatomische, chymische und botanische Abhandlungen der königlichen Akademie der Wissenschaften zu Berlin von den Jahren 1692-1737. - Bd.1-9 (1692-1737) 1749-1762. Breslau: W. G. Korn, 1749-1762. Translated by Wolfgang Balthasar Adolph von Steinwehr. **TRANSLATION OF:** Histoire de l'Académie royale des sciences et des belles-lettres de Berlin. **SOURCES:** G487h, S2437a **SUBJECT:** Science - Societies - Berlin - Translations.

149   Anatomische, chymische und botanische Abhandlungen der königlichen Akademie der Wissenschaften zu Paris. - T.1 (1692-3, 1699-1701)- t.9, 1749-(1737)1760. Breslau: Johann Jacob Korn, 1749-1760. Translated by Wolfgang Balthasar Adolf von Steinwehr. **SOURCES:** G487h, K3204, S2438 **LOCATIONS:** DSG, ZDB   **SUBJECT:** Science   Societies - Paris - Translations.

150   Anatomische merckwürdige Nachrichten von der grossen Spiese-Safts-Röhre in der Brust etc. - v.1, 1740. Frankfurt: Johann Gottfriaed Conradi, 1740. A translation of six articles from Selecta medica Francofurtensia. **TRANSLATION OF:** Selecta medica Francofurtensia. **SOURCES:** OCLC

LOCATIONS: DSG   SUBJECT: Medicine - Periodicals - Germany - Translations.

151   Anleitingen / Naturforschende Gesellschaft, Zürich. - Zürich: 1771-1800. Title varies. Prize papers published as a result of the agricultural prize essay competition were published under this title and printed in large numbers but not issued in series or collected (Ref.39). SUBJECT: Agriculture - Societies - Zürich - Prize essays.

152   Annalen aller Verhandlungen und Arbeiten der ökonomisch-patriotischen Societät der Furstentheimer Schweidnitz und Jauer. - Jauer: 1785-1803. SOURCES: (Ref.40)   SUBJECT: Economics - Societies - Jauer.

153   Annalen der Akademie der Künste und mechanischen Wissenschaften. - Heft 1, 1791. Berlin: 1791. CONTINUATION OF: Monatsschrift der Akademie der Künste und mechanischen Wissenschaften zu Berlin. EDITOR: Carl Phillip Moritz. SOURCES: K4093 LOCATIONS: ULS,ZDB SUBJECT: Technology - Societies - Berlin.

154   Annalen der Arzneimittellehre. - v.1-2, no.1, 1795-1799. Leipzig: Schäfer, 1795-1799. EDITORS: Johann Jacob Römer and Carl Gottlob Kühn. CONTINUED BY: Instructions et observations sur les maladies des animaux domestiques. SOURCES: G218, K3714 LOCATIONS: BUC SUBJECT: Medicine - Periodicals - Germany.

155   Annalen der Botanik. - Heft. 1-24, of which 3 make up a vol. 1791-1800. Zürich: Orell, 1791-1800. CONTINUATION OF: Magazin für die Botanik. Imprint varies. Heft 7-24 also carries the title Neue Annalen. OTHER TITLE: Neue Annalen der Botanik. EDITOR: Paul Usteri. MICROFORMS: IDC(8). SOURCES: B2778, S2191 LOCATIONS: BUC, CCF, CCN, ULS, ZDB SUBJECT: Botany - Periodicals - Germany.

156   Annalen der deutschen Akademien. St.1-2, 1790-1791. Stuttgart: Erhard, 1790-1791. EDITORS: Friedrich Christian Franz and Eugen Karl Ludwig von Scheler. SOURCES: B308, K1223,S3239 LOCATIONS: CCF SUBJECT: General - Periodicals - Germany.

157   Annalen der Entbindungs-Lehranstalt auf der Universität Göttingen nebst einer Anzeige und Beurtheilung neuer Schriften für Geburtshelfer. Bd.1, St.1 (1800), St.2(1801), Bd.2, St.1(1801), St.2(1804). Göttingen: Johann Christian Dietrich, 1800-1804. EDITOR: Friedrich Benjamin Osiander. SOURCES: G173 LOCATIONS: CCF, ULS, ZDB SUBJECT: Obstetrics - Periodicals - Germany.

158   Annalen der Gärtnerei, nebst einem allgemeinen intelligenzblatt für Garten-und Blumen-freunde. - 12 parts in 2 vol. 1795-1800. Erfurt: Keyser, 1795-1800. Subtitle varies. EDITOR: Neuenhahn den Jüngern. INDEXES: Index to St. 1-6 in 6, and 7-12 in 12. SOURCES: B310 LOCATIONS: BUC, CCN, ULS, ZDB SUBJECT: Gardening - Periodicals - Germany.

159   Annalen der Geburtshülfe, Frauenzimmer und Kinderkrankheiten. v.1-2 für die
      Jahre 1790-1791, 1793-1794. Winterthur: Steiner, 1793-1794. **EDITOR:**
      Johann Jacob Römer. **SOURCES:** G112, K3678 **LOCATIONS:** BUC, ULS
      SUBJECT: Obstetrics - Periodicals - Germany - Reviews. **SUBJECT:**
      Pediatrics - Periodicals - Germany - Reviews.

160   Annalen der Geographie und Statistik. Bd, 1-3, 1790-1792. Braunschweig:
      Crusius, 1790-1792. Imprint varies. **OTHER TITLE:** Annalen der
      geographischen und statistishen Wissenschaften. **EDITOR:** Eberhard August
      Wilhelm Zimmermann. **CONTINUED BY:** Allgemeine geographische
      Ephemeriden. **SOURCES:** B311, K1224 **LOCATIONS:** BUC, CCF, ULS,
      ZDB **SUBJECT:** Geography - Periodicals - Germany. **SUBJECT:** Statistics
      - Periodicals - Germany.

161   Annalen der leidenden Menschheit. - Heft 1-10, 1795-1801. Altona: Hammerich,
      1795-1801. **EDITOR:** August Adolph Friedrich Hennings. Kraus reprint,
      1972. **SOURCES:** G250, K1287, OCLC **LOCATIONS:** BUC, CCF, ULS,
      ZDB **SUBJECT:** Social sciences - Periodicals - Germany.

162   Annalen der Märkischen ökonomischen Gesellschaft zu Potsdam. - Bd.1-3(each
      with 3 Hefte) 1792-1802. Potsdam: Hovath, 1792-1802. **SOURCES:** K2984
      LOCATIONS: CCF, ULS, ZDB   **SUBJECT:** Economics - Societies -
      Potsdam.

163   Annalen der Naturgeschichte. - Bd.1, 1791. Göttingen: Dieterich, 1791.
      EDITOR: Heinrich Friedrich Link. **CONTINUED BY:** Beiträge zur
      Naturgeschichte. **SOURCES:** B318, G526, K3344 **LOCATIONS:** BUC,
      CCF, ULS **SUBJECT:** Natural history - Periodicals - Germany.

164   Annalen der neuesten englischen und französischen Chirurgie und Geburtshülfe.
      - Heft 1-3, 1799-1800. Erlangen: Johann Christian Schubart, 1799-1800.
      EDITORS: Bernhard Nathaniel Gottlob Schreyer and Johann Christian
      Friedrich Harless. **CONTINUED BY:** Journal der ausländischen medicinischen
      Literatur. **SOURCES:** G260, K3721 **LOCATIONS:** CCF, ULS, ZDB
      SUBJECT: Surgery - Periodicals - Germany - Reviews. **SUBJECT:** Obstetrics
      - Periodicals - Germany - Reviews.

165   Annalen der niedersächsischen Landwirthschaft / Königliche Churfürstliche
      Landwirthschaft- Gesellschaft zu Celle. - Bd.1-6, 1799-1804. Celle: Schulze,
      1799-1804. **CONTINUATION OF:** Nachrichten von Verbesserung der
      Landwirthschaft und des Gewerbes. **CONTINUATION OF:** Neue
      Abhandlungen und Nachrichten. **EDITORS:** Albrecht Daniel Thaer and J.C.
      Benecke. **CONTINUED BY:** Annalen des Ackerbaues. **SOURCES:** B5226,
      K3021,OCLC **LOCATIONS:** BUC, CCF, CCN, ULS, ZDB **SUBJECT:**
      Agriculture - Societies - Celle.

166   Annalen der Ökonomie, Cameralistik und anderer dahin einschlagender
      Wissenschaften. - Heft 1-2, 1787. Leipzig: Böhme, 1787. **SOURCES:** K2749
      LOCATIONS: CCF **SUBJECT:** Economics - Periodicals - Germany.

167   Annalen der Physik. - v.1-76 in 3 series, 1799-1824. Halle: Renger, 1799-1808;
      Leipzig: Barth, 1809-1818. **CONTINUATION OF:** Journal der Physik.
      CONTINUATION OF: Neues Journal der Physik. Ser. 1: Jahrg.1-10 in 30
      Bde., 1799-1808, Supplement to Bd.12; Ser. 2: Jahrg 11-20, Bd.31-60, also as
      neue Folge Jahrg.1-30, Bd.1-30; Ser. 3: Jahrg 21-25 and Jahrg 25 Jan-Apr.
      Bd.61-76, also as Jahrg.11-15 or Bde. 31-46. **EDITORS:** Friedrich Albrecht
      Gren and Ludwig Wilhelm Gilbert. **CONTINUED BY:** Annalen der Physik
      und der physikalischen Chemie. **INDEXES:** Index to v.1-76 (1799-1824) in:
      Vollständige und systematisch geordneters Sach - und Namenregister angefertigt
      von Heinrich Müller. Leipzig, 1826 (Ref.4). **MICROFORMS:** IDC(8), UM(1),
      PMC(1,5,6). **SOURCES:** B2410c, G599, K3376, OCLC, S2704
      LOCATIONS: BUC, CCF, CCN, ULS, ZDB   **SUBJECT:** Physics -
      Periodicals Germany. **SUBJECT:** Science - Periodicals - Germany.

168   Annalen der Rostockschen Akademie. v.1-13, 1788-1807. Rostock: Koppe,
      1788-1807. Annual. **EDITOR:** Johann Christian Eschenbach. **SOURCES:**
      K415, S3205   **LOCATIONS:** CCF, ULS, ZDB   **SUBJECT:** General -
      Periodicals - Germany.

169   Annalen der Staats-arzneykunde. - St.1-3. 1790-1791. Zülichau; Jena: Frommann,
      1790-1791. **CONTINUATION OF:** Medicinisch-gerichtliche Bibliothek.
      CONTINUATION OF: Bibliothek für Physiker. Formed by the union of
      Bibliothek für Physiker and Medicinisch-gerichtliche Bibliothek. **EDITOR:**
      Johann Daniel Metzger. **CONTINUED BY:** Materialen für die
      Staatsarzneikunde und Jurisprudenz. **SOURCES:** G89, K3659 **LOCATIONS:**
      BUC, CCF, ULS, ZDB   **SUBJECT:** Medical jurisprudence - Periodicals -
      Germany.

170   Annalen des englishen Ackerbaues, und anderer nützlicher Künste. - Bd.1-3,
      1790-1802 (Ref.32:307). Leipzg: Crusius, 1790-1802. Translated by S.
      Hahnemann and Johann Riem. **OTHER TITLE:** Annalen des Ackerbaues.
      TRANSLATION OF: Annals of agriculture. **SOURCES:** K3056 **SUBJECT:**
      Agriculture - Periodicals - England - Translations.

171   Annalen des Handels, der Schiffahrt und Gewerbe in und ausser Deutschland,
      oder, Bibliothek des Wissenswürdigen und Nützlichen für Nährstand überhaupt.-
      Bd.1, 1793. Frankfurt und Leipizg: Perrenonische Buchhandlung, 1793.
      SOURCES: K2762a   **SUBJECT:** Commerce - Periodicals - Germany.

172   Annalen des klinischen Instituts in dem Juliushospital zu Würzburg. - Bd.1-2
      (1800-1801)1803-1804.Würzburg: 1803-1804.**CONTINUATION OF:** Annales
      Instituti medico-clinici Wirceburgensis. **OTHER TITLE:** Annalen des
      klinischen Anstalt, Juliushospital, Würzburg. **EDITOR:** Joseph Nicolas
      Thomann. **LOCATIONS:** BUC, ULS, ZDB   **SUBJECT:** Medicine -
      Societies - Würzburg.

173   Annalen des klnischen Institutes zu Berlin. Heft 1-3, 1791-1794. Berlin:
      Heinrich August Rothman, 1791-1794. **EDITOR:** Johann Friedrich Fritze.
      LOCATIONS: CCF, ZDB **SUBJECT:** Medicine - Societies - Berlin.

174   Annalen zur Geschichte der Klinik nach dem Laufe der Zeiten. - Th.1-2, 1797-1803. Prague: 1797-1803. **EDITORS:** Johann Anton Sebald and Johann Ambrosi. **SOURCES:** G141, OCLC   **LOCATIONS:** ULS   **SUBJECT:** Medicine - Periodicals - Czechoslovakia.

175   Annales avec des notices statistiques concernant l'agriculture et la médecine / Observatoire de l'académie de Turin. - v.1-5, 1787-1818. Turin: 1787-1818. **INDEXES:** Table général des publications de la Société, 1759-1813/14, in v.13 of its Mémoires. **SOURCES:** S2069   **SUBJECT:** Agriculture - Societies - Turin. **SUBJECT:** Medicine - Societies - Turin.

176   Annales de chimie, ou, Recueil de mémoires concernant la chimie et les arts qui en dependant. T.1-96, 1789-1815. Paris: Cuchet, 1789-1815. Three vols. a year to 1793, then irregular. **TRANSLATED AS:** Annals of chemistry. **TRANSLATED AS:** Auswahl vorzüglicher Abhandlungen aus den sämmtlichen Bänden der französischen Annalen der Chemie. After 1800 "et specialement la pharmacie" is added to the title. Some of the earlier volumes are to be found in other editions. T.1-3 was reprinted at Paris in 1830. A Spanish translation has been cited (Ref.5). Publisher varies. Suspended Oct. 1793-1796 inclusive. **OTHER TITLE:** Recueil de mémoires concernant la chimie et les arts qui en dépendant. **EDITORS:** Pierre Auguste Adet (from v.7), Louis Bernard Guyton Morveau (from v.9, 1791), Antoine Laurent Lavoisier (to v.18, 1793), Claude Louis Berthollet (to v.73, 1810), Antoine François de Fourcroy (to v.73, 1810), Gaspard Monge (to v.18, 1810), Jean Henri Hassenfratz (to v.18, 1793, then from v.24, 1797), Baron Philipp Friedrich Dietrich (to v.17, 1792) (Ref.93). **CONTINUED BY:** Journal de pharmacie. **CONTINUED BY:** Annales de chimie et de physique. **INDEXES:** Three index volumes were published: 1. Table générale des matieès contenues dans les trente premiers volumes. Paris, Fuchs, 1801; 2. dans les volumes 31 jusqu'a 60. Paris, Bernard, 1807; 3. dans les tomes 61-96. Paris Cuchard, 1816. **MICROFORMS:** PMC(1,5,6), RM(3). **SOURCES:** B346, G600, OCLC, S1292   **LOCATIONS:** BUC, CCF, ULS, ZDB   **SUBJECT:** Chemistry - Periodicals - France.

177   Annales de la Société des amis réunis de Strasbourg: ou, Cures que des membres de cette Société ont operées par le magnétisme animal. - v.1-3, 1786-1789. Strasbourg: Société Harmonique, 1786-1789. Title of v.2: Suite des cures faites par differents magnetiseurs. Publisher varies. **OTHER TITLE:** Suite des cures faites par differents magnetiseurs. **SOURCES:** G235 **LOCATIONS:** ULS   **SUBJECT:** Animal magnetism - Societies - Strasbourg.

178   Annales de l'agriculture française, contenant des observations et des mémoires sur toutes les parties de l'agriculture. - T.1-79, 1796-1817, ser. 2, 1-36, 1818-1873. Paris: Madame Husar, 1796-1873. Title and imprint vary slightly. **EDITOR:** Henri Alexander Tessier. **SOURCES:** NUC   **LOCATIONS:** BUC, CCF, CCN, ULS   **SUBJECT:** Agriculture - Periodicals - France.

179   Annales de l'observatoire imperial de Paris. - v.1-61 (1800-1907) 1855-1867. Paris: 1907. Imprint varies. None published for 1833-1836 and 1894-96.

SOURCES: OCLC   LOCATIONS: CCF, ZDB   SUBJECT: Astronomy - Societies - Paris.

180   Annales Instituti medico-clinici Wirceburgensis. - v.1-2, 1799-1802. Würzburg: 1799-1802. EDITOR: Joseph Nicolas Thomann. CONTINUED BY: Annalen des klinischen Instituts in dem Juliushospital zu Würzburg. SOURCES: G160, K3722   LOCATIONS: CCF, CCN   SUBJECT: Medicine - Societies - Würzburg.

181   Annali di chimica. - v.1-24, 1790-1805. Pavia: Baldassiare Comini, 1790-1805. Subtitle from v.4: e storia naturale, ovvero raccolta di memorie sulle scienze, arti e manufatture ad esse relative. EDITOR: Luigi Gaspero Brugnatelli. INDEXES: Index to v.1-21, in v.21.   SOURCES: B403, S1990 LOCATIONS: BUC, CCF, ULS   SUBJECT: Chemistry - Periodicals - Italy.

182   Annals of agriculture and other useful arts. - v.1-46, 1784-1815. London; Bury St. Edmunds: Arthur Young, 1784-1815. Frequency varies. TRANSLATED AS: Annalen des englischen Ackerbaues und anderer nützlichen künste (Ref.6,p.307). Printer varies. v.1-46 also numbered 1-271. OTHER TITLE: Young's annals. EDITOR: Arthur Young. INDEXES: Index to v.1-10 in v.10; v.1-20 in v.20; 1-30 in v.30; v.31-40 in v.40. SOURCES: B412, OCLC LOCATIONS: BUC, CCF, ULS   SUBJECT: Agriculture - Periodicals - England.

183   Annals of Chemistry. - v.1, 1791. London: 1791. A translation of v.5 of the Annales de chimie. TRANSLATION OF: Annales de chimie, ou, Recueil de mémoires concernant la chimie et les arts qui en dependent.   SOURCES: B418, G600a, S188   SUBJECT: Chemistry - Periodicals - France - Translations.

184   Annals of medicine. - v.1-8, 1796-1804. Edinburgh: G. Muddie; London: A.J Robinson, 1796-1804. TRANSLATED AS: Medicinische Annalen Englischen Aerzte von 1796. CONTINUATION OF: Medical and philosophical commentaries. CONTINUATION OF: Essays and observations, physical and literary. CONTINUATION OF: Medical commentaries. In two series. EDITORS: Andrew Duncan, Sr. and Andrew Duncan, Jr. CONTINUED BY: Edinburgh medical and surgical journal. INDEXES: Index to v.1-8 in Edinburgh medical journal 1804. SOURCES: G132, OCLC LOCATIONS: BUC, CCF, ULS, ZDB   SUBJECT: Medicine - Periodicals - Scotland.

185   Annals of philosophy, natural history, chemistry, literature, agriculture, and the mechanical and fine arts. - v.1-3(1800-1802)1800-1804. London: 1801-1804. EDITOR: Thomas Garnett. SOURCES: B426, S195   LOCATIONS: BUC, CCF, ULS   SUBJECT: General - Periodicals - England.

186   Année rurale ou calendrier à l'usage des cultivateurs de la Géneralité de Paris. - T.1-2, 1787-1788. Paris: 1787-1788. EDITOR: P.M.A. Brussonet. LOCATIONS: BUC   SUBJECT: Agriculture - Periodicals - France.

28   SCIENTIFIC AND TECHNICAL PERIODICALS

187   Anno médico. - v.1, 1706. Porto: Malleu Filhos, 1796. **EDITOR:** José Bento Lopes (Ref.34:35). **SOURCES:** OCLC **SUBJECT:** Medicine - Periodicals - Portugal.

188   Annuaire de l'Institut nationale des sciences, lettres et arts de France. - 81 v., 1796-1876. Paris: 1796-1876. **CONTINUATION OF:** Histoire et mémoires de l'Academie royale des sciences. **CONTINUATION OF:** Historia in quae praeter ipsius academiae originem. **CONTINUATION OF:** Mémoires de mathematiques et de physique tirez des registres de l'Académie royale des sciences. Formed in 1795 by the amalgamation of five independent societies including the Académie des science, Paris. **SOURCES:** S1396b **LOCATIONS:** BUC, ULS, ZDB **SUBJECT:** Science - Societies - Paris.

189   Annuaire du bureau des longitudes, augmenté de notices scientifiques. - T.1-98, 1796-1895. Paris: 1796-1895. Title varies. **OTHER TITLE:** Annuaire de la République Francaise, presenté par le Bureau des longitudes; Annuaire présenté au Gouvernement par le Bureau des longitudes. **SOURCES:** B477, S1354A **LOCATIONS:** BUC, CCF **SUBJECT:** Astronomy - Societies - Paris.

190   Annuaire du Lycée des arts. - Paris: 1795-1798. **CONTINUATION OF:** Journal des inventions et découvertes. **CONTINUED BY:** Annuaire de l'Athenée des arts. **LOCATIONS:** CCF **SUBJECT:** Technology - Societies - Paris

191   Annus medicus quo sistuntur observationes circa morbos acutos et chronicos, adjiciunturque eorum curationes, et quaedam anatomicae cadavrerum sectiones. - Vienna: Johann T.Trattner, 1759-1762. A German translation in 1774 is cited (Ref.41,v.5:436). **EDITOR:** Anton Störck. **SOURCES:** OCLC **SUBJECT:** Medicine - Periodicals - Germany.

192   Antichità de Ercolano esposte / Reale Accademia Ercolanese di Archeologia, Napoli. - v.1-8, 1752-1792. Naples: 1752-1792. **SOURCES:** G490 **SUBJECT:** Archeology - Societies - Naples.

193   Anzeige für alle Aerzte und Wundaerzte in Schleissien und Sudpreussen, die Freunde ihrer Wissenschaft and des Vaterland sind. - By an organization of physicians. **SOURCES:** (Ref.137) **SUBJECT:** Medicine - Periodicals - Germany.

194   Anzeige und Rezensionen der neuesten erscheinenden Werke der Militär-Litteratur. - Heft 1-8, 1781. Vienna: 1781. **SOURCES:** K3966 **SUBJECT:** Military art and science - Periodicals - Austria - Reviews.

195   Anzeige von den Sammlungen einer Privatgesellchaft in der Oberlausitz / Görlitz. - St.1-12, 1780-1795. Görlitz: 1780-1795. Appeared in manuscript for members only (K344). **CONTINUED BY:** Anzeigen der Oberlausitzischen Gesellschaft der Wissenschaften. **SOURCES:** K344 **SUBJECT:** General - Societies - Görlitz.

196 Anzeigen der Oberlausitzischen Gesellschaft der Wissenschaften / Goörlitz. - Lauban: Scharf, 1796-1797; Görlitz: Burghart, 1798-1800. **CONTINUATION OF:** Anzeige von den Sammlungen einer Privatgessechaft in der Oberlausitz, Görlitz. **SOURCES:** K450 **SUBJECT:** General - Societies - Görlitz.

197 Anzeigen von der Leipziger ökonomischen Societät, nebst Auszüge aus den bey derselben eingelaufenen halfjahrigen Nachrichten. - Heft 1-62, 1764-1815. Dresden: Walther, 1764-1815. Imprint and subtitle vary. 1791-1797 as Neue Sammlung. **OTHER TITLE:** Neue Sammlung vermischte ökonomische Schriften. **CONTINUED BY:** Schriften der ökonomische Societät zu Leipizg. **SOURCES:** K2878 **LOCATIONS:** CCF, ULS, ZDB **SUBJECT:** Economics - Societies - Leipzig.

198 Apologie des Brownschen Systems der Heilkunde auf Vernunft und Erfahrungen gegründet. - Bd.1-2, 1795-1800. Vienna: 1796-1800. Contributions by authors on the Brunonian doctrine and reviews of relevant works. **EDITOR:** Carl Werner. **SOURCES:** G258,OCLC **SUBJECT:** Medicine - Periodicals - Austria. **SUBJECT:** Brunonian doctrine - Periodicals - Austria.

199 Archaeologia, or, miscellaneous tracts relating to antiquity / Society of antiquaries of London. v.1-50, 1770-1887, ser 2, 1- 1888- . London: The Society, 1700-. v.51-68 also called ser, v.1-v.18. **OTHER TITLE:** Miscellaneous tracts relating to antiquity. **INDEXES:** v.1-v.15, 1809, v.17-v.30, 1844. **MICROFORMS:** DA(1), LOC(1),UM(1),IDC(8). **SOURCES:** B543, OCLC **LOCATIONS:** BUC, CCF, CCN, ULS, ZDB **SUBJECT:** Archeology - Societies - London.

200 Archaeologia Scotica, or, Transactions of the Society of Antiquaries of Scotland. v.1-v.5, 1792-1890. Edinburgh: William Creech, 1792-1890. There were several editions of some of the volumes. **OTHER TITLE:** Transactions of the Society of Antiquaries of Scotland. **CONTINUED BY:** Proceedings, Society of Antiquaries of Scotland. **INDEXES:** Indexes to v.1-v.3 in v.3. **SOURCES:** B546, OCLC, S128a **LOCATIONS:** BUC, CCF, ULS, ZDB **SUBJECT:** Archeology - Societies - Edinburgh.

201 Archiv der Ärzte und Seelsorger wider die Pockennoth. Stück 1-Stück 7, 1796-1799. Leipzig: Weygand, 1796-1799. **EDITOR:** Johann Christian Wilhelm Juncker. **SOURCES:** G221, K3699 **LOCATIONS:** BUC, ULS, ZDB **SUBJECT:** Smallpox - Periodicals - Germany. **SUBJECT:** Medicine - Periodicals - Germany.

202 Archiv der Insectengeschichte. - Heft.1-8, 1781-1786. Zürich; Winterthur: H. Steiner, 1781-1786. **TRANSLATED AS:** Archives de l'histoire des insectes. **TRANSLATED AS:** Archives of entomology. Sold in sheets and plates. It was planned to be issued in 10 Hefte, but discontinued after 8. The pages are not numbered (Ref.8). **EDITORS:** Johann Caspar Fuessli and Johann Friedrich Wilhelm Herbst. **MICROFORMS:** RM(3),IDC(8). **SOURCES:** B571, 577, OCLC, S2194 **LOCATIONS:** BUC, CCF, ULS, ZDB **SUBJECT:** Entomology - Periodicals - Switzerland.

203   Archiv der interessanten und nützlichen Aufsätze für Landwirthe und Haushaltungen. Bd.1 (St.1-St.6) 1794-1795. Leipzig: 1794-1795. **OTHER TITLE:** Archiv für Landwirthschaft und Haushaltungen. **EDITOR:** Georg Heinrich Piependring. **SOURCES:** K2995   **SUBJECT:** Agriculture - Periodicals - Germany. **SUBJECT:** Home economics - Periodicals - Germany.

204   Archiv der medicinischen Polizey und der gemeinnützigen Arzneykunde. - Bd.1-6, 1783-1787. Leipzig: Weygand, 1783-1787. **EDITOR:** Johann Christian Friedrich Scherf. **CONTINUED BY:** Beiträge zum Archiv der medicinischen Polizei. **CONTINUED BY:** Allgemeines Archiv der Gesundheitspolizei. **INDEXES:** Cumulative subject index for v.1-2 in v.3. **SOURCES:** G58, K3599, OCLC   **LOCATIONS:** BUC, CCF, CCN, ULS, ZDB   **SUBJECT:** Medical jurisprudence - Periodicals - Germany.

205   Archiv der practischen Heilkunde für Schlesien und Südpreussen. Bd.1-Bd.3, Bd.4, St.1, 1799-1804.   Breslau: Hirschberg und Lissa, 1799-1804. **EDITORS:** Abraham Zadig (August Theodor Zanth) and Friedrich Gotthelf Friese. **SOURCES:** G164,K3723   **LOCATIONS:** BUC, CCF, ULS, ZDB   **SUBJECT:** Medicine - Periodicals - Germany.

206   Archiv der practischen Arzneykunst für Aerzte, Wundärzte und Apotheker. - Bd.1-Bd.3, 1785-1787. Leipzig: Weygand, 1785-1787. **EDITOR:** Philipp Friedrich Theodor Meckel. **CONTINUED BY:** Neues Archiv der praktischen Arzneykunst. **SOURCES:** G65, K3614   **LOCATIONS:** BUC, CCF, CCN, ULS, ZDB   **SUBJECT:** Medicine - Periodicals - Germany. **SUBJECT:** Surgery - Periodicals - Germany. **SUBJECT:** Pharmacy - Periodicals - Germany.

207   Archiv der reinen und angewandten Mathematik. - Bd.1-Bd.3 (Hefte 1-11), 1794-1800. Leipzig: Schäfer, 1795-1800. **CONTINUATION OF:** Leipziger Magazin für reine und angewandte Mathematik. The 11th issue announced that the periodical would cease with the 12th issue, but it does not seem to have appeared. **EDITORS:** Johann Bernouilli and Carl Friedrich Hindenburg. **MICROFORMS:** RM(3). **SOURCES:** B574, K3360,S 2891   **LOCATIONS:** BUC, CCF, CCN, ULS, ZDB   **SUBJECT:** Mathematics - Periodicals - Germany.

208   Archiv der Schwarmärmerei und Aufklärung. Bd.1-Bd.3 (6 issues each), Bd.4 (2 issues), 1787-1791. Altona und Hamburg: Matthiesen, 1787-1791. **EDITOR:** Friedrich Wilhelm von Schütz. **CONTINUED BY:** Neues Archiv der Schwärmerei und Aufklärung. **SOURCES:** K4562   **LOCATIONS:** ZDB   **SUBJECT:** General - Periodicals - Germany.

209   Archiv der Verhandlungen einer Gesellschaft von Aerzten zur Gründung einer durchaus zweckmässigen Volksarneikunde für Ärzte. Bd.1, St.1, 1796. Neustrelitz: Albanus, 1796. **EDITOR:** Adolf Friedrich Nolde. **SOURCES:** G130, K3700   **SUBJECT:** Medicine - Periodicals - Germany.

210 Archiv für Magnetismus und Somnambulismus. - Bd.1-8, 1787-1788. Strassburg: Academische Buchhandlung, 1787-1788. EDITOR: Johann Lorenz Böckmann. SOURCES: B592, G234, K3638 LOCATIONS: CCF, ULS, ZDB SUBJECT: Animal magnetism - Periodicals - Germany.

211 Archiv für den practischen Arzt. St. 1-St.3, 1794-1796. Marburg: Krieger, 1794-1796. SOURCES: G121, K3687 LOCATIONS: ZDB SUBJECT: Medicine - Periodicals - Germany.

212 Archiv für ältere und neuere vorzüglich deutsche Geschichte, Staatsklugheit und Erdkunde. Bd.1-Bd.2, 1790-1792. Memmingen: Seyler, 1790-1792. EDITOR: Ernst Ludwig Posselt. SOURCES: B579, K1227, S3089 LOCATIONS: BUC, CCF, ULS, ZDB SUBJECT: Geography - Periodicals - Germany.

213 Archiv für die allgemeine Heilkunde. Bd.1-Bd.2, 1790-1792. Berlin: C.F. Himburg, 1790-1792. EDITOR: August Friedrich Hecker. CONTINUED BY: Neues Archiv für die allgemeine Heilkunde. SOURCES: G91, K3661 LOCATIONS: CCF, ULS, ZDB SUBJECT: Medicine - Periodicals - Germany.

214 Archiv für die Botanik. - Bd.1-3, 1796-1805. Leipzig: Kühn, 1796-1805. Publisher varies. EDITOR: Johann Jacob Römer. MICROFORMS: IDC(8). SOURCES: B602, G553, K3359, OCLC, S2889 LOCATIONS: BUC, CCF, CCN, ULS, ZDB SUBJECT: Botany - Periodicals - Germany.

215 Archiv für die Geburtshülfe, Frauenzimmer und neugebohrner Kinder-Krankheiten. Bd.1-Bd.6, 1787-1797. Jena: Akademische Buchhandlung, 1787-1797. Two isssues a year. EDITOR: Johann Christian Stark. CONTINUED BY: Neues Archiv für die Geburtshülfe. INDEXES: Subject index at the end of the year. SOURCES: G76, K3633, OCLC LOCATIONS: CCF, CCN, ULS, ZDB SUBJECT: Obstetrics - Periodicals - Germany. SUBJECT: Pediatrics - Periodicals - Germany.

216 Archiv für die Geschichte der Arzneykunde in ihrem ganzen Umfang. Bd.1, St.1, 1790. Nürnberg: Ernst Grattenauer, 1790. EDITOR: Philipp Ludwig Wittwer. SOURCES: G86, K3660 LOCATIONS: ULS, ZDB SUBJECT: Medicine - Periodicals - Germany.

217 Archiv für die Physiologie. Bd.1- Bd.12, 1796-1815. Halle: Kurt, 1796-1815. Three issues to a volume. The issues are paged separately in the first volume, but consecutively thereafter. EDITORS: Johann Christian Reil, 1795-1805; Johann Heinrich Ferdinand Autenrieth, 1806-1815. CONTINUED BY: Deutsches Archiv für die Physiologie. CONTINUED BY: Archiv für Anatomie und Physiologie. MICROFORMS: RM(3). SOURCES: B611, G133 K3693, OCLC, S2705 LOCATIONS: BUC, CCF, CCN, ULS, ZDB SUBJECT: Physiology - Periodicals - Germany.

217a Archiv für die theoretische Chemie - Heft 1-Heft 4, 1800-1802. Jena: J. G. Voigt, 1800-1802. Imprint varies. EDITOR: Alexander Nicolaus Scherer.

SOURCES: B614, K3379, S2810.   LOCATIONS: ZDB   SUBJECT:
Chemistry - Periodicals - Germany.

218   Archiv für die thierische Chemie. - Heft 1-2, 1800-1801. Halle; Stemmerde:
      Schwetschke, 1800-1801. EDITOR: Johann Horkel. SOURCES: B615,
      K3380, S2706 LOCATIONS: BUC, ULS, ZDB SUBJECT: Biological
      chemistry - Periodicals - Germany. SUBJECT: Chemistry - Periodicals -
      Germany.

219   Archiv für Freymäurer und Rosenkreuzer. Bd.1-Bd.2, 1783-1785. Berlin: August
      Mylius, 1783-1785. EDITOR: Conrad Friedrich Uden. SOURCES: B587,
      GXX, K6595, OCLC LOCATIONS: CCF, ULS, ZDB SUBJECT: Alchemy
      - Periodicals - Germany.

220   Archiv für medicinische Länderkunde. St.1-St.2, 1800-1801. Coburg und Leipzig:
      J.C.D.Sinner, 1800-1801. EDITOR: Winckler. SOURCES: G169, K3734
      LOCATIONS: ZDB SUBJECT: Medicine - Periodicals - Germany.

221   Archiv für Rossärzte und Pferdeliebhaber. - Bd.1-4, 1788-1796. Marburg:
      Academische Buchhandlung, 1788-1796. EDITORS: Johann David Busch and
      Heinrich Daum. INDEXES: There is a modern index (Ref.7).   SOURCES:
      G227, K3644 LOCATIONS: CCN, ULS, ZDB SUBJECT: Veterinary
      medicine - Periodicals - Germany.

222   Archiv für Zoologie und Zootomie. Bd.1-Bd.5 (2 issues each), 1800-1806. Berlin
      und Braunschweig: Karl Reichard, 1800-1806. EDITOR: Christian Rudolf
      Wilhelm Wiedemann. CONTINUED BY: Neues Archiv für Zoologie und
      Zootomie. SOURCES: B599, G570, K3381, OCLC, S2281 LOCATIONS:
      BUC, CCF, CCN, ULS, ZDB SUBJECT: Zoology - Periodicals - Germany.

223   Archiv gemeinnützger physischer und medicinischer Kenntnisse. Bd.1-Bd.6,
      1787-1791. Zürich: J. C. Füessli, 1787-1791. CONTINUATION OF: Gazette
      de Santé, oder, gemeinnütziges medizinisches Magazin für Leser aus allen
      Ständen. EDITOR: Johann Heinrich Rahn. CONTINUED BY: Magazin für
      gemeinnnützige Arzneikunde und medicinische Polizei. SOURCES: B616,
      G73, K3634, S2193 LOCATIONS: CCF, CCN, ULS, ZDB   SUBJECT:
      Medicine - Periodicals - Switzerland.

223a Archiv kleiner zerstreuter Reisebeschreibungen der merckwürdigsten Gegenden
      der Sweiz. v.1-2, 1796-1802. St. Gallen: 1796-1802. EDITOR: Georg
      Leonhard Hartmann. SOURCES: B617, S2176. SUBJECT: Geography -
      Periodicals - Switzerland.

224   Archiv nützlicher Erfindungen und wichtiger Entdeckungen in Künsten und
      Wissenschaften. Leipzig: 1792. EDITOR: H. E. Vollbeding. SOURCES:
      K3914 LOCATIONS: ZDB SUBJECT: Science - Periodicals - Germany -
      Popular.

225 Archiv zur neuern Geschichte, Geographie, Natur- und Menschenkenntnisse. Bd.1-Bd.8, 1785-1788. Leipzig: Georg Emanuel Beer, 1785-1788. EDITOR: Johann Bernoulli. SOURCES: B621, K1160, S2890 LOCATIONS: BUC, CCF, ZDB SUBJECT: Geography - Periodicals - Germany.

226 Archives de l'histoire des insectes, publiées en allemand, traduites en français avec des augmentations et corrections du traducteur. - 1 v.1794. Winterthur: Ziegler, 1794. EDITOR: Johann Caspar Fuessli. TRANSLATION OF: Archiv der Insectengeschichte. MICROFORMS: RM(3). SOURCES: B637, G577a, OCLC, S2186 LOCATIONS: BUC, CCF, ULS, ZDB SUBJECT: Entomology - Periodicals - France.

227 Archives mythohermétiques, ouvrages périodiques. - No. 1-2, 1779. Paris: 1779. TRANSLATED AS: Mythohermtisches Archiv. EDITOR: Clavier Duplessis. LOCATIONS: BUC, CCF SUBJECT: Alchemy - Periodicals - France. SUBJECT: Astrology - Periodicals - France.

228 Archives of entomology, containing the history, or ascertaining the characters and classes of insects not hitherto described, imperfectly known, or erroneously classified. - 1 v.1795. London: J. Johnson, 1795. Translated from the German of J.C. Fuessli; with notes, and the original plates, fifty-one in number, coloured, to which is added the French translation. Some plates unnumbered; 37-42 omitted in numbering. EDITOR: Johann Caspar Fuessli. TRANSLATION OF: Archiv der Insectengeschichte. MICROFORMS: RM(3). SOURCES: B642, G577b, OCLC, S204a LOCATIONS: BUC, ULS SUBJECT: Entomology - Periodicals - England.

229 Arithmetisches Wochenblatt. Qu.1- Qu.4, 1796. Berlin: Felisch, 1796. SOURCES: K3367 SUBJECT: Mathematics - Periodicals - Germany.

230 Artis medicae principes, Hippocrates, Aretaeus, Alexander, Aurelianus, Celsus, Rhazeus. Recensuit, praefatus est Albertus de Haller. v.1-v.11, 1769-1774. Lausanne: Grasset, 1769-1774. Issued under Albrecht von Haller's direction. A 2d ed. was issued in Lausanne, 1784-1787 in 11 v. OTHER TITLE: Principes artis medicae. EDITOR: Philippe Rodolphe Vicat. SOURCES: G319 LOCATIONS: NUC SUBJECT: Medicine - Periodicals - Switzerland.

231 De Artz of Geneesheer, en aangename spectatorialle vertogen op eene klaare en einvoudige wyze leerende, wat men moet doen, om gezond, lang, en gelukkig. v.1-v.12, 1765-1771. Amsterdam: Kornelis von Tongerlo, 1765-1771. A 2d ed. was published in Amsterdam in 6 v.1767-1770. TRANSLATION OF: Arzt der Frauenzimmer. CONTINUED BY: Naalezing van den Artz of Geneesheer. SOURCES: G21 LOCATIONS: BUC, CCF, CCN, ULS SUBJECT: Medicine - Periodicals - Netherlands - Popular - Translations.

232 Arzneikunde Annalen. Heft 1-Heft 13, 1787-1792. Copenhagen: Johann Gottlob Rothe, 1787-1792. CONTINUATION OF: Medicinisch-chirurgische Bibliothek. EDITOR: Johann Clemens Tode. CONTINUED BY:

Medicinisch-chirurgisches Journal. **SOURCES: G74, K3632 LOCATIONS:**
BUC, CCF, ULS SUBJECT: Medicine - Periodicals - Denmark.
SUBJECT: Surgery - Periodicals - Denmark.

233 Arzneiwissenschaftliche Aufsätze böhmischer Gelehrten. Prague and Dresden:
Walther, 1798. **EDITOR:** Johann Dionis John. **SOURCES: G320, K3717**
SUBJECT: Medicine - Periodicals - Czechoslovakia.

234 Arzneyen, eine physikalisch-medicinische Monatsschrift, zum Unterricht alle
denen, welche den Schaden des Quacksalbers nicht kennen. Bd.1-Bd.2,
1766-1767. Langensalza: Johann Christian Martini, 1766-1767. **EDITOR:**
Ernest Gottfried Baldinger. **CONTINUED BY:** Neue Arzneyen wider die
medicinischen Vorurtheile. **SOURCES: G20, K3546 LOCATIONS:** CCF,
ULS **SUBJECT:** Medicine - Periodicals - Germany - Popular.

235 Arzneyen wider physikalische, ökonomische und diätetische Vorurtheile. Jg.1-Jg.
2(each with 12 issues) 1774-1776. Heilbronn: Allinger, 1774-1776. **EDITOR:**
Friedrich August Weber. **SOURCES: K5069 SUBJECT:** Medicine -
Periodicals - Germany - Popular. **SUBJECT:** Nutrition - Periodicals - Germany
- Popular.

236 Arzneykundige Abhandlungen, hrsg. von dem Collegio der Aerzte in London.
Bd.1-Bd.3, 1768-1787. Leipzig: Fritsch, 1768-1787. A translation by Carl
Christian Krause of the Medical transactions. **TRANSLATION OF:** Medical
transactions published by the Royal College of Physicians of London.
SOURCES: G26a, K3549 LOCATIONS: CCF, ZDB **SUBJECT:** Medicine
- Societies - London - Translations.

237 Der Arzt. Heft 1-Heft 4, 1795-1796. Eisenach und Halle: 1795-1796. Forms
Abtheilung 7 of the Compendiöse Bibliothek. **OTHER TITLE:** Compendiöse
Bibliothek der gemeinnützigen Kenntnisse für alle Stände. **EDITOR:** Christian
Carl André. **SOURCES: G126 SUBJECT:** Medicine - Periodicals -
Germany.

238 Der Arzt der Frauenzimmer. Bd.1-Bd.3, 1771-1773. Leipizg: J. G. Müller,
1771-1773. **TRANSLATED AS:** Arzt, of genees-heer voor vrouwelyk geslacht.
EDITOR: Johann Georg Friedrich Franz. **MICROFORMS:** RP(1).
SOURCES: G27, OCLC LOCATIONS: BUC, ULS, ZDB **SUBJECT:**
Gynecology - Periodicals - Germany. **SUBJECT:** Women - Periodicals -
Germany.

239 Der Arzt, eine medicinische Wochenschrift. T.1-t.12, 1759-1764. Hamburg:
G.C. Grund, 1759-1764. Weekly. **TRANSLATED AS:** Laeger et medicinst
ugeskrift. It was translated into Swedish in Stockholm by Pfeifer and also into
Dutch and Danish. There were several editions of some of the earlier volumes,
as well as a new edition in 6 v. in Hamburg in 1769. It also appeared in a
reformatted version under the title: Johann Unzer's medicinisches Handbuch
nach den Grundsätzen seiner medicnischen Wochenscrift Der Arzt, in Lüneberg
and Hamburg: Gotthelf Christian Berth, 1770. **EDITOR:** Johann August

Unzer. **INDEXES:** There is an index to the first 12 v.of the 1st ed in v.12, and in v.6 of the 1769 ed. **SOURCES:** GIII:13, K3540, OCLC **LOCATIONS:** BUC, CCF, ULS, ZDB **SUBJECT:** Medicine - Periodicals - Germany - Popular.

240 De Arzt, of genees-heer voor vrouwelyk geslacht in't gemeen in spectatoriale vertoogen. Deel 1. stuk 1, 1772. The Hague: Pieter van Cleef, 1772. **TRANSLATION OF:** Arzt der Frauenzimmer. **SOURCES:** OCLC **LOCATIONS:** BUC **SUBJECT:** Gynecology - Periodicals - Germany - Translations. **SUBJECT:** Women - Periodicals - Germany - Translations.

241 The Asiatic miscellany, consisting of translations, citations, fugitive pieces, original productions and extracts from curious. v.1-v.2, 1785-1787. Calcutta: Daniel Stuart, 1785-1787. Reprinted in London for J. Wallis in 1787 and in Dublin in 1793. Reprinted for J. Murray under the title: Dissertations and miscellaneous pieces relating to the history and antiquities, arts, sciences and literature of Asia, in 2 v. which also included material from the Asiatic Researches. **OTHER TITLE:** Dissertations and miscellaneous pieces. **MICROFORMS:** IDC(8),RP(1). **SOURCES:** B1476, G635, OCLC, S257a **LOCATIONS:** BUC,ULS **SUBJECT:** General - Periodicals - India.

242 Asiatic researches, or, transactions of the Society instituted in Bengal for inquiring into the history and antiquities, the arts, sciences and literature of Asia. - v.1-20, 1788-1839. Calcutta: Manuel Cantopher, 1788-1839. **TRANSLATED AS:** Abhandlungen über die Geschichte und Alterthümer, die Künste, Wissenschaften und Literatur Asiens. **TRANSLATED AS:** Recherches asiatiques, ou mémoires de la Société établie à Bengale pour faire des recherches sur l'histoire et les antiquitées, les arts, les sciences et la litérature de l'Asie. There were reprints in London in quarto and octavo in 12 v.1799-1817. **INDEXES:** Index to the publications of the Asiatic Society, 1788-1953. Compiled by Sibdas Chaudin, published as a Supplement to the Journal of the Asiatic Society, Ser. 3, v.22, 1956. **MICROFORMS:** IDC(8). **SOURCES:** G458,S3790 **LOCATIONS:** BUC, CCF, CCN, ULS, ZDB **SUBJECT:** General - Societies - Calcutta.

243 Assemblée publique de la Société royale des sciences, Montpellier. Montpellier: Jean Martel, 1706-1788. Nos.1-15 issued irregularly 1706-1749, nos.16-28 annually 1772-1784, no.28 in 1788. Others may also have been issued (Ref.43:316-9) **SOURCES:** G514, OCLC, S1241a **LOCATIONS:** BUC, CCF, ULS, ZDB **SUBJECT:** Science - Societies - Montpellier.

244 Astrologer's magazine and philosophical miscellany, consisting of an easy introduction to the celestial science of astrology...chymical secrets, medical prescriptions, extracts in interesting subjects from foreign and domestic philosophical transactions. v.1, August 1793-February 1794. London: W. Locke, 1793-1794. **CONTINUATION OF:** Conjuror's magazine and physiolognomical mirror. **SOURCES:** B5434 **LOCATIONS:** BUC, ULS **SUBJECT:** Alchemy - Periodicals - England. **SUBJECT:** Astrology - Periodicals - England. **SUBJECT:** Medicine - Periodicals - England.

245   Astronomical observations made at the Royal observatory at Greenwich. Ser.1:
      v.1-2; ser2: v.1-4; ser.3:v.1-5; ser.4: v.1-12; ser.5: v.1-3; (1750/62- 1838)
      1750-1924. London and Oxford: 1750-1924. The first series contains the
      observations of James Bradley and were published privately as were the
      observations made earlier by Flamsteed and Halley. OTHER TITLE:
      Astronomical and magnetical and meteorological observations. SOURCES:
      B695, OCLC, S401b   LOCATIONS: BUC, CCF, CCN, ULS, ZDB
      SUBJECT: Astronomy - Societies - London.   SUBJECT: Meteorology -
      Societies - London.

246   Astronomisches Jahrbuch, oder, Ephemeriden für das Jahr nebst einer Sammlung
      der neuesten in die astronomischen Wissenschaften einschlagenden
      Beobachtungen, Nachrichten, Bemerkungen und Abhandlungen. - Bd.1-49,
      (1776-1824) 1774-1821. Berlin: Georg Jacob Decker, 1776-1829. Annual.
      Contains astronomical articles as well as tabular meteorological and
      astronomical data projected 2-3 years ahead. EDITORS: J.H. Lambert and
      Johann Elert Bode. CONTINUED BY: Berliner astronomischer Jahrbuch.
      Sammlung astronomischen Abhandlungen, Beobacthungen und Nachrichten.
      INDEXES: Namen-und Sach-register der Berliner astronomischer Jahrbucher
      von 1776 bis 1829. Berlin, Dummler, 1829 (Ref.126). SOURCES: B716,
      S2370b   LOCATIONS: BUC, CCF, CCN, ULS, ZDB   SUBJECT:
      Astronomy - Societies - Berlin.   SUBJECT: Meteorology - Societies - Berlin.

247   Asuntos various sobre ciencias y artes, obra periodica al Rey N. Sr. No.1-no.12,
      October 1772-January 1773. Mexico: Imprenta de la Biblioteca Mexicana, 1772.
      Weekly.   EDITOR: José Antonio de Alzate y Ramirez (Ref.44).
      LOCATIONS: ULS SUBJECT: Science - Periodicals - Mexico.

248   Atti della reale accademia delle scienze e belle lettere di Napoli. v.1 (1778-1787)
      1788. Naples: 1788. SOURCES: G492, S1933 LOCATIONS: BUC, CCF,
      CCN, ULS SUBJECT: Science - Societies - Naples.

249   Atti della reale accademia di scienze, lettere e belle arte de Palermo. v.1-v.2,
      1775-1780. Palermo: 1775-1780. SOURCES: G495 LOCATIONS: ULS
      SUBJECT: General - Societies - Palermo.

250   Atti della reale società economica di Firenze ossia de georgofili. v.1-v.8,
      1791-1817. Florence: Accademia dei georgofili, 1792-1817. LOCATIONS:
      BUC, CCF, ULS SUBJECT: Agriculture - Societies - Florence. SUBJECT:
      Economics - Societies - Florence.

251   Atti della società patriotica di Milano, diretta all'avanzamento dell'agricoltura,
      della arte e delle manufatture. v.1-v.3, (1778-1783) 1783-1793.   Milan:
      1783-1793.   SOURCES: S1916   LOCATIONS: BUC, CCF, ULS
      SUBJECT: Economics - Societies - Milan. SUBJECT: Agriculture - Societies
      - Milan.

252 Atti delle reale accademia de' fisiocritici, Siena. v.1-v.10, 1760-1841. Siena: 1760-1841. Publishers vary. SOURCES: OCLC, S2050a LOCATIONS: BUC, CCF, CCN, ULS, ZDB SUBJECT: Science - Societies - Siena.

253 Atti e memorie inediti dell'Accademia del cimento e notizie aneddoti dei progressi delle scienze in Toscana. v.1-v.3, 1780. Florence: Guiseppe Tofani, 1780. CONTINUATION OF: Saggi di naturali esperienze fatte nell Accademia del cimento. Published in the same year under the title: Notizie degli aggrandimento... by Guiseppe Bouchard in Florence. OTHER TITLE: Notizie degli aggrandimento delle scienze fisiche accaduti in Toscano. EDITOR: Giovanni Targioni-Tozetti. MICROFORMS:RM(3). SOURCES: GII:5c, OCLC, S1821a LOCATIONS: BUC, ULS SUBJECT: Science - Societies - Florence.

254 Aufklärung der Arzneiwissenschaft. v.1 (St.1-St.3), 1793. Weimar: Industrie Comptoir, 1793. Translated by Christoph Wilhelm Hufeland and Johann Friedrich August Göttling. TRANSLATION OF: Médecine éclairée par les sciences physiques. SOURCES: G98a, K3681 LOCATIONS: ULS SUBJECT: Medicine - Periodicals - France - Translations.

255 Der aufrichtige Medikus. No.1-no.17, 1726-1727. Nürnberg, 1726-1727. SOURCES: K3519 LOCATIONS: ZDB SUBJECT: Medicine - Periodicals - Germany.

256 Aufsätze und Beobachtungen aus der gerichtlichen Arzneywissenschaft. Bd.1-Bd.8, 1783-1793. Berlin: 1783-1793. A 2d ed. was published in Berlin: Mylius, 1803-1815, in 3 v. EDITOR: Johann Theodor Pyl. SOURCES: G321, K3600 LOCATIONS: ULS, ZDB SUBJECT: Medical jurisprudence - Periodicals - Germany.

257 Der aus dem Reiche der Wissenschaften wohlversuchte Referendarius, oder, auserlesene Sammlungen von allerhand wichtiger Abhandlungen, Schriften un Versuchen aus der Naturlehre, Arzneiwissenschaft, natürlichen Theologie und Rechtsgelehrsamkeit. Bd.1-Bd.6 (Th. 1-12), 1750-1769. Augusburg: 1750-1769. EDITOR: Johann Andreas Erdmann Maschenbauer. TRANSLATION OF: Handlingar, Kungliga Svenska Vetenskapsakademien. SOURCES: B736, S2251 LOCATIONS: ZDB SUBJECT: Science - Periodicals - Germany - Popular.

258 Auserlesene Abhandlungen der Naturgeschichte, Physik und Arzneywissenschaft, aus dem Lateinischen übersetzt von Höpfner. - Bd.1-Bd.3, 1776-1778. Leipzig: 1776-1778. SOURCES: B745, G601, S2894 SUBJECT: Science - Periodicals - Germany - Translations.

259 Auserlesene Abhandlungen für Aerzte, Naturforscher und Psychologen aus den Schriften der literarisch-philosophischen Gesellschaft zu Manchester. Bd.1, 1795. Leipizg: 1795. Translated and annotated by August Wilhelm Schwenger. TRANSLATION OF: Memoirs of the literary and philosophical society of

Manchester.   **SOURCES:** B467, G480b, S2964a   **LOCATIONS:** ZDB
SUBJECT: Science - Societies - Manchester - Translations.

260 Auserlesene Abhandlungen praktischen und chirurgischen Inhalts aus den
     Philosophischen Transactionen. Bd.1-Bd.5, (1699-1765) 1774-1778. Lübeck und
     Leipizg: 1774-1778. A translation by Nathanael Gottfried Leske of selected
     surgical items from the Lowthorp abridgement of the Philosophical transactions
     of the Royal Society of London. **TRANSLATION OF:** Philosophical
     transactions. **SOURCES:** GII:41, K3563, OCLC   **LOCATIONS:** ZDB
     SUBJECT: Science - Societies - London - Translations. **SUBJECT:** Surgery
     - Periodicals - Germany.

261 Auserlesene Abhandlungen über Gegenstände der Policey, des Finanzen und
     Oekonomie gezogen aus der Hannoverischen Magazin. - v.1-3, 1786-1788.
     Hannover: Helwig, 1786-1788. **CONTINUATION OF:** Hannoverisches
     Magazin.   **LOCATIONS:** ZDB   **SUBJECT:** Economics - Periodicals -
     Germany.

262 Auserlesene Abhandlungen, welche der Kaiserlichen Akademie der Wissenschaften
     zu Paris von einigen Gelehrten eingesendet, in ihren Versammlungen abgelesen
     und von ihr herausgegeben worden. - 2 parts in 1 v.1752-1754.   Leipizg:
     Gleditsch, 1752-1754. Translated from the Mémoires de mathématiques et de
     physique, Académie royale des sciences, Paris (1750ff) by Ferdinand Wilhelm
     Beer. **TRANSLATION OF:** Mémoires de mathématiques et de physique.
     SOURCES: B747, GII:17m, K3211, S2881   **SUBJECT:** Science - Societies
     - Paris - Translations.

263 Auserlesene Beobachtungen der wetteifernden medicinischen Gesellschaft in Paris.
     - Bd.1-Bd.3, 1802-1803.   Leipzig: Johann Ambroise Barth, 1802-1803.
     TRANSLATION OF: Mémoires de la société médical d'émulation.
     SOURCES: G145a **LOCATIONS:** ZDB **SUBJECT:** Medicine - Societies -
     Paris - Translations.

264 Auserlesene Beyträge zur Thierarzneykunst. Bd.1-Bd.4, 1786-1788. Leipzig:
     Weidmann und Reich, 1786-1788.   **OTHER TITLE:** Beiträge zur
     Thierarzneykunst. **EDITOR:** Christian Friedrich Ludwig. **INDEXES:** Indexed
     in Wimmel (Ref.7). **SOURCES:** G226, K3626   **LOCATIONS:** ZDB
     SUBJECT: Veterinary medicine - Periodicals - Germany.

265 Auserlesene chirurgische Disputationen von Albrecht von Haller in einen Auszug.
     Bd.1-Bd.5, 1777-1787. Leipzig: Christian G. Hertel, 1777-1787. Translated
     with annotations by Friedrich August Weiz (Waitz). **EDITOR:** Albrecht von
     Haller.   **TRANSLATION OF:** Disputationes chirurgicae selectae.
     SOURCES: G352b   **SUBJECT:** Surgery - Periodicals - Switzerland -
     Dissertations - Translations.

266 Auserlesene chirurgische Wahrnehmungen. - Bd.1-Bd.12, 1791-1806. Frankfurt
     A.M.: 1791-1806. Translated by F. Dörner. **TRANSLATION OF:** Journal

de chirurgie. **SOURCES:** G97a   **SUBJECT:** Surgery - Periodicals - France
- Translations.

267   Auserlesene medicinisch-chirurgisch-anatomisch-chymische und botanische
Abhandlungen / Römisch-Kaiserlichen Akademie der Naturforscher. - Bd.1-20,
1755-1771. (Ref.21 v.23:953g), Nürnberg: Stein, 1755-1771. Publisher varies.
TRANSLATION OF: Miscellanea curiosa, Academia naturae curiosorum.
MICROFORMS: RP(1).   **SOURCES:** B749, GII:7a, K3219, S3148
LOCATIONS: BL, ULS, ZDB   **SUBJECT:** Science - Societies - Halle -
Translations.

268   Auserlesene ökonomische Bibliothek. - Bd.1-Bd.2, 1799-1800. Vienna: Döll,
1799-1800. **EDITOR:** Christian Friedrich Berger.   **SOURCES:** K3023
SUBJECT: Economics - Periodicals - Austria.

269   Auserlesene Sammlung vermischter ökonomischer Schriften, oder, neue Zugaben
zur practisch-ökonomischen Encyclopädie von Johann Riem. - Bd.1-Bd.2,
1790-1792. Dresden: Gaerlach, 1790-1792.   **EDITOR:** Johann Riem.
CONTINUED BY: Neue Sammlung vermischter ökonomischer Schriften.
CONTINUED BY: Neufortgesetzte Sammlung ökonomischer und
Bienenschriften. Two Beilage issued in 1791.   **SOURCES:** K2976
LOCATIONS: BUC, NUC   **SUBJECT:** Economics - Periodicals - Germany.

270   Auserlesene Sammlung zum vortheil der Staats-wirthschaft, der Naturforschung,
des Feldbaues, mit Beifall der löbl. ökonomischen Gesellschaft in Bern.
Bd.1-Bd.2, 1763-1769. Basel: Emanuel Thurneysen, 1763-1769. Translated
from the proceedings of the Svenska Vetenskapsakademien, Stockholm, by
Gottlieb Sigmund Gruner with a preface by Albrecht von Haller. **SOURCES:**
B750, K2877, S2121   **LOCATIONS:** BL   **SUBJECT:** Science - Societies -
Stockholm - Translations.

271   Auserlesene Sammlung zur Geschichte und Ausübung des Blasenschnitts gehöriger
Abhandlungen. - Bd.1, 1784. Leipzig: Weygand, 1784. A collection of
translations of works on lithotomy. **SOURCES:** G322   **SUBJECT:** Medicine
- Periodicals - Germany - Translations.   **SUBJECT:** Lithotomy - Periodicals
- Germany - Translations.

272   Ausgesuchte Beiträge für die Entbindungskunst. St.1-St.2, 1789. Leipizg:
Weygand, 1789. **CONTINUATION OF:** Sammlung auserlesensten und
neuesten Abhandlungen für Wundärzte. Articles on obstetrics extracted from:
Sammlung auserlesensten und neuesten Abhandlungen für Wundärzte
(1778-1781). **SOURCES:** G323, K3651 **LOCATIONS:** ZDB **SUBJECT:**
Obstetrics - Periodicals - Germany.

273   Auswahl aller eigenthümlicher Abhandlungen und Beobachtungen aus den neuesten
Enteckungen in der Chemie mit einigen Verbesserungen und Zusätzen. -
Bd.1-Bd.4, 1786-1787. Leipizg: Weygand, 1786-1787. **CONTINUATION
OF:** Chemische Journal für die Freunde der Naturlehre. **CONTINUATION
OF:** Neuesten Entdeckungen in der Chemie. Contains Crell's original

contributions to the Entdeckungen with corrections and annotations. Bd.4 consists of papers not previously printed. **EDITOR:** Lorenz Florenz Friedrich von Crell. **INDEXES:** Indexed by Engelhardt (Ref.51). **SOURCES:** B756, S2896 **LOCATIONS:** CCN, ULS, ZDB **SUBJECT:** Chemistry - Periodicals - Germany.

274   Auswahl der besten Aufsätze und Beobachtungen für Wundärzte, aus italienischen Zeitschriften. St.1-St.2, 1783. Leipizg: Weygand, 1783. **EDITOR:** Christian Gotthold Eschenbach. **SOURCES:** G324, K3601 **SUBJECT:** Surgery - Periodicals - Italy - Translations.

275   Auswahl der besten ausländischen geographischen und statistischen Nachrichten. Bd.1-Bd.4, 1784-1800. Halle: Renger, 1794-1800. **CONTINUATION OF:** Beiträge zur Völker-und Länderkunde. **CONTINUATION OF:** Neue Beiträge zur Völker-und Länderkunde. **EDITOR:** Matthias Christian Sprengel. **SOURCES:** K2765 **LOCATIONS:** CCF, ZDB **SUBJECT:** Geography - Periodicals - Germany.

276   Auswahl der medicinischen Aufsätze und Beobachtungen aus den Nürnbergischen gelehrten Unterhandlungen. Bd.1-Bd.3, 1787-1788. Halle: Francke, 1787-1788. A translation by August Friedrich Hecker of the first four years of the Commercium litterarium. **TRANSLATION OF:** Commercium litterarium ad rei medicae et scientiae naturalis incrementum institutum. **SOURCES:** G345a, K3636 **SUBJECT:** Medicine - Periodicals - Germany - Translations.

277   Auswahl der neuesten Abhandlungen und Beobachtungen auswärtiger Gelehrten über Gegenstände der Physik, Chemie und Mineralogie, aus mehreren Sprachen übersetzt und durch die neueren deutsche Entdeckungen bereichert. - Bd.1-Bd.2, 1790. Quedlinburg: Ernst, 1790. **EDITOR:** Friedrich Wolff. **SOURCES:** B759, S3187 **SUBJECT:** Science - Periodicals - Germany.

278   Auswahl ökonomischer Abhandlungen welche die freie ökonomische Gesellschaft in St. Petersburg in teutscher Sprache erhalten hat. - Bd.1-4, 1790-1793. St. Petersburg: 1790-1793. **CONTINUATION OF:** Trudy Vol'nago ekonomecheskago obtchestva. **LOCATIONS:** BUC **SUBJECT:** Economics - Societies - St. Petersburg.

279   Auswahl vorzuglicher Abhandlungen aus den sämmtlichen Banden der französischen Annalen der Chemie. - Bd.1, 1800. Helmstädt: 1800. Translated by Lorenz Florenz Friedrich von Crell. **TRANSLATION OF:** Annales de chimie. **LOCATIONS:** ULS **SUBJECT:** Chemistry - Periodicals - France - Translations.

280   Auszug aus den Transaktionen der Societät zu London zur Aufmunterung der Künste, der Manufakturen und der Handlung. - Bd.1-Bd.3, 1795-1798. Translated and edited by Johann Gottlieb Geissler. **OTHER TITLE:** Auszüge aus den Transkaten der Societät zu London. **TRANSLATION OF:** Transactions of the Society instituted at London for the encouragement of the arts, manufactures and commerce. **SOURCES:** B764, S2499 **SUBJECT:**

Technology - Societies - London - Translations. **SUBJECT:** Economics - Societies - London - Translations. **SUBJECT:** Agriculture - Societies - London - Translations.

281 Auszüge aus den besten französischen periodischen medicinischen, chirurgischen, pharmaceutischen Schriften. - Bd.1-Bd.5, 1780-1784. Leipizg: Adam Friedrich Böhme, 1780-1784. **EDITOR:** Christian Friedrich Held. **TRANSLATION OF:** Journal de médecine, chirurgie et de pharmacie. **SOURCES:** G262, K3582 **LOCATIONS:** BUC, ULS, ZDB **SUBJECT:** Medicine - Periodicals - France - Translations.

282 Auszüge aus den besten und neuesten Englischen medicinischen Streitschriften. Bd.1, 1792. Heidelberg und Leipzig: Pfähler, 1792. Translations by Heinrich Tabor primarily of dissertations of the University of Edinburgh from 1779-1789. **SOURCES:** G264 **SUBJECT:** Medicine - Periodicals - Germany - Dissertations.

283 Auszüge aus den neuesten Dissertationen über die Naturlehre, Arzneywissenschaft, Chemie und Physik gehören. Bd.1 (St.1-St.6), 1769-1772. Berlin und Stralsund: Gottlob August Lange, 1769-1772. Title varies. **EDITOR:** Ernst Gottfried Baldinger and later Johann Carl Weber. **SOURCES:** B763, K3554, S2283 **LOCATIONS:** DSG **SUBJECT:** Science - Periodicals - Germany - Dissertations.

284 Auszüge aus den neuesten medicinischen Probe-und Einladungsschriften. - Bd.1 (St.1-St.4) 1796-1797. Altona: 1796-1797. **EDITORS:** Friedrich Georg August Bouchholz and Johann Hermann Becker. **SOURCES:** G266 **LOCATIONS:** ZDB **SUBJECT:** Medicine - Periodicals - Germany - Dissertations.

285 Auszüge aus den neuesten medicinischen und chirurgischen Disputationen. Bd.1-Bd,4, 1749-1755. **LOCATIONS:** ZDB **SUBJECT:** Medicine - Periodicals - Germany - Dissertations - Translations.

286 Auszüge der besten medicinischen Probeschriften des 16ten und 17ten Jahrhunderts. Th.1-Th.4, 1771-1786. Altenburg: Richter, 1771-1786. **EDITOR:** Christoph Jakob Mellin. **CONTINUED BY:** Fortgesetzte Auszüge aus den besten alten medicinischen Probeschriften. **SOURCES:** G263 **LOCATIONS:** NUC **SUBJECT:** Medicine - Periodicals - Germany - Dissertations.

287 Auszüge medicinisch-chirurgischer Beobachtungen. - Theil 1-2, 1789-1790. Nürnberg: Johann Adam Stein, 1789-1790. A translation of medical and surgical papers by Georg Albrecht Weinrich from the Acta Physico-medica (1727-1754) of the Academia Caesarea-Leopoldinae-Carolina naturae curiosorum. **TRANSLATION OF:** Acta physico-medica, Academia Caesarea Leopoldina-Carolina naturae curiosorum. **SOURCES:** GII:7e **LOCATIONS:** ULS **SUBJECT:** Science - Societies - Halle - Translations. **SUBJECT:** Medicine - Societies - Halle - Translations. **SUBJECT:** Surgery - Societies - Halle - Translations.

288   Auszüge medicinischer Probe-und Einladungsschriften. - St.1-St.2, 1790-1791.
Schwerin und Wismar: Bodmer, 1790-1791. EDITOR: Johann Georg Reyher.
SOURCES: G265   SUBJECT: Medicine - Periodicals - Germany -
Dissertations.

289   Auszüge verscheidener arzneiwissenschaftliche Abhandlungen aus den
wöchentlichen Hallischen Anzeigen. Bd.1 (1729-56) 1788,-Bd.2 (1761-84)
1789. Halle: Renger, 1788-1789. CONTINUATION OF: Wöchentliche
Hallische Anzeigen. Selected medical articles from the Wöchentliche Hallische
Anzeigen. EDITOR: August Gottlob Weber. SOURCES: G267 SUBJECT:
Medicine - Periodicals - Germany.

289a   Avantcoureur, feuille hebdomadaire où sont annoncés les objets particuliers des
sciences et des arts. Paris: Michel Lambert, 1759-1773. CONTINUATION
OF: Feuille nécessaire. Weekly. Included some of Lavoisier's early work
(Ref.47). Publisher varies. EDITORS: Meussnier de Querlon and others.
LOCATIONS: CCF, ULS   SUBJECT: General - Periodicals - France.

290   Avanzamenti della medicina e fisica. - v.17-20, 1795-1796. Pavia: 1795-1796.
CONTINUATION OF: Biblioteca fisica d'Europa. CONTINUATION OF:
Giornale fisico-medico. EDITOR: Luigi Gaspero Brugnatelli. LOCATIONS:
CCF, ULS   SUBJECT: Science - Periodicals - Italy.

291   Avvisi sopra la salute umana. - v.1-v.10, 1776-1785. Florence: Gaetana
Cambiagi, 1776-1785. Weekly. LOCATIONS: DSG SUBJECT: Medicine
- Periodicals - Italy - Popular.

292   Baierisch-ökonomische Hausvater zum Nutzen und Vergnügen, oder, gesammelte
und vermehrte Schriften der Churfürstlicher Gesellschaft sittlich und
landwirthschaftlicher Wissenschaften in Burghausen. - Bd.1-Bd.8, 1778-1786.
Munich: Stein, 1778-1786.   EDITOR: Aloysius Friedrich Wilhelm von
Hillesheim.   SOURCES: K2920   SUBJECT: Economics - Periodicals -
Germany - Popular.   SUBJECT: Agriculture - Periodicals - Germany.

293   Beiträge zur practischen und gerichtlichen Arzneikunde. - Bd.1, 1799. Stendal:
Franzen, 1799. EDITOR: Johann Christoph Fahner.   SOURCES: G328,
K3725   LOCATIONS: ZDB SUBJECT: Medical jurisprudence - Periodicals
- Germany.

294   Beiträge für die Zergliederungskunst. - Bd.1-Bd.2, 1800-1803. Leipzig: Karl
Tauschnitz, 1800-1803.   EDITOR: Johann Christian Rosenmüller.
SOURCES: B799, G329, S2904   LOCATIONS: BUC, ZDB   SUBJECT:
Anatomy - Periodicals - Germany.

295   Beiträge zu den chemischen Annalen. - Bd.1-Bd.6 (each with 4 issues) 1786-1799.
Leipizg: J.G.Müller, 1786-1799. CONTINUATION OF: Chemische Annalen
für die Freunde der Naturlehre. Published as a supplement to the Chemische
Annalen. v.4 also under the title: Beiträge zur Erweiterung der Chemie.
OTHER TITLE: Beiträge zur Erweiterung der Chemie. EDITOR: Lorenz

Florenz Friedrich von Crell. **INDEXES:** Indexed in Geuss,(Ref.53, v.4). **SOURCES:** G603 K3309, S2800 **LOCATIONS:** CCF, CCN, ULS, ZDB **SUBJECT:** Chemistry - Periodicals - Germany.

296 Beiträge zum Archiv der medicinischen Polizei und der Volksarzneikunde. - Bd.1-Bd.8, 1789-1799. Leipzig: Weygand, 1789-1799. **CONTINUATION OF:** Archiv der medicinischen Polizei. **EDITOR:** Johann Christian Friedrich Scherf. **SOURCES:** G84, K3650, OCLC **LOCATIONS:** BUC, CCF, CCN, ULS, ZDB **SUBJECT:** Medical jurisprudence - Periodicals - Germany.

297 Beiträge zur Arznei-und Apothekerkunde. - Bd.1, 1791. Göttingen: 1791. **EDITOR:** Georg Heinrich Piepenbring. **SOURCES:** K3671 **SUBJECT:** Medicine - Periodicals - Germany. **SUBJECT:** Pharmacy - Periodicals - Germany.

298 Beiträge zur Arzneiwissenschaft, Wundarznei - und Entbindungskunst. - Bd.1-Bd.2, 1796-1802. Münster: Theysing, 1796-1802. **EDITOR:** Wilhelm Anton Fischer. **SOURCES:** K3701 **LOCATIONS:** ZDB **SUBJECT:** Medicine - Periodicals - Germany. **SUBJECT:** Surgery - Periodicals - Germany. **SUBJECT:** Obstetrics - Periodicals - Germany.

299 Beiträge zur Aufhebung der Gemeinheiten und Verbesserung der Landwirthschaft, von einer ökonomischen Gesellschaft im Magdeburgischen etc. - Samml.1-Samml.4, 1775-1780. Brandenburg: Johann Wendelin, 1775-1780. **SOURCES:** K2915 **LOCATIONS:** CCF **SUBJECT:** Agriculture - Societies - Magdeburg.

300 Beiträge zur Aufnahme der theoretischen Mathematik. - St.1-St.4, 1758-1761. Greifswald und Leipzig: Cnobloch, 1758-1761. **EDITOR:** Wenzeslaus Johann Gustav Karsten. **SOURCES:** K3228 **LOCATIONS:** CCN, ZDB **SUBJECT:** Mathematics - Periodicals - Germany.

301 Beiträge zur Beförderung der Geschichte und Heilung der Krankheiten. aus dessen Sammlung practischen Streitschriften, in einem vollstandigen Auszug gebracht und mit Anmerkungen versehen von Lorenz von Crell. - v.1-v.6, 1781-1784. Berlin und Stettin: Friedrich Nicolai, 1781-1784. **EDITOR:** Albrecht von Haller. **TRANSLATION OF:** Disputationes ad morborum historiam et curationem facientes. **SOURCES:** G353b **LOCATIONS:** NUC **SUBJECT:** Medicine - Periodicals - Switzerland - Dissertations.

302 Beiträge zur Beförderung der Naturkunde. - Bd.1 (St.1-25). 1774. Halle: Johann Gottfried Trampe, 1774. **EDITOR:** Friedrich Wilhelm von Leysser. **SOURCES:** B943, K3254, S2707 **SUBJECT:** Natural history - Periodicals - Germany.

303 Beiträge zur Botanik. - Bd.1-Bd.2, 1782-1783.Bremen: Müller, 1782-1783. There was a new edition in 1892 with the title: Neue Beiträge zur Botanik. **EDITOR:** Albrecht Wilhelm Roth. **CONTINUED BY:** Neue Beiträge zur Botanik.

MICROFORMS: IDC(8). SOURCES: B806, K3285, OCLC, S2428 LOCATIONS: BUC, ZDB SUBJECT: Botany - Periodicals - Germany.

304 Beiträge zur bürglichen Geschichte der Cultur, zur Naturgeschichte, Naturlehre und dem Feldbau. - Bd.1, 1783. Leipzig: Kleefeld, 1783. Publisher varies. EDITOR: Johann Christoph Adelung. TRANSLATION OF: Mémoires de l'Académie royale des sciences, des lettres et des beaux arts de Bruxelles. SOURCES: B807, G443a, K368, S2879 SUBJECT: General - Societies - Brussels - Translations.

305 Beiträge zur Chemie in Übersetzungen oder vollständigen Auszügen neuer chemischer Abhandlungen, sammt einigen neuen Aufsätzen. - Bd.1, 1791. Vienna: Hörling, 1791. EDITOR: Franz Xaver August von Wasserberg. SOURCES: B808, G602, S3569 SUBJECT: Chemistry - Periodicals - Austria.

306 Beiträge zur chemischen Kenntniss der Mineralkörper. - Bd.1-Bd.6, 1795-1815. Posen: Decker, 1795-1815. TRANSLATED AS: Analytical essays towards promoting the chemical knowledge of mineral substances. Imprint varies. Vol. 6 has added title page: Chemische Abhandlungen gemischten inhalts. OTHER TITLE: Chemische Abhandlungen gemischten Inhalts. EDITOR: Martin Heinrich Klaproth. MICROFORMS: RM(3). SOURCES: B809 SUBJECT: Chemistry - Periodicals - Germany. SUBJECT: Mineralogy - Periodicals - Germany.

307 Beiträge zur Erweiterung und Berichtigung der Chemie. - Bd.1-Bd.2, 1799-1802. Erfurt: Beyer und Maring, 1799-1802. EDITOR: Christian Friedrich Bucholz. SOURCES: B814, S2551 LOCATIONS: CCF, ULS SUBJECT: Chemistry - Periodicals - Germany.

308 Beiträge zur Geographie, Geschichte und Staatenkunde. - Bd.1-Bd.2, 1794-1796. Nürnberg: Schneider und Weigels, 1794-1796. CONTINUATION OF: Geographisches Magazin. CONTINUATION OF: Neues geographisches Magazin. CONTINUATION OF: Historische und geographische Monatsschrift. EDITOR: Johann Ernst Fabri. CONTINUED BY: Magazin für die Geographie, Staatenkunde und Geschichte. SOURCES: B1779a, K1273 LOCATIONS: BUC, ZDB SUBJECT: Geography - Periodicals - Germany.

309 Beiträge zur gerichtlichen Arzneigelahrtheit. - Theil 1, 1787. Frankfurt und Leipizg: 1787. Reissued in Strassbourg by König in 1789 under the title Medicinisch-gerichtliche Beobachtungen. OTHER TITLE: Medicinisch-gerichtliche Beobachtungen nebst ihrer Beurtheilung. EDITOR: Christian Ludwig Schweikhard. SOURCES: G325 SUBJECT: Medical jurisprudence - Periodicals - Germany.

310 Beiträge zur gerichtlichen Arzneigelartheit und zur medicinischen Polizei. - Theil1-Th.4, 1782-1792. Weimar: Carl Ludolf Hoffman, 1782-1792. Kirchner (Ref.130) classes it as a monograph. EDITOR: Wilhelm Heinrich Sebastien

Bucholtz.   **SOURCES:** G326, OCLC   **LOCATIONS:** CCF   **SUBJECT:** Medical jurisprudence - Periodicals - Germany.

311   Beiträge zur Geschichte der Erfindungen. - Bd.1-Bd.5 (each with four issues), 1780-1805. Leipzig: Paul Gotthelf Kummer, 1786-1805. **TRANSLATED AS:** History of inventions and discoveries. Some of the volumes appear in other editions. **EDITOR:** Johann Beckmann. **MICROFORMS:** RM(3). **SOURCES:** G604, K2729, OCLC, S2899 **LOCATIONS:** ULS **SUBJECT:** Economics - Periodicals - Germany. **SUBJECT:** Manufactures - Periodicals - Germany.

312   Beiträge zur Geschichte der Medizin. - Bd.1 (St.1-3), 1794-1796. Halle: Renger, 1794-1796. **EDITOR:** Curt Sprengel. **SOURCES:** G122 **LOCATIONS:** BUC, CCF, ULS, ZDB **SUBJECT:** Medicine - Periodicals - Germany - History.

313   Beiträge zur Ingenieurwissenschaft. - St.1, 1776. Prague: 1776. **EDITOR:** Franz Joseph von Kinsky. **SOURCES:** K3912 **SUBJECT:** Engineering - Periodicals - Germany.

314   Beiträge zur Insektengeschichte. - Bd.1-Bd.3, 1780-1793. Frankfurt: Varrentrapp und Wenner, 1790-1793. **EDITOR:** Ludwig Gottlieb Scriba. **MICROFORMS:** RM(3). **SOURCES:** B820, G578, K3334, S2582 **LOCATIONS:** BUC, ULS **SUBJECT:** Entomology - Periodicals - Germany.

315   Beiträge zur Insektengeschichte. - Bd.1-Bd.3, 1781-1783. Leipzig: Schweikert, 1781-1783. **OTHER TITLE:** Neue Beiträge zur Insektengeschichte. **EDITOR:** August Wilhelm Knoch. **SOURCES:** K3275 **LOCATIONS:** ULS, ZDB **SUBJECT:** Entomology - Periodicals - Germany.

316   Beiträge zur Kriegskunst und Geschichte des Kriegs von 1756-63. - Bd.1-Bd.6, 1775-1786. Freiberg: Craz, 1775-1786. **EDITOR:** Johann Gottlieb Tielke. **SOURCES:** K3960, OCLC **SUBJECT:** Military art and science - Periodicals - Germany.

317   Beiträge zur Kriegswissenschaft. - Bd.1-Bd.3, 1794-1796.Frankfurt: Jäger, 1794-1796. **EDITOR:** Johann Christoph Jäger. **SOURCES:** K3979a **SUBJECT:** Military art and science - Periodicals - Germany.

318   Beiträge zur medicinischen Gelehrsamkeit. - Bd.1-Bd.3, 1749-1755. Halle: Renger, 1749-1755. **EDITOR:** Christian Friedrich Daniel. **SOURCES:** K3529 **LOCATIONS:** BN, ZDB **SUBJECT:** Medicine - Periodicals - Germany.

319   Beiträge zur näheren Kenntniss des Galvanismus und der Resultate seiner Untersuchung. - B.1-Bd.2 (each with 4 issues), 1800-1805. Jena: Frommann, 1800-1805. **EDITOR:** Johann Wilhelm Ritter. **CONTINUED BY:** Neue Beiträge zur näheren Kenntniss des Galvanismus (1808). **SOURCES:** B827, G327, K3382, S2811 **SUBJECT:** Electricity - Periodicals - Germany.

320   Beiträge zur natürlichen, ökonomischen und politischen Geschichte der Ober-und
      Niederlausitz. - Zittau: Schöps, 1791. **EDITOR**: Christian August Pescheck.
      LOCATIONS: ZDB **SUBJECT**: General - Periodicals - Germany.

321   Beiträge zur Natur oder Insektengeschichte. - T.1-t.37. 1761. Nürnberg: 1761.
      Published in expanded form in Nürnberg, 1792-1793 with the assistance of
      Christian Schwarz. **OTHER TITLE**: Beiträge zur Natur und Insektenlehre;
      Beiträge zur Natur und Insecten-Geschichte. **EDITOR**: Christian Friedrich Carl
      Kleemann. **SOURCES**: K3233, OCLC **LOCATIONS**: BL, CCF
      SUBJECT: Natural history - Periodicals - Germany. **SUBJECT**: Entomology
      - Periodicals - Germany.

322   Beiträge zur Naturgeschichte. - Bd.1-Bd.2, 1794-1801. Rostock: 1794-1801.
      CONTINUATION OF: Annalen der Naturgeschichte. **EDITOR**: Heinrich
      Friedrich Link. **SOURCES**: B828, G528, S3201 **SUBJECT**: Natural history
      - Periodicals - Germany.

323   Beiträge zur Naturgeschichte des Schweizerlandes. - St.1-3, 1773-1774. Bern:
      A.Wagner, 1775. There was another edition in 1775. **EDITOR**: Jacob Samuel
      Wyttenbach. **SOURCES**: B829, G529, K3258, OCLC, S2131 **LOCATIONS**:
      NUC **SUBJECT**: Natural history - Periodicals - Switzerland.

324   Beiträge zur Naturgeschichte, sonderlich des Mineralreichs aus ungedruckten
      Briefen gelehrten Naturforscher und aufmerksamer Freunde der Natur. -
      Bd.1-Bd.2, 1774-1776. Altenburg: Richter, 1774-1776. **EDITOR**: Johann
      Samuel Schröter. **SOURCES**: K3255 **LOCATIONS**: BL, ULS **SUBJECT**:
      Natural history - Periodicals - Germany. **SUBJECT**: Mineralogy - Periodicals
      - Germany.

325   Beiträge zur Naturgeschichte und Bergpolizeywissenschaft. - Pt 1-pt.4, 1780.
      Leipizg: W. Vogel, 1780. **EDITOR**: Friedrich Gottlob Gläser. **SOURCES**:
      B831, S2901 **SUBJECT**: Natural history - Periodicals - Germany.
      SUBJECT: Mineral industries - Periodicals - Germany.

326   Beiträge zur Naturkunde und den damit verwandten Wissenschaften, besonders der
      Botanik, Chemie, Haus- und Landwirthschaft. - Th.1-Th.7, 1787-1792.
      Hannover und Osnabrück: Schmidt, 1797-1792. Publisher varies. **EDITOR**:
      Friedrich Ehrhardt. **MICROFORMS**: IDC(8). **SOURCES**: B915, G606,
      OCLC, S2777 **LOCATIONS**: CCN, ULS **SUBJECT**: Science - Periodicals
      - Germany.

327   Beiträge zur öffentlichen und gerichtlichen Arzneikunde. - St.1-St.2, 1798-1802.
      Braunschweig und Frankfurt: Schulbuchhandlung, 1798- 1802. **EDITOR**:
      Theodor Georg August Roose. **SOURCES**: G154, K3718 **LOCATIONS**:
      CCF, ULS, ZDB **SUBJECT**: Medical jurisprudence - Periodicals - Germany.

328   Beiträge zur Ökonomie, Technologie, Polizey-und Cameralwissenschaften. -
      Bd.1-Bd.12, 1779-1791. Göttingen: Vandenhöck, 1779-1791. **EDITOR**:
      Johann Beckmann. **SOURCES**: K2923 **LOCATIONS**: BUC, CCF, CCN,

ZDB   **SUBJECT:** Economics - Periodicals - Germany.   **SUBJECT:**
Technology - Periodicals - Germany.

329   Beiträge zur Pflanzenantomie, Pflanzenphysiologie und zu einer neuen
      Characteristik der Bäume und Sträucher. - Heft 1-Heft 7, 1799-1800. Leipzig:
      Heinrich Gräff, 1799-1800. **EDITOR:** Friedrich Casimir Medicus.
      SOURCES: B841, G554, K3377, S2902 **LOCATIONS:** ZDB **SUBJECT:**
      Botany - Periodicals - Germany.

330   Beiträge zur philosophischen Anthropologie und der damit verwandten
      Wissenschaften. - Bd.1-Bd.2, 1794-1796. Vienna: Schaumburg, 1794-1796.
      EDITOR: Johann Michael Wagner. **SOURCES:** B843, G249, K3354, S3571
      LOCATIONS: ZDB **SUBJECT:** Anthropology - Periodicals - Germany.

331   Beiträge zur Physik, Ökonomie, Mineralogie, Technologie und Statistik, besonders
      der russischen und angränzenden Länder. - Bd.1-Bd.3, 1786-1789. Berlin und
      Stettin: Nicolai, 1786-1789. **EDITOR:** Benedict Franz Johann Hermann.
      SOURCES: B917, 5522, K3310, S2299 **SUBJECT:** Science - Periodicals -
      Germany.

332   Beiträge zur practischen Anwendung des thierischen Magnetismus, nebst Anhang.
      - Strasburg: Academische Buchhandlung, Leipzig: Hinrichs, 1786. **SOURCES:**
      G:VIII **LOCATIONS:** DSG **SUBJECT:** Animal magnetism - Periodicals -
      Germany.

      Beiträge zur practischen und gerichtlichen Arzneikunde, see no. 293.

333   Beiträge zur Rechtsgelehrsamkeit, Ökonomie, Polizey und Cameralwissenschaft
      wie auch zur Naturwissenschaft, besonders von Hessen. - Bd.1 (St.1-4), 1769.
      Frankfurt: 1769. **EDITOR:** Georg Conrad Stockhausen. **SOURCES:** K2721,
      S2583 **SUBJECT:** General - Periodicals - Germany.

334   Beiträge zur Sittenlehre, Ökonomie, Arzneywissenschaft, Naturlehre und
      Geschichte in ihrem allgemeinen Umfange, aus den westlichen Gegenden
      Deutschlands. - St.1-St.2, 1770-1772. Mannheim: Schwann, 1770-1772.
      SOURCES: B846, G604, K5350, S3069 **LOCATIONS:** BUC **SUBJECT:**
      General - Periodicals - Germany.

335   Beiträge zur theoretischen und practischen Geburtshülfe und zur Kenntniss und
      Cur einiger Kinderkrankheiten. - Heft 1-Heft 3, 1798-1808. Hamburg:
      Bachmann und Gundermann, 1798-1808. **EDITOR:** Justus Heinrich Wigand.
      SOURCES: G156 **LOCATIONS:** ULS **SUBJECT:** Obstetrics - Periodicals
      - Germany. **SUBJECT:** Pediatrics - Periodicals - Germany.

336   Beiträge zur theoretischen und praktischen Elektrizitätslehre. - St.1-St.4,
      1793-1795. Stuttgart: Metzler, 1793-1795. **EDITOR:** Gabriel Christoph
      Bohnenberger. **SOURCES:** K3349 **SUBJECT:** Electricity - Periodicals -
      Germany. **SUBJECT:** Physics - Periodicals - Germany.

337   Beiträge zur vaterländischen Historie, Geographie, Statistik und Landwirthschaft sammt einer Übersicht der schönen Litteratur. - Bd.1-Bd.10, 1785-1818. Munich: Lindauer, 1785-1818. EDITOR: Lorenz von Westenreider. CONTINUED BY: Neue Beiträge zur vaterdischen Historie, Geographie, Statistik. SOURCES: K1163 LOCATIONS: BUC, CCF, CCN, ZDB SUBJECT: General - Periodicals - Germany.

338   Beiträge zur Verbesserung der Landwirthschaft. - Jahrgang 1-Jahrgang 3 (Bd.1-Bd.12), 1795-1798,1808- 1810. Prague: Herl, 1795-1810. None issued between 1798 and 1808. EDITOR: Franz Fuss. SOURCES: K3003 SUBJECT: Agriculture - Periodicals - Germany.

339   Beiträge zur Völker-und Länderkunde. Bd.1-Bd.14, 1781-1790. Leipzig: Weygand, 1781-1790. OTHER TITLE: Beiträge zur Erweiterung der Länder und Völkerkunde; Auswahl kleiner Reisebeschreibungen und anderer statistischer und geographischer Nachrichten. EDITORS: Johann Reinhold Forster and Matthias Christian Sprengel. CONTINUED BY: Neue Beiträge zur Völker-und Länderkunde. CONTINUED BY: Auswahl der besten ausländischen geographischen und statistischen Nachrichten. SOURCES: B813, G239, K1113, OCLC, S2898 LOCATIONS: BUC, CCF, ULS SUBJECT: Geography - Periodicals - Germany.

340   Bekjendgiöreise / Kongelige Danske videnskabs selskab, Copenhagen. - Nos. 1-14, June 22, 1794-July 23, 1813. Copenhagen: 1794-1813. EDITOR: Thomas Buggel. INDEXES: Fortegnelse over de Koneglige danske Videnkabernes selskabs publicationer, 1742-1930. Ed. Asger Lomholt. Copenhagen, 1930. SOURCES: G475f, S669a LOCATIONS: BUC SUBJECT: Science - Societies - Copenhagen - Prize essays.

341   Belehrende Nachrichten für den Nahrungstand, besonders für Landwirthe, Hausmutter, Handwerker, Künstler, Manufakturisten und Kaufleuten. - Jahrgang 1 (Quartel 1-4), 1783. Leipzig: Breitkopf und Härtel, 1785. SOURCES: K2938 SUBJECT: Medicine - Periodicals - Germany - Popular. SUBJECT: Nutrition - Periodicals - Germany - Popular. SUBJECT: Economics - Periodicals - Germany - Popular.

342   Bemerkungen der Cuhrpfälzischen physikalisch- ökonomischen Gesellschaft, Mannheim. - Bd.1-Bd.16, 1770-1785. Mannheim: C.F. Schwan; Lautern: Akademische Buchhandlung, 1770-1785. OTHER TITLE: Physikalisch-ökonomische und Bienengesellschaft zu Lautern (Ref.51,p.48). CONTINUED BY: Vorlesungen der Churfpälzischen physicalisch-ökonomischen Gesellschaft. SOURCES: K2901, S3075a LOCATIONS: BUC, CCF, CCN, ULS, ZDB SUBJECT: Science - Societies - Mannheim. SUBJECT: Bee culture - Societies - Mannheim.

343   Beobachtungen des meteorologischen observatoriums auf dem Hohenspeisenberg. - v.1-14, 1792-1884. Munich: 1792-1884. Issued as supplement to Annalen der Münchener Sternwarts. CONTINUED BY: Annalen der Königlichen

Sternwaarts bei München. **LOCATIONS:** CCF, CCN **SUBJECT:** Meteorology - Periodicals - Germany.

344 Beobachtungen verschiedener chirurgischen Vorfälle. - Bd.1-Bd.2, 1780-1783. Vienna: 1780-1783. A collection of medical case studies by Joseph Mohrenheim. **SOURCES:** G330 **LOCATIONS:** BN **SUBJECT:** Medicine - Periodicals - Austria.

345 Beobachtungen, Versuche und Erfahrungen über des Salpeters. - Tübingen: J.G. Cotta, 1783-1786. **SOURCES:** OCLC **LOCATIONS:** ZDB **SUBJECT:** Chemistry - Periodicals - Germany.

346 Bergbaukunde / Societät der Bergbaukunde, Zellerfeld. - Bd.1-2, 1789-1790. Leipzig: B.J.Goeschen, 1789-1790. **EDITORS:** Ignaz Edler von Born and Friedrich Wilhelm Heinrich von Trebra. **SOURCES:** NUC **LOCATIONS:** BUC, ULS **SUBJECT:** Mineral industries - Societies - Zellerfeld.

347 Bergmännischer Kalender. - v.1-v.2, 1790-1791. Freiberg und Annaberg: Craz, 1790-1791. **OTHER TITLE:** Bergmaännisches Taschenbuch. **EDITOR:** Alexander Wilhelm Köhler. **SOURCES:** B863, K3338 **LOCATIONS:** ZDB **SUBJECT:** Mineral industries - Periodicals - Germany.

348 Bergmännisches Journal. - Jg. 1-6 (v.1-12), 1788-1794. Freiburg: Craz, 1788-1794. Monthly. **EDITORS:** Alexander Wilhelm Köhler and Christian August Siegfried Hoffmann. **CONTINUED BY:** Neues bergännisches Journal. **SOURCES:** B864, K3326, OCLC, S2617 **LOCATIONS:** BUC, CCF, ULS, ZDB **SUBJECT:** Mineral industries - Periodicals - Germany.

349 Bergs Journal. - v.1, 1787-1788. Stockholm: J.G. Lange, 1787-1788. **SOURCES:** (Ref.54:285) **SUBJECT:** Mineral industries - Periodicals - Sweden.

350 Berliner Beiträge zur Landwirthschaftswissenschaft. - Bd.1-8, 1771-1791. Berlin: Pauli, 1771-1791. Each volume consists of 12 parts. **EDITORS:** Carl Friedrich von Benekendorf (v.1-7), Gottfried Ludolf Grassmann (v.8). **CONTINUED BY:** Neue Berliner Beiträge zur Landwirthschaftswissenschaft. **SOURCES:** B895, K2904, OCLC **LOCATIONS:** CCF, CCN, ULS, ZDB **SUBJECT:** Agriculture - Periodicals - Germany.

351 Berlinische Blätter. - Bd.1-Bd.4, 1797-1798. Berlin: Nicolai, 1797-1798. **CONTINUATION OF:** Berlinische Monatsschrift. **EDITOR:** Johann Erich Biester. **CONTINUED BY:** Neue Berlinische Monatsschrift. **INDEXES:** Indexed in Hocks (Ref.55). **MICROFORMS:** GO(9). **SOURCES:** G608, K4613 **LOCATIONS:** BUC, ZDB **SUBJECT:** General - Periodicals - Germany.

352 Berlinische Monatsschrift. - Bd.1-Bd.28, 1783-1796. Berlin: Haude und Spener, 1783-1796. Monthly. **EDITORS:** Friedrich Gedike and Johann Erich Biester. **CONTINUED BY:** Berlinische Blätter. **CONTINUED BY:** Neue Berlinische

Monatsschrift.   **INDEXES:** Indexed in Hocks (Ref.55), and Allgemeines Sachregister (q.v.). **MICROFORMS:** GO(9), IDC(8).   **SOURCES:** G610, K369, OCLC **LOCATIONS:** BUC, CCF, CCN, ULS, ZDB **SUBJECT:** General - Periodicals - Germany.

353   Berlinische Sammlung zur Beförderung der Arzneiwissenschaft, der Naturgeschichte, der Haushaltungskunst, Cameralwissenschaft und der dahin einschlagenden Literatur. - Bd.1-Bd.10 (6 issues each), 1768-1779. Berlin: Joachim Pauli, 1768-1779. **CONTINUATION OF:** Berlinisches Magazin. **EDITOR:** Friedrich Heinrich Wilhelm Martini. **INDEXES:** Index Bd.1-10 in 10. **SOURCES:** B901, G611, K3245, OCLC, S2296 **LOCATIONS:** BUC, CCF, ULS, ZDB **SUBJECT:** Science - Periodicals - Germany.

354   Berlinisches Jahrbuch für die Pharmacie und für die damit verbundenen Wissenschaften. - Bd.1-Bd.35, 1795-1835.   Berlin: Ferdinand Oehmigke, 1795-1835. Bd.9-15 issued under the title: Neues Jahrbuch für die Pharmacie, Bd.16-35 under the title: Deutsches Jahrbuch für die Pharmacie. **OTHER TITLE:** Neues Jahrbuch für die Pharmacie; Deutsches Jahrbuch für die Pharmacie. **EDITORS:** Georg Heinrich Stoltze, Adolph Ferdinand Gehlen and others. **INDEXES:** Index to Bd.1-8 in v.8, to Bd.9-14 in v.14.   **SOURCES:** B902, G219, K3695, OCLC **LOCATIONS:** BUC, CCF, CCN, ULS, ZDB **SUBJECT:** Pharmacy - Periodicals - Germany.

355   Berlinisches Magazin der Wissenschaften und Künste. - Jahrgang 1 (St.1-4)-Jahrgang 2(St.1) 1782-1784.   Berlin: J.F. Unger, 1782-1784. **EDITOR:** Wilhelm Jakob Wippel. **SOURCES:** B903, G614, K364, S2295 **LOCATIONS:** CCF, ULS, ZDB **SUBJECT:** General - Periodicals - Germany.

356   Berlinisches Magazin, oder, gesammlete Schriften und Nachrichten für die Liebhaber der Arnzeywissenschaft, Naturgeschichte und der angenehmen Wissenschaften überhaupt.- Bd.1-Bd.4 (6 issues each), 1765-1769. Berlin: Arnold Wever, 1765-1769. **EDITOR:** Friedrich Heinrich Wilhelm Martini. **CONTINUED BY:** Berlinische Sammlung zur Beförderung der Arzneiwissenschaft, der Naturgeschichte, der Haushaltungskunst, Cameralwissenschaft und der dahin einschlagenden Literatur. **SOURCES:** B904, G613, K3237, S2294 **LOCATIONS:** BUC, CCF, ULS, ZDB **SUBJECT:** Science - Periodicals - Germany.

357   Berlinisches Taschenbuch für Freunde der Gesundheit. - Jahrgang 1-2, 1783-1784. Berlin: Nicolai, 1783-1784. **EDITOR:** Conrad Friedrich Uden. **SOURCES:** G180, K3604 **SUBJECT:** Medicine - Periodicals - Germany - Popular.

358   Bernisches Magazin der Natur, Kunst und Wissenschaften. - Bd.1-2, 3(St.1), 1775-1779. Bern: Typographisches Gesellschaft, 1775-1779. Part two did not appear until 1777. It consists of translations and extracts, largely from French journals. **OTHER TITLE:** Bernerisches Magazin der Natur, Kunst und Wissenschaften. **EDITOR:** Jacob Samuel Wyttenbach. **SOURCES:** B908,

K315, S2133 LOCATIONS: CCF, ULS, ZDB SUBJECT: Science - Periodicals - Switzerland.

359 Beschäftigungen der Berlinischen Gesellschaft naturforschender Freunde. - Bd.1-Bd.4, 1775-1779. Berlin: Joachim Pauli, 1775-1779. The volume for 1778 was not published. CONTINUED BY: Schriften der Berlinischen Gesellschaft Naturforschender Freunde. CONTINUED BY: Neue Schriften der Gesellschaft naturforschender Freunde zu Berlin. INDEXES: There is a modern index (Ref.53, v.3). MICROFORMS: RM(3). SOURCES: B909, G469, K3259, OCLC, S2292a LOCATIONS: BUC, CCF, CCN, ULS SUBJECT: Natural history - Societies - Berlin.

360 Der besorgte Forstmann, eine Zeitschrift über das Verderbniss der Wälder durch Thiere und vorzüglich Insecten überhaupt, besonders aber durch die Kiefer-, Fichten-, Tannen- und Birkenraupen. - St. 1-4, 1798-1799. Weimar: Industrie Comptoir, 1798-1799. EDITOR: Johann J. Lynker von Lützenwich. SOURCES: B912, K3017, S3298 LOCATIONS: ULS, ZDB SUBJECT: Forestry management - Periodicals - Germany.

361 Bibliographie analytique de médecine, ou, journal abbréviateur des meilleurs ouvrages nouveaux, latins ou français de médecine clinique, d'hygiène et de médecine preservatoire. - v.1-3, 1799-1801. Paris et Tours: Croullebois, 1799-1801. Imprint varies. EDITOR: Laurent Bodin. SOURCES: G268 SUBJECT: Medicine - Periodicals - France - Indexes.

362 Biblioteca Browniana germanica. - v.1-13, 1797-1800. Florence: Guglielmo Piatti, 1797-1800. CONTINUATION OF: Biblioteca medica Browniana. EDITORS: Guiseppe Belluomini and Luigi Giobbe. SOURCES: G254a SUBJECT: Brunonianism - Periodicals - Italy. SUBJECT: Medicine - Periodicals - Italy.

363 Biblioteca fisica d'Europa, ossia raccolta di osservazioni sopra la fisica, matematica, chimica, storia naturale, medicina ed arti. - v.1-20, 1788-1791. Pavia: Monastero di Salvatore, 1788-1791. Subtitle varies. EDITOR: Luigi Gaspero Brugnatelli. CONTINUED BY: Giornale fisico-medico. CONTINUED BY: Avanzamenti della medicina e fisica. INDEXES: v.1-12 in v.12; v.13-19 in v.20. SOURCES: B921, G616, OCLC, S1991 LOCATIONS: BUC, CCF, ULS SUBJECT: Science - Periodicals - Italy.

364 Biblioteca italiana, ou, Tableau des progrès des sciences et des arts en Italie. - v.1-9, 1793-1801. Turin: Bocca and Ballino; Paris: Huzard, 1793-1801. EDITORS: Gioberti and others. SOURCES: B924, G617, S2057 SUBJECT: Science - Periodicals - Italy.

365 Biblioteca medica Browniana. - v.1-6, 1797-1798. Florence; Milan: Guiseppi Maopero, 1797-1798. EDITOR: Luigi Frank. CONTINUED BY: Biblioteca Browniana germanica. SOURCES: G254 LOCATIONS: ULS SUBJECT: Brunonianism - Periodicals - Italy. SUBJECT: Medicine - Periodicals - Italy.

366 Bibliotheca della più recente letteratura medico-chirurgica. - Pavia: Galiazzi, 1790-1792. Augmented and annotated by Tommaso Volpi. **EDITORS:** Johann Neopomuk Hunczowski and Johann Adam Schmidt. **TRANSLATION OF:** Bibliothek der neuesten medicinisch-chirurgischen Literatur. **SOURCES:** G269a **SUBJECT:** Medicine - Periodicals - Austria - Translations. **SUBJECT:** Medicine, Military - Periodicals - Austria - Translations.

367 Bibliotheca physico-medica. - Bd.1-2, 1776-1777. Breslau: Wilhelm Gottlieb Korn, 1776-1777. **SOURCES:** K3576 **LOCATIONS:** BUC, ZDB **SUBJECT:** Medicine - Periodicals - Germany. **SUBJECT:** Science - Periodicals - Germany.

368 Bibliothek auslandischer Chemisten, Mineralogen und mit Mineralien beschäftigter Fabrikanten, nebst derlei biographischen Nachrichten. - Bd.1-4, 1781-1784. Nürnberg: Johann Abraham Stein, 1781-1784. **EDITOR:** Johann Hermann Pfingsten. **LOCATIONS:** BL, ZDB **SUBJECT:** Mineralogy - Periodicals - Germany. **SUBJECT:** Chemistry - Periodicals - Germany.

369 Bibliothek der ausländischen neuesten Litteratur in Auszügen aus den besten Wochen- und Monatschriften. - v.1-2 (1770-1771) 1772. Hanau: Schulze, 1772. **CONTINUATION OF:** Neue Auszüge aus den besten ausländischen Wochen- und Monatschriften. **EDITOR:** Joseph Christoph Stockhausen. **SOURCES:** K4464 **SUBJECT:** General - Periodicals - Germany - Reviews.

370 Bibliothek der gesammten Naturgeschichte. - Bd.1-2, 1789-1791. Frankfurt; Mainz: Varrentrapp und Wenner, 1789-1791. **EDITORS:** Johann Fibig and Bernhard Sebastian von Nau. **MICROFORMS:** RM(3). **SOURCES:** B925, G527, K3335, OCLC, S2584 **LOCATIONS:** BUC, CCF, ULS, ZDB **SUBJECT:** Science - Periodicals - Germany - Reviews.

371 Bibliothek der neuesten medicinisch-chirurgischen Literatur für die Kais. Königl. Feldchirurgen. - Bd.1-3 (4 issues each), Bd.4 (St.1) 1789-1793. Vienna: Rudolph Graffer, 1789-1793. **TRANSLATED AS:** Bibliotheca della più recente letteratura medico-chirurgica. Primarily abstracts and translations. **EDITORS:** Johann Neopomuk Hunczowski and Johann Adam Schmidt. **SOURCES:** G269, K3652 **LOCATIONS:** BUC, ULS, ZDB **SUBJECT:** Medicine - Periodicals - Austria - Reviews. **SUBJECT:** Medicine, Military - Periodicals - Austria - Reviews.

372 Bibliothek der neuesten physisch-chemischen, metallurgischen, technologischen und pharmaceutischen Literatur. - Bd.1-4, 1788-1795. Berlin: 1788-1795. **EDITOR:** Sigismund Friedrich Hermbstädt. **CONTINUED BY:** Annalen der chemischen Literatur. **CONTINUED BY:** Bulletin des Neuesten und Wissenswürdigsten aus der Naturwissenschaften. **CONTINUED BY:** Museum des Neuesten und Wissenswürdigsten aus der Naturwissenschaften. **SOURCES:** B926, G621, K3323, S2300 **LOCATIONS:** ULS, ZDB **SUBJECT:** Chemistry - Periodicals - Germany - Reviews. **SUBJECT:** Pharmacy - Periodicals - Germany - Reviews. **SUBJECT:** Technology - Periodicals - Germany - Reviews.

373   Bibliothek der neuesten und wichtigsten Reisebeschreibungen und geographischen Nachrichten zur Erweiterung der Erdkunde. - Bd.1-50, 1800-1814. Weimar: Industrie Comptoir, 1800-1814. A series of monographs on travel and geography. EDITOR: Matthias Christian Sprengel. CONTINUED BY: Neue Bibliothek der wichtigsten Reisebeschreibungen. INDEXES: Sprengel, M.C. and T.F. Ehrmann. Register über die ersten vier und zwanzig Bände. Weimar, 1806. SOURCES: S3299 LOCATIONS: CCN, ULS, ZDB SUBJECT: Geography - Periodicals - Germany.

374   Bibliothek der practischen Heilkunde. - Bd.1-76, 1799-1836. Jena: Reimer, 1799-1836. Monthly. Frequency varies; Imprint varies. Includes six supplements. Bd.21-74, also as new series. EDITOR: Christian Wilhelm Hufeland. SOURCES: G270 LOCATIONS: BUC, CCF, CCN, ULS, ZDB SUBJECT: Medicine - Periodicals - Germany - Reviews.

375   Bibliothek der wichtigsten practischen Ärzte des 17. Jahrhunderts. - Bd.1-2, 1785-1786. Leipzig: Weygand, 1785-1786. A collection of earlier medical authors. SOURCES: G271, OCLC LOCATIONS: ZDB SUBJECT: Medicine - Periodicals - Germany.

376   Bibliothek for physik, medicin og oeconomie. - v.13-18, 1798-1800. Copenhagen: C.L. Buchs and Sebastien Papp, 1798-1800. CONTINUATION OF: Physicalsk, ökonomisk og medico-chirurgisk Bibliothek for Danmark og Norge. EDITOR: Ole Hieronymus Mynster (Ref.42). CONTINUED BY: Nyt Bibiliothek for physik, medicin og oeconomie. SOURCES: B3660a, S645.2 LOCATIONS: ULS SUBJECT: Science - Periodicals - Denmark.

377   Bibliothek für Chirurgie und practische Medicin. - Bd.1 (St.1-3) 1790-1793. Göttingen: Vandenhoeck und Ruprecht, 1790-1793. EDITOR: Justus Arnemann. CONTINUED BY: Bibliothek für die Medicin, Chirurgie und Geburtshülfe. SOURCES: G92, K3662 LOCATIONS: BUC, CCN, ZDB SUBJECT: Medicine - Periodicals - Germany - Reviews. SUBJECT: Surgery - Periodicals - Germany - Reviews.

378   Bibliothek für das Merkwürdigste aus der Natur- und Völkergeschichte. - Bd.1, 1796. Leipzig: 1796. SOURCES: B928, G531, S2906 SUBJECT: Natural history - Periodicals - Germany.

379   Bibliothek für die Medicin, Chirurgie und Geburtshülfe. - Bd.1 (St.1-2) 1799. Göttingen: Dieterich, 1799. CONTINUATION OF: Bibliothek für Chirurgie und practische Medicin. EDITOR: Justus Arnemann. SOURCES: G157, K3726 LOCATIONS: BUC, ZDB SUBJECT: Medicine - Periodicals - Germany - Reviews. SUBJECT: Obstetrics - Periodicals - Germany - Reviews. SUBJECT: Surgery - Periodicals - Germany - Reviews.

380   Bibliothek für Kinderärzte. - Bd.1,1782; Bd.2, 1798. Vienna: Hörling, 1782-1798. EDITOR: Paula Joseph Carl von Aborner von Ahornrain. SOURCES: G333 SUBJECT: Pediatrics - Periodicals - Austria.

381 Bibliothek für Physiker. - St.1-4, 1787-1789. Königsberg: Hartung, 1787-1789. **CONTINUATION OF:** Medicinisch-gerichtliche Bibliothek. **EDITOR:** Johann Daniel Metzger. **CONTINUED BY:** Annalen der Staats-arzneikunde. **CONTINUED BY:** Materialen für die Staatsarzneikunde und Jurisprudenz. **SOURCES:** G71,K637 **LOCATIONS:** CCF, ULS **SUBJECT:** Medical jurisprudence - Periodicals - Germany.

382 Bibliothek für Thierärzte, Landwirthe und Liebhaber der Thierarzneikunde. - Theil 1-2, 1794-1796. Marburg: Academische Buchhandlung, 1794-1795. **EDITOR:** Johann David Busch. **SOURCES:** G230 **SUBJECT:** Veterinary medicine - Periodicals - Germany.

383 Bibliothek von Anzeigen und Auszügen kleiner meist akademischer Schriften theologischen, philosophischen, mathematischen, historischen und philologischen Inhalts. - Bd.1-3 (each 4 issues) 1790-1793. Jena: Cuno, 1790-1792. **EDITOR:** Heinrich Eberthard Gottlob Paulus. **SOURCES:** B931, K431, S2812 **LOCATIONS:** BUC, CCF, ULS, ZDB **SUBJECT:** General - Periodicals - Germany - Dissertations.

384 Bibliothèque ancienne et moderne, pour servir de suite aux bibliothèque universelles et choisies. - v.1-29, 1714-1729. Amsterdam: 1714-1729. **CONTINUATION OF:** Bibliothèque universelle et historique. **CONTINUATION OF:** Bibliothèque choisie, pour servir de suite à la bibliotheèque universelle. Reprinted by Slatkine in 1968. There were several other editions. **EDITOR:** Jean LeClerc. **CONTINUED BY:** Bibliothèque raisonée des ouvrages des savans de l'Europe. **INDEXES:** Index to v.1-28 in v.29. **MICROFORMS:** ACR(1),DA(1). **SOURCES:** B940b, OCLC **LOCATIONS:** BUC, CCF, CCN, ULS, ZDB **SUBJECT:** General - Periodicals - Netherlands.

385 Bibliothèque angloise, ou, histoire litteraire de la Grand- Bretagne. - T.1-15, 1717-1728. Amsterdam: Paul Marret, 1717-1728. Reprinted by Slatkine Reprints 1968. **EDITORS:** Michel de la Roche and Armand de La Chapelle (Ref.56,p.14). **MICROFORMS:** ACR(1). **SOURCES:** OCLC **LOCATIONS:** BUC, NNF, CCN, ULS, ZDB **SUBJECT:** General - Periodicals - Netherlands - Reviews.

386 Bibliothèque britannique, ou, histoire des ouvrages des savans de la Grande Bretagne. - T.1-25, 1733-1747. Lay Haye: Pierre de Hondt, 1733-1747. Monthly. Each volume is divided into two parts each with its own title page. Paged continuously except for v.20 and 25 (Ref.9) Slatkine reprint (1969). **OTHER TITLE:** Histoire des ouvrages des savans de la Grand-Bretagne. **EDITORS:** Kemp, P. Desmaiseaux, Sthoelin, J. Bernard, B. Duval, L. de Beaufort, J. Rousset de Missy and T.Lédiard. **CONTINUED BY:** Journal Britannique. **INDEXES:** v.25 is a general alphabetical index to the whole series. **MICROFORMS:** ACR(1),RM(3). **SOURCES:** B932, OCLC, S816 **LOCATIONS:** BUC, CCF, CCN, ULS, ZDB **SUBJECT:** General - Periodicals - Netherlands - Reviews.

387 Bibliothèque britannique, ou, Recueil extrait des ouvrages anglais périodique et autres. - T.1-60, 1796-1815. Geneva: Imprimerie de la Bibliothèque britannique, 1796-1815. **EDITORS:** Marc Auguste Pictet, Charles Pictet de Rochemont and Fréderic Guillaume Maurice (Ref.57&58). **CONTINUED BY:** Bibliothèque universelle des sciences, belles-lettres et arts. **CONTINUED BY:** Bibliothèque universelle de Genève. **CONTINUED BY:** Archives des sciences physiques et naturelles. **MICROFORMS:** RM(3). **SOURCES:** B933, S2154.1 **LOCATIONS:** BUC, CCF, CCN, ULS, ZDB **SUBJECT:** General - Periodicals - Switzerland - Reviews.

388 Bibliothèque choisie de médecine, tirée des ouvrages périodiques, tant françois qu'étrangers. - T.1-10, 1748-1770. Paris: D'Houry, 1748-1770. Published in an octavo ed. in 31 v. and a quarto ed. in 10 v. A medical encyclopedia arranged alphabetically under topics. Cites the journal literature. **EDITOR:** François Planque and with v.10 by Jean Goulin. **MICROFORMS:** DA(1) **SOURCES:** G273, OCLC **LOCATIONS:** BUC, ULS **SUBJECT:** Medicine - Periodicals - France.

389 Bibliothèque choisie, pour servir de suite à la bibliothèque universelle. - v.1-28, 1703-1728. Amsterdam: H. Schelte, 1703-1728. **CONTINUATION OF:** Bibliothèque universelle et historique. Publisher varies. Reprinted in 1718 (Amsterdam, Wetstein) and in 1968 (Slatkine Reprints). **EDITOR:** Jean LeClerc. **CONTINUED BY:** Bibliothèque ancienne et moderne, pour servir de suite aux bibliothèque universelles et choisies. **CONTINUED BY:** Bibliothèque raisonée des ouvrages des savans de l'Europe. **INDEXES:** Index to v.1-27 in v.28. **SOURCES:** B940a OCLC **LOCATIONS:** BUC, CCF, CCN, ULS, ZDB **SUBJECT:** General - Periodicals - Netherlands.

390 Bibliothèque de chirurgie du Nord, ou, extrait des meilleurs ouvrages de chirurgie, publiées dans le nord. - T.1, pars 1-2, 1788-1789. Bonn: J.F. Abshoven; Paris: Barrois, 1788-1789. Consists primarily of translations from German review journals. Editor, Joseph Claude Rougemont. **SOURCES:** G272, OCLC **LOCATIONS:** NUC **SUBJECT:** Surgery - Periodicals - France - Reviews.

391 Bibliothèque de physique et d'histoire naturelle, contenant la physique générale, la physique particuliere, la méchanique, la chymie, l'anatomie, la botanique, la médecine, l'histoire naturelle des insectes, des animaux & des coquillages. - T.1-4, 1758-1765. Paris: 1758-1765. There was another ed., 6 vol. in 7 vol., Paris: David, 1765-1769. An anthology drawn from a wide variety of literary and scientific journals. **EDITOR:** Claude François Lambert. **MICROFORMS:** RM(3) . **SOURCES:** B934, G618, OCLC, S1343 **SUBJECT:** Science - Periodicals - France.

392 Bibliothèque des sciences et des beaux arts. - T.1 (Jan.-Mar. 1754)-T.50 (1780). La Haye: Pierre Gosse, 1754-1780. **EDITORS:** C. Chais, Elie de Joncourt, J. de la Fite, Marie Elisabeth de la Fite, C.G.F. Dumas, H. Hop, J. Guiot, et al. **INDEXES:** v.25 (1754-1765) and v.50 (1766-1778) are indexes to the previous 24 volumes. **MICROFORMS:** ACR(1),RM(3). **SOURCES:** B935, G620,

OCLC, S817   **LOCATIONS:** BUC, CCF, CCN, ULS, ZDB   **SUBJECT:** Science - Periodicals - Netherlands - Reviews.

393   Bibliothèque du Nord, ouvrage destinée à faire connoitre en France ce que l'Allemagne produit d'interessant, d'agreeable& d'utile dans tous les genres de sciences, de litterature & d'arts. - Jan. 1778-Jan. 1780. Paris: 1778-1780. Published monthly by the Société de Hesse-Homburg pour l'encouragement des connaissances des moeurs (Ref.59,p.63).   **LOCATIONS:** CCF, ULS, ZDB   **SUBJECT:** General - Societies - Hesse-Homburg.

394   Bibliothèque germanique médico-chirurugicale, ou, extraits des meilleurs ouvrages de médecine et de chirurgie publiés en Allemagne. - T.1-8, 1798-1802. Paris: Huzard; Strasbourg: König, 1798-1802.   **EDITORS:** C. Brewer and Michel De La Roche.   **INDEXES:** Index v.1-8 in v.8.   **MICROFORMS:** ACR(1).   **SOURCES:** G274, OCLC   **LOCATIONS:** BUC, CCP, ULS   **SUBJECT:** Medicine - Periodicals - France - Reviews.   **SUBJECT:** Surgery - Periodicals - France - Reviews.

395   Bibliothèque impartiale. - T.1-18 (3 issues each) 1750-1758. Göttingen und Leyden: Elie Luzac, 1750-1758.   Analyzed by Marx (Ref.29:89-107).   **EDITOR:** Jean Henri Samuel Formey.   **SOURCES:** K169, OCLC   **LOCATIONS:** BUC, CCP, CCN, ULS, ZDB   **SUBJECT:** General - Periodicals - Netherlands.

396   Bibliothèque médico-physique du Nord, ou, recueil périodique de ce qu'il y à d'éssentiel, d'interessant et de plus nouveau surtout en fait d'observations et déscouvertes, dans les collections académiques, et dans les autres ouvrages des savants du nord, soit en médecine, pharmacie, histoire naturelle et de physique. - T.1-3, 1783-1784. Lausanne: François Grasset, 1783-1784.   A collection of extracts from various German journals. Each volume has an individual title.   **OTHER TITLE:** Bibliothèque de médecine, de chirurgie et de pharmacie; Bibliothèque d'histoire naturelle et de physique; Bibliothèque de chymie.   **EDITORS:** Philippe Rodolphe Vicat and M.H. Struve.   **SOURCES:** B937, G275, OCLC, S2167   **LOCATIONS:** BUC, CCF, ULS   **SUBJECT:** Science - Periodicals - Switzerland.

397   Bibliothèque physico-économique, instructive et amusante, contenant des mémoires et observations pratiques sur l'économie rustique, sur les nouvelles découvertes les plus intéressantes, la description des nouvelles machines et instruments inventés pour la perfection des arts utiles, agreeable etc. - T.1-82, 1782-1831. Paris: 1782-1831. 1786-1792 and 1795ff published in two vols. each. Subtitle varies. Suspended 1798-1801. There were at least 5 editions.   **EDITORS:** Antoine Augustin Parmentier, Nicolas Deyeux, Charles Nicolas Sigisbert Sonnini de Manoncourt, Jean Baptiste Bory de St. Vincent and Jean Simon Étienne Julia de Fontenelle.   **SOURCES:** B938, S1341   **LOCATIONS:** BUC, CCF, ULS, ZDB   **SUBJECT:** Science - Periodicals - France - Popular.

398   Bibliothèque raisonée des ouvrages des savans de l'Europe. - v.1-50, 1728-1753. Amsterdam: Wetstein & Smith, 1728-1753.   **CONTINUATION OF:**

Bibliothèque universelle et historique. **CONTINUATION OF:** Bibliothèque choisie, pour servir de suite à la bibliothèque universelle. **CONTINUATION OF:** Bibliothèque ancienne et moderne, pour servir de suite aux bibliothèque universelles et choisies.Reprint by Slatkine in 1969. **EDITORS:** Pierre Massuet, Willem Jacob Gravesande, Jean Rousset and Louis de Jaucourt. **INDEXES:** There are indexes to v.1-25 and v.26-50. **MICROFORMS:** ACR(1). **SOURCES:** B939, G619,OCLC **LOCATIONS:** BUC, CCF, CCN, ULS, ZDB **SUBJECT:** General - Periodicals - Netherlands.

399   Bibliothèque salutaire, ou, recueil choisie d'observations sur la physique, la médecine, la chirurgie, l'histoire naturelle et l'économie animale; extrait des mémoires de toutes les compagnies savantes de l'Europe. - T.1-4, 1787. Paris: Leory, 1787. **SOURCES:** G276 **LOCATIONS:** **SUBJECT:** Science - Periodicals - France - Reviews. **SUBJECT:** Medicine - Periodicals - France - Reviews.

400   Bibliothèque universelle et historique. - v.1-25, 1686-1694. Amsterdam: Wetstein, 1686-1694. Monthly. Imprint varies. There are several editions and reprints. Analyzed in Reesink (Ref.13). **EDITORS:** Jean LeClerc, Jean Cornand de La Crose and Jacques Bernard. **CONTINUED BY:** Bibliothèque choisie, pour servir de suite à la bibliothèque universelle. **CONTINUED BY:** Bibliothèque ancienne et moderne, pour servir de suite aux bibliothèque universelles et choisies. **CONTINUED BY:** Bibliothèque raisonée des ouvrages des savans de l'Europe. **INDEXES:** Indexes to v.1-25 in v.26. Index multiplex à la Bibliothèque universelle et historique, 1686-1694, ed. Jean Pierre Lobies. Osnabrück, Biblio Verlag, 1968-1972, in 7v. **MICROFORMS:** RP(1). **SOURCES:** B940, OCLC **LOCATIONS:** BUC, CCF, CCN, ULS, ZDB **SUBJECT:** General - Periodicals - Netherlands.

401   Der Bienenstock, eine ökonomische Wochenschrift. - Bd.1-3, 1768-1771. Vienna: Trattner, 1768-1771. Publisher varies. **EDITOR:** Johann Georg Wolf. **SOURCES:** K2891 **LOCATIONS:** CCP, ZDB **SUBJECT:** Economics - Periodicals - Germany.

402   Bijdragen ter bevordering der schoone kunsten en wetenschappen. - Deel 1-13, 1793-1796. Amsterdam: Johannes Allert, 1793-1796. Continued for two issues in 1825. **EDITORS:** Rhijnuis Feith and Jacobius Kantelaar. **SOURCES:** S899.1 **LOCATIONS:** BUC, CCN **SUBJECT:** General - Periodicals - Netherlands.

403   Blicke in die Natur, periodisches Werk. - Bd.1, 1792. Prague: 1792. **EDITOR:** G.J. Wenzel. **SOURCES:** K3347 **SUBJECT:** Natural history - Periodicals - Czechoslovakia.

404   Blumenlese nützlicher und angenehmer Kenntnisse aus der Natur. - Quartal 1-5, 1787-1788. Leipzig: 1787-1788. **EDITOR:** Christian Gottlieb Weise. **SOURCES:** B985, K3317, S2907 **SUBJECT:** Natural history - Periodicals - Germany.

405 Boekzaal der geleerde wereld of tijdschrift voor leteerkundigen. - Amsterdam: B. Bos, 1705-1708. **CONTINUATION OF:** Boekzaal van Europe. **CONTINUATION OF:** Twee-maandellijke uitreksel. Continued under this title from 1811-1863 following the Maandelijke uittreksels. **EDITORS:** Pieter Rabus, Willem Sewel and Joan Van Gavern. **CONTINUED BY:** Maandelijke uitreksels of Boekzaal der geleerde wereld. **INDEXES:** Register der boeken, uytgetrokken in de Boekzaalen, of, Tweemandelyke uytreksels, vorrheenen gesticht in de jaaren 1692 tot 1708 van de heeren P. Rabus, W. Sewel, en J. van Gavern: en nu in ordre gebracht door J. Le Long. Amsterdam, 1716. **SOURCES:LOCATIONS:** BUC,CCF,CCN **SUBJECT:** General - Periodicals - Netherlands - Reviews.

406 De Boekzaal van Europe. - Deel 1-19, 1692-1791. Rotterdam: Van der Staart, 1692-1791. Bimonthly. **EDITOR:** Pieter Rabus. **CONTINUED BY:** Twee-maandellijke uitreksels. **CONTINUED BY:** Boekzaal der geleerde wereld of tijdschrift voor leteerkundigen. **CONTINUED BY:** Maandelijke uitreksels of Boekzaal der geleerde wereld. **INDEXES:** Index to Deel 1-19 in 19. See also: Register der boeken, uytgetrokken in de Boekzaalen of Tweemandelyke uytreksels, vorrheenen gesticht in de jaaren 1692 tot 1708 van de heeren P. Rabus, W. Sewel, en J. van Gavern: en nu in ordre gebracht door J. Le Long. Amsterdam, 1716. **SOURCES:** OCLC **LOCATIONS:** BUC, CCF, CCN **SUBJECT:** General - Periodicals - Netherlands - Reviews.

407 Le Bon jardinier almanach. - v.1-46, 1753-1798. Paris: Audot, 1753-1798. **CONTINUED BY:** Jardinier portatif. **CONTINUED BY:** Almanach du bon jardinier. **CONTINUED BY:** Bon jardinier. **SOURCES:** B1011 **LOCATIONS:** BUC, ULS **SUBJECT:** Gardening - Periodicals - France.

408 De Bononiensi scientiarum et artium instituto atque academia commentarii / Reale Accademia delle Scienze dell' Istituo di Bologna. - v.1 (1731), v.2 in 3 pts.(1745-1747), v.3(1755), v.4(1757), v.5 in 2 pts.(1767), v.6(1783), v.7(1791) (Ref.28,p.311ff). T.1-7, 1731-1791. **TRANSLATED AS:** Abhandlungen zur Naturgeschichte, Chemie, Anatomie, Medizin und Physik aus den Schriften des Instituts der Künste und Wissenschaften zu Bologna. Other editions of some of the volumes are cited. Translated into French in Collection Académique, v.10. **OTHER TITLE:** Commentarii societatis regiae scientiarum, Bologna. **CONTINUED BY:** Novi commentarii societatis regiae scientiarum: Reale Accademia delle scienze dell' Istituto di Bologna. **MICROFORMS:** RM(3). **SOURCES:** G493, S1807 **LOCATIONS:** BUC, CCF, ULS, ZDB **SUBJECT:** Science - Societies - Bologna.

409 The Botanical magazine, or, flower garden displayed. - v.1-14, 1787-1800. London: W. Couchman, 1787-1904. Supplements issued 1788-1790 and 1835-36 under the title: Companion to the Botanical Magazine (B:1330). **OTHER TITLE:** Curtis's botanical magazine. **EDITOR:** William Curtis. **INDEXES:** Curtis, S. General indexes to plates contained in the first fifty-three volumes of the Botanical magazine. London, 1828. Tonks, E. General index to the Latin names and synonyms of the plants depicted in the first hundred and seven volumes. London, 1883. Hemsley, W. Batting. A new and complete index to

the Botanical magazine from its commencment in 1787 to the end of 1904, including the first, second and third series. London, 1906. **MICROFORMS: IDC(8). SOURCES:** B1025, G555, OCLC **LOCATIONS: BUC, CCF, CCN, ULS, ZDB SUBJECT:** Botany - Periodicals - England.

410 Botanical review. - v.1, 1790. London: 1790. **EDITOR:** Edward O. Donovan. **SOURCES:** B1028 **LOCATIONS:** ULS **SUBJECT:** Botany - Periodicals - England.

411 Der Botaniker, oder, compendiöse Bibliothek aller Wissenswürdigen aus dem Gebiete der Botanik. - 15 parts in 1 vol. Gotha: 1793-1795. Imprint and subtitle vary. **SOURCES:** B1029, G556, S2659 **LOCATIONS:** CCP,ZDB **SUBJECT:** Botany - Periodicals - Germany.

412 Botanische Beobachtungen. - Mannheim: Neue Hof und Akademische Buchhandlung, 1783-1784. **LOCATIONS:** ZDB **SUBJECT:** Botany - Periodicals - Germany.

413 Botanische Unterhaltungen mit jungen Freunden der Kräuterkunde auf Spatziergängen. - St.1-22, 1784-1786. Munich: Johann Bapt.Stobl, 1784-1786. Essays in the form of dialogues. **EDITOR:** Georg Anton Weizenbeck. **SOURCES:** B1033, G557, K3304, S3113 **LOCATIONS: BUC SUBJECT:** Botany - Periodicals - Germany - Popular.

414 Botanisches Jahrbuch für Jedermann. - Bd.1, 1799.Lüneberg: 1799. **EDITOR:** J.J. Schmidt. **SOURCES:** B1040, S3059 **SUBJECT:** Botany - Periodicals - Germany.

415 Botanisches Taschenbuch für die Anfänger dieser Wissenschaft und der Apothekerkunst. - Bd.1-15, 1790-1804. Regensburg: Montags Erben, 1790-1804. Imprint and subtitle vary. **EDITOR:** David Heinrich Hoppe. **CONTINUED BY:** Neues botanisches Taschenbuch für die Anfänger diesser Wissenschaft und der Apothekerkunst. **SOURCES:** B1041, G558, K3341, OCLC, S3192 **LOCATIONS:** CCF, ULS, ZDB **SUBJECT:** Botany - Periodicals - Germany. **SUBJECT:** Pharmacy - Periodicals - Germany.

416 Botanist's repository for new and rare plants. - v.1-10, 1797-1815. London: J. White, 1797-1815. **EDITOR:** Henry Charles Andrews. **SOURCES:** B1049, G559 **LOCATIONS:** BUC, ULS **SUBJECT:** Botany - Periodicals - England.

417 Bremisches Magazin zur Ausbreitung der Wissenschaften, Künste und Tugend, von einigen Liebhabern derselben, mehrentheils aus den englischen Monatsschriften gesammelt und herausgegeben. - Bd.1-7 (3 issues each) 1757-1765. Hannover: Nicolai Förster, 1757-1765. Imprint varies. **EDITOR:** Johann Philip Cassel. **CONTINUED BY:** Neues Bremisches Magazin zur Ausbreitung der Wissenschaften, Künste und Tugend. **MICROFORMS:** RM(3). **SOURCES:** B1066, K5253, OCLC, S2780 **LOCATIONS:** BUC, CCF, ULS, ZDB **SUBJECT:** Science - Periodicals - Germany.

418 Briefe an das schöne Geschlecht über verschiedene Gegenstände aus dem Reiche der Natur. - Bd.1-3, 1770-1771. Jena: 1770-1771. **SOURCES:** K3250 **SUBJECT:** Natural history - Periodicals - Germany - Popular.

419 Britische Bibliothek. - Bd.1-6, Bd.7, St.1, 1756-1767. Leipzig: Johann Wendler, 1756-1767. **EDITOR:** Karl Wilhelm Müller. **SOURCES:** K208 **LOCATIONS:** BUC, CCF **SUBJECT:** General - Periodicals - Germany - Reviews

420 British diary. - v.1-9, 1787-1796. Birmingham: Thomas Pearson, 1787-1796. Contained mathematical questions (Ref.16,53:451). **OTHER TITLE:** Diaria Britannica. **EDITORS:** John Cotes and George Taylor, v.1-7; John Cotes and Patrick, v.8-9. **SOURCES:** (Ref.16,v.53:451). **SUBJECT:** Mathematics - Periodicals - England - Popular. **SUBJECT:** General - Periodicals - England.

421 The British merchant, or commerce preserv'd. - Nos.1-103, 1713-1714. London: A. Baldwin, 1713-1714. Translated into French as Négociant anglois. Dresden, 1753. **OTHER TITLE:** Négociant anglois. **EDITOR:** Charles King. Microforms RP(1) (Goldsmith Kress Library). **LOCATIONS:** BUC, ULS **SUBJECT:** Economics - Periodicals - England.

422 The British physician, treating of our diet and common nourishment, of air, of medicines, of physical and chirurgical writers &c. - No. 1-5, 1716. London: Mayhew, 1716. **OTHER TITLE:** Great Britain's rules of health (nos.3-5). **SOURCES:** GIII:2, OCLC **LOCATIONS:** BUC **SUBJECT:** Medicine - Periodicals - England - Popular.

423 The British telescope, being an ephemeris of the coelestial motions, with an almanack-for the year. - No.1-26, 1724-1749. London: Henry Parker, 1724-1749. **EDITOR:** Edmund Weaver. **SOURCES:** B1087, OCLC, S229 **LOCATIONS:** BUC **SUBJECT:** Astronomy - Periodicals - England. **SUBJECT:** Astrology - Periodicals - England.

424 Builder's magazine, or, monthly companion for architects, carpenters, masons, bricklayers &c. By a society of architects. - v.1, 1774. London: Newberry, 1774. **MICROFORMS:** RP(1) (Goldsmiths'-Kress Library). **LOCATIONS:** BUC, ULS **SUBJECT:** Building - Periodicals - England.

425 Bulletin de la Société libre d'emulation de Rouen. - Rouen: 1797-1899. Published also under the titles: Séances publique, Bulletin des travaux,, Reports des travaux and Procés verbal de la séance. **INDEXES:** Société libre d'emulation et de l'industrie de la Seine- Inferieure. Table générale du Bulletin publié par la Société de 1797 à 1899. Rouen, E.Cognard, 1900. **SOURCES:** G511, S1646a **SUBJECT:** Science - Societies - Rouen.

426 Bulletin de la Société Philomathique de Paris. - T.1-3 (Nos. 1-96) 1789-1805. Paris: 1789-1805. The first 15 nos. appeared monthly in manuscript and were reprinted in 1802 (Ref.61). **CONTINUED BY:** Bulletin des sciences par la Société Philomathique de Paris. **CONTINUED BY:** Nouveau bulletin des

sciences, Société Philomathique de Paris. **MICROFORMS:** Micro editions Hachete, MHA(9). **SOURCES:** G512, OCLC, S1594b **LOCATIONS:** BUC, CCF, ULS, ZDB **SUBJECT:** Science - Societies - Paris.

427 Bulletin universel des sciences, des lettres et des arts. - No.1-14, 1800-1801. Paris: 1800-1801. **SOURCES:** B1122, G622, S1351 **LOCATIONS:** CCF **SUBJECT:** General - Periodicals - France.

428 Cahiers périodiques du cours d'agriculture. - Pars 1-26, 1789-1790. Paris: 1789-1790. **EDITOR:** Sarcey de Sutières. **SOURCES:** B1179 **LOCATIONS:** NUC **SUBJECT:** Agriculture - Periodicals - France.

429 Calendario reale georgico, ossia almanacco d'agricultura ad uso principalemente degli agronomi piemontesi. - Turin: 1792-1802. **SOURCES:** (Ref.76) **SUBJECT:** Agriculture - Periodicals - Italy.

430 Calendarium astrologicum. - No.1-79, 1669-1747. London: 1669-1747. **EDITOR:** Trigge. **SOURCES:** B1186, S230 **LOCATIONS:** BL **SUBJECT:** Astrology - Periodicals - England.

431 Calendarium medicum ad usum saluberrimae Facultatis parisiensis, in quo habentur quaestiones propositae, tem en thesibus, quam in actibus vesperiarum, doctoratus et pastellariae (Ref.35). - Paris: 1757-1785. **LOCATIONS:** BN **SUBJECT:** Medicine - Societies - Paris.

432 Calendrier à l'usage des membres du Collège de Pharmacie et de leurs éleves (Ref.36). - Paris, 1782-1801. **CONTINUED BY:** Annuaire à l'usage des membres du Collège de Pharmacie. **LOCATIONS:** CCF **SUBJECT:** Pharmacy - Societies - Paris.

433 Calendrier à l'usage du Collège de Chirurgie de Paris. - Paris: 1793. **SOURCES:** G1793 **SUBJECT:** Surgery - Societies - Paris.

434 Calendrier de l'almanach sous-verre, avec l'indication des articles de la grand notice des associés. - Paris: 1769-1832.Notice de l'almanach sous-verre des associeés, issued as supplement. **LOCATIONS:** BUC **SUBJECT:** Science - Periodicals - France.

435 Calendrier des jardiniers, qui enseigne ce qu'il faut faire dans le potager, dans les pépinieres, dans les serres & dans le jardin de fleurs tout les mois de l'année. - Paris: Piget et Durand, 1738-1785, (Ref.35:175). Translated from the English of Richard Bradley. Imprint varies. There was another translation by Philip Miller with a variant title from the 6th English edition published in Brussels in 1789. **SOURCES:** OCLC **SUBJECT:** Gardening - Periodicals - France.

436 Calendrier des jardiniers. (Duplicates 435)

437 Calendrier intéresant, ou, almanach physico-économique. - Paris: 1772-1773. **LOCATIONS:** BN **SUBJECT:** Science - Periodicals - France.

438   Calendrier intéressant pour l'année, ou, almanach physico-économique. - Bouillon: Société typographique, 1783. **SOURCES: OCLC   SUBJECT:** Science - Periodicals - Belgium.

439   Carlsruher nützliche Sammlungen, oder, Abhandlungen aus allen Theilen derer Wissenschaften, besonders dem Staats-und Lehnrechte, denen Geschichten der Naturlehre, dem Policei-, Cameral-, Handlungs- und Fabrikwesen, wie auch der Haus-und Landwirthschaft. - Bd.1 (St.1-52) 1758. Carlsruhe: Michael Macklot, 1758. **SOURCES:** B1213, K218, S2841 **LOCATIONS:** ZDB **SUBJECT:** Economics - Periodicals - Germany. **SUBJECT:** Commerce - Periodicals - Germany.

440   Cartas criticas periódicos destinadas a la Facultad de Medicina. - Madrid: 1788. Appeared in an edition of 500. **SOURCES:** (Ref.62,p.26). **SUBJECT:** Medicine - Societies - Madrid.

441   Cases and observations / Medical Society of New Haven County. - v.1, 1788. New Haven: J. Meigs, 1788. The first volume of medical transactions to be published in the United States (Ref.63). Published in facsimile in the Yale Journal of Biology and Medicine, 1933/4, 6:213-298. **SOURCES:** G79, OCLC **LOCATIONS:** ULS **SUBJECT:** Medicine - Societies - New Haven.

442   Cases, medical, chirurgical, and anatomical, with observations, translated from the History and memoirs of the Royal Academy from 1666 to the Present time, by Loftus Wood. - v.1, 1776. **LOCATIONS: NUC   SUBJECT:** Medicine - Societies - Paris - Translations.

443   Der Cederbaum, ökonomische Wochenschrift. - Vienna: 1768. **SOURCES:** K2897 **SUBJECT:** Economics - Periodicals - Austria.

444   Le Censeur universel, revue générale de toutes les manufacture, le commerec &c. - Paris: Guillot, 1785-1787. Issued daily in 1785, it became a weekly in 1786. **LOCATIONS: CCF   SUBJECT:** General - Periodicals - France - Reviews.

445   Centurie des questions traitées ès conférences du bureau d'adresse. - T.1-5 (1633-1642) 1636-1655. Paris: 1636-1655. **TRANSLATED AS:** General collection of discourses of the virtuosi of France. Reprinted under the title: Recueil générale des questions traictées ès conférences du Bureau d'adresse in 5 vols. Paris: 1658-1660, and in Lyon (1668) in 2 vols. under the title: La Bibliothèque des sciences contenant les plus belles questions de la philosophie. A translation in English by G. Havers was published in London in 1664 under the title: A General collection of the discourses of the virtuosi of France. Several of the individual conferences were also translated into English from 1640 (Ref.64). **OTHER TITLE:** Recueil générale des questions traictées ès conférences du Bureau d'adresse; Bibliothèque des sciences contenant les plus belles questions de la philosophie; Conférences du Bureau d'adresse. **EDITORS:** Théophraste Renaudot, and Eusèbe Renaudot. **MICROFORMS:** RP(1).   **SOURCES:** S1357 **LOCATIONS:** CCF **SUBJECT:** General - Societies - Paris.

446   Chemische Annalen für die Freunde der Naturlehre, Arzneygelahrtheit,
      Haushaltungskunst und Manufacturen. - Bd.1-Bd.40, 1784-1804. Helmstädt und
      Leipzig: C.G.Fleckeisen, 1784-1803. TRANSLATED AS: Crell's chemical
      journal. EDITOR: Lorenz Florenz Friedrich von Crell. CONTINUED BY:
      Beiträge zu den chemischen Annalen. Supplemented with: Beiträge zu den
      chemischen Annalen. INDEXES: Indexes for 1784-1791 in v.2 of 1791, for
      1792-1799 in v.2 of 1799, and for 1800-1803 in v.2 of 1804. Also indexed in
      Geuss (Ref.53, v.4). SOURCES: B1254, G623, K3297, OCLC, S2801
      LOCATIONS: BUC, CCF, CCN, ULS, ZDB   SUBJECT: Chemistry -
      Periodicals - Germany.

447   Chemische en phijsische oefeningen voor de bimnnaars der schei en naturkunde.
      - Deel 1-3, 1792-1797. Amsterdam; Leyden: Honkoop en van Tiffelen,
      1792-1797. Imprint varies. EDITORS: Petrus Johannes Kastelyn, Nicolas
      Bondt and J.R. Deiman. CONTINUED BY: Nieuwe chemische en phijsische
      oefeningen. SOURCES: B1256, OCLC, S728   SUBJECT: Science -
      Periodicals - Netherlands.

448   Chemisches Archiv.- Bd.1-Bd.2, 1783. Leipizg: Weygand, 1783. EDITOR:
      Lorenz Florenz Friedrich von Crell. CONTINUED BY: Neues chemisches
      Archiv. CONTINUED BY: Neuestes chemisches Archiv. INDEXES: Indexed
      by Engelhardt (Ref.51). SOURCES: B1257, G624, K3293, OCLC, S2912
      LOCATIONS: BUC, CCF, CCN, ULS, ZDB   SUBJECT: Chemistry -
      Periodicals - Germany.

449   Chemisches Journal für die Freunde der Naturlehre, Arzneygelahrtheit,
      Haushaltungskunst und Manufakturen. - Theil 1-Theil 6, 1778-1781. Lemgo:
      Meyer, 1778-1781.   EDITOR: Lorenz Florenz Friedrich von Crell.
      CONTINUED BY: Neuesten Entdeckungen in der Chemie. CONTINUED
      BY: Auswahl aller eigenthümlicher Abhandlungen und Beobachtungen in der
      Chemie. INDEXES: Indexed by Engelhardt (Ref.51). MICROFORMS:
      RM(3). SOURCES: B1256, G625, K3265, OCLC, S3054 LOCATIONS:
      BUC, CCF, CCN, ULS, ZDB   SUBJECT: Chemistry - Periodicals -
      Germany.

450   Chirurgische Bibliothek. - Bd.1-15, 1771-1796. Göttingen: Johann Christian
      Dieterich, 1771-1797. EDITOR: August Gottlieb Richter. CONTINUED BY:
      Bibliothek für die Chirurgie. INDEXES: Index to Bd.1-6, and Bd.7-12, by C.F.
      Witting. Göttingen, Dieterich, 1794 and 1796.   SOURCES: G28, K3557,
      OCLC LOCATIONS: BUC, ULS, ZDB   SUBJECT: Surgery - Periodicals
      - Germany - Reviews.

451   Choix des mémoires et abrégé d'histoire / Académie des sciences et belles-lettres,
      Berlin. - v.1-4, 1768. Berlin: 1768. A 2nd edition in Paris, 1770.
      TRANSLATION OF: Miscellanea Berolinensis.   LOCATIONS: BUC
      SUBJECT: Science - Societies - Berlin - Translations.

452   Choix économique et moral. - T.1-2, 1777.Geneva: Téron, 1777. Bimonthly 10
      July -25 December, 1777. **EDITOR:** Jean Téron. **SOURCES:**
      (Ref.29,p.36-7). **SUBJECT:** Economics - Periodicals - Switzerland.

453   Christianias physicalske aarbog. - v.1-3, 1783-1785. Christiania: 1783-1785.
      EDITOR: Christian Ernst Wildberg Schulze. **CONTINUED BY:** Physicalske
      aarbog. **SOURCES:** B1282, S579   **SUBJECT:** Science - Periodicals -
      Denmark.

454   Christian's scholar's and farmer's magazine, calculated, in an eminent degree, to
      promote religion, to disseminate useful knowledge, to afford literary pleasure
      and amusement, and to advance the interest of agriculture. - v.1, no.1,
      Apr./May, 1789-v.2, no.6, Feb./Mar. 1791. Elizabeth-Town, N.J.: Shepard
      Kollock, 1789-1791. Bimonthly. **MICROFORMS:** UM(1). **SOURCES:**
      OCLC **LOCATIONS:** BUC, ULS, ZDB **SUBJECT:** General - Periodicals
      - United States. **SUBJECT:** Agriculture - Periodicals - United States.

455   Chursächsische Bergwerks-calender. - Bd.1-11, 1773-1783.   Marienberg:
      1773-1783. **SOURCES:** B1287   **SUBJECT:** Mineral industries - Periodicals
      - Germany.

456   Der Chursächsische Landphysikus, eine medicinisch-physikalische Monatsschrift
      zum Besten des Landmannes. - Bd.1-3, 1771-1773. Naumburg: 1771-1773.
      Monthly. Imprint varies. **EDITOR:** Friedrich August Weiz (also known as
      Waitz). **SOURCES:** K3559 **LOCATIONS:** NUC **SUBJECT:** Medicine -
      Periodcals - Germany - Popular.

457   Chymische Experimente einer Gesellschaft im Erzgebürge. - Nr.1-6, 1753-1756.
      Berlin: Johann Jacob Schütze, 1753-1759. It may be the first periodical by a
      chemical society (Ref.33,p.36-7). Imprint varies.   **SOURCES:** K3216
      LOCATIONS: ZDB   **SUBJECT:** Mineral industries - Societies - Berlin.
      SUBJECT: Chemistry - Societies - Berlin.

458   Chymische Nebenstunden. - St.1-2, 1762-1768. St. Petersburg: 1762-1768.
      TRANSLATED AS: Recreations physiques, économiques et chimiques.
      EDITOR: Johann Georg Model. **SOURCES:** K3234 **LOCATIONS:** BN,
      ZDB **SUBJECT:** Chemistry - Periodicals - Germany.

459   Der Chymische Warsager, oder, eine zwanzig Wochenblättern herausgegebene
      Unterweisung. - Bd.1-20, 1747-1748. Hamburg: J.C. Bohn, 1747-1748.
      Weekly. An alchemical journal (Ref.65). **MICROFORMS:** UWM(1).
      SOURCES: K4913 **SUBJECT:** Alchemy - Periodicals - Germany.

460   Collectanea austriaca ad botanicam, chemiam et historiam naturalem spectantium.
      - v.1-4, 1786-1790. Vienna: Wappler, 1786-1790. **CONTINUATION OF:**
      Miscellanea austriaca ad botanicam, chemiam et historiam naturalem
      spectantiam. **EDITOR:** Nikolaus Joseph Jacquin. A supplement to the series
      was published in 1796.   **SOURCES:** B1307, G626, OCLC, S3579
      LOCATIONS: ULS **SUBJECT:** Science - Periodicals - Austria.

461 Collectanea chirurgica. - v.1-2 (1721-1722) 1722-1723. Hannover: 1722-1723. SOURCES: K3517 LOCATIONS: ZDB SUBJECT: Surgery - Periodicals - Germany.

462 Collectanea Hibernica medica. - Dublin: J. Exshaw, 1783. EDITOR: Richard Harris. SOURCES: (Ref.66). SUBJECT: Medicine - Periodicals - Ireland.

463 Collectanea medico-physica, oder, Holländisch Jahr-Register sonderbahrer Anmerkungen, die sowohl in der Artzney-Kunst als Wissenschaft der Natur in gantz Europa vorgefallen. - v.1-9, 1680-1690. Leipzig: Weidmann, 1690. EDITOR: Steven Blankaart. TRANSLATION OF: Collectanea medico-physica oft Hollands jaar-register der genees-en-natuurkundige aanmerkingen van gantsch Europa. MICROFORMS: RP(1). SOURCES: GI:5a, K3514 LOCATIONS: CCF SUBJECT: Science - Periodicals - Netherlands - Translations. SUBJECT: Medicine - Periodicals - Netherlands - Translations.

464 Collectanea medico-physica, oft, Holland jaar-register der genees-en-natuurkundige aanmerkingen van gantsch Europa. - Deel 1-4, 1680-1688. Amsterdam: Johan ten Hoorn, 1680-1688. TRANSLATED AS: Collectanea medico-physica oder Holländisch Jahr-Register. EDITOR: Steven Blankaart. SOURCES: B1398, GI:5, OCLC, S729 LOCATIONS: CCN, ULS SUBJECT: Science - Periodicals - Netherlands. SUBJECT: Medicine - Periodicals - Netherlands.

465 Collectanea societatis medicae Havniensis. - v.1-2, 1774-1775. Copenhagen: J.G. Rothe, 1774-1775. CONTINUATION OF: Acta medica et philosophica Hafniensia. CONTINUED BY: Acta societatis medicae Havniensis. CONTINUED BY: Acta regiae societatis medicae Havniensis. CONTINUED BY: Acta nova regiae societatis medicae Hafniensis. SOURCES: GI3 LOCATIONS: BUC, CCF, CCN, ULS, ZDB SUBJECT: Medicine - Societies - Copenhagen.

466 Collectanea zum nützlichen Vergnügen bei müssigen Gartenstunden. - St.1-20, 1768. Hamburg: Schröder, 1768. SOURCES: K5325 SUBJECT: Botany - Periodicals - Germany - Popular.

467 Collection académique, composées des mémoires, actes ou journaux des plus célèbres académies et sociétés. - v.1-13, 1755-1779. Dijon: F. Desventes; Paris: Pancoucke, 1755-1779. Various editions are cited (Ref.21,v.23:1503a). Imprint varies. Also published as part of: Recueil des mémoires, partie étrangères. Includes translations of proceedings of various scientific societies and periodicals (Ref.26,p.212-15). OTHER TITLE: Collection académique, partie étrangères; Recueil des mémoires, ou, collection des pièces académiques. Edited by Jean Berryat and others. INDEXES: Indexed in Rozier's Nouvelle table des articles contenue dans les Collection Académique. MICROFORMS: IDC(8). SOURCES: B1310, G465, S1139 LOCATIONS: BUC, CCF, CCN, ULS SUBJECT: Science - Periodicals - France - Abstracts. SUBJECT: Science - Societies - France - Abstracts.

468   Collection de différentes pièces, concernant la chirurgie, l'anatomie & la médecine pratique, extraites principalement des ouvrages étrangers. - v.1, 1761. Paris: Le Breton, 1761. Compiled by Jean François Simon. **SOURCES:** G343, OCLC **LOCATIONS:** ULS **SUBJECT:** Medicine - Periodicals - France - Reviews.

469   Collection de thèses médico-chirurgicales sur les points les plus important de la chirurgie publiées par Haller. T.1-T.5, 1757-1760. Paris: 1757-1760. Translated by Henri Jacques Macquart. **EDITOR:** Albrecht von Haller. **TRANSLATION OF:** Disputationes chirurgicae selectae. **SOURCES:** G352a **SUBJECT:** Surgery - Periodicals - Switzerland - Dissertations - Translations.

470   A Collection of letters for the improvement of husbandry and trade. - London: J. Gain, 1681-1703. Weekly. Originally published in 583 weekly folio numbers in two series, 1681-1684 and 1692-1703. Issued in a revised edition in London, 1727-1728 in 4 v. by Richard Bradley under the title: Husbandry and trade improved. Reprinted by Gregg International Publishers in 1969. **OTHER TITLE:** Husbandry and trade improved. **EDITORS:** John Houghton and Richard Bradley. **MICROFORMS:** RP(1) (Goldsmiths' Kress Library). **SOURCES:** OCLC **LOCATIONS:** BUC, ULS **SUBJECT:** Agriculture - Periodicals - England. **SUBJECT:** Economics - Periodicals - England.

471   A Collection of mathematical problems and aenigmas. - 5 v., 1741-1745. London: J. Fuller, 1741-1745. **OTHER TITLE:** Gentlemen's diary, or, The mathematical repository, supplement. **EDITOR:** Thomas Peat (Ref.12, p.383). Supplement to Gentlemen's diary; or, The mathematical repository. **SOURCES:** OCLC **SUBJECT:** Mathematics - Periodicals - England.

472   Collegium experimentale, sive, curiosum in quo primaria seculi superioris inventa. - Nürnberg: W.M. Endteri, 1676-1701. Imprint varies. **OTHER TITLE:** Tentamium Collegi curiosi. Edited by Johann Christophorus Sturm. Supplement issued as Tentaminum Collegii curiosi in 1676. **MICROFORMS:** RP(1). **SOURCES:** GII:8, OCLC **SUBJECT:** Science - Societies - Nuremberg.

473   Cometicae observationes / Academia physico-mathematica. - 1 v.1681. Rome: 1681. **SOURCES:** GII:10 **LOCATIONS:** BUC **SUBJECT:** Science - Societies - Rome.

474   Commentarien der neuern Arzneikunde. - v.1-6, 1793-1800. Tübingen: 1793-1800. **EDITOR:** Christian Gottlob Hopf. **SOURCES:** G115 **LOCATIONS:** BUC, ULS, ZDB **SUBJECT:** Medicine - Periodicals - Germany.

475   Commentarii Academia scientiarum imperialis Petropolitana. - T.1-14 (1726-1746) 1728-1751. St. Petersburg: Academiae Scientiarum; Leipzig: Hartknoch, 1728-1751. A new edition published in 14 v. in Bologna beginning with 1740 is cited (Ref.25.v.1:20). **CONTINUED BY:** Novi commentarii Academiae scientiarum imperialis Petropolitanae. **CONTINUED BY:** Acta Academia

scientiarum imperialis Petropolitanae. **CONTINUED BY:** Nova acta Academiae scientiarum imperialis Petropolitanae. **INDEXES:** Cumulative index (Ref.115). **MICROFORMS:** IDC(8). **SOURCES:** G442, S3706 **LOCATIONS:** BUC, CCF, CCN, ULS, ZDB **SUBJECT:** Science - Societies - St. Petersburg.

476   Commentarii de rebus in scientia naturali et medicina gestis. - v.1-37, 1752-1808. Leipzig: Fredericum Gleditsch, 1782-1808. **EDITORS:** Christian Gottlieb Ludwig, Johann Daniel Reichel, and Nathanael Gottfried Leske. 4 supplements were issued to each of the first three series of ten volumes and one to the fourth. **INDEXES:** Each of the first three series have their own indexes. **SOURCES:** B1325, G627, K3534, OCLC **LOCATIONS:** BUC, CCF, CCN, ULS, ZDB **SUBJECT:** Science - Periodicals - Germany - Reviews.

477   Commentarii medici, opera periodici. - v.1-3, 1795-1800. Pavia: 1795-1800. **EDITORS:** Luigi Gaspero Brugnatelli and Valeriano Ludwig Brera. **SOURCES:** G128 **LOCATIONS:** ZDB **SUBJECT:** Medicine - Periodicals - Italy.

478   Commentarii Societatis regiae scientiarum gottingensis. - v.1-4 (1751-1754) 1752-1754. Göttingen: Vandenhoeck, 1752-1754. Publisher varies. In a format similar to that of the Histoire et mémoires of the Académie des Sciences, Paris. Kraus reprint in 1972. **CONTINUED BY:** Commentationes regiae societatis Gottingensis. **CONTINUED BY:** Novi commentarii societatis regiae scientiarum Gottingensis. **CONTINUED BY:** Commentationes societatis regiae scientiarum Gottingensis. **CONTINUED BY:** Deutsche Schriften der königlichen Societät der Wissenschaften zu Göttingen. **INDEXES:** Index to the society's publications from 1751-1808 in v.16 of its Commentarii for 1804/1808. Also indexed in Vezenyi (Ref.1). **MICROFORMS:** IDC(8), RM(3). **SOURCES:** G472, K181, OCLC, S2694 **LOCATIONS:** BUC, CCF, CCN, ULS, ZDB **SUBJECT:** General - Societies - Göttingen.

479   Commentationes chemicae / Academia Electoralis Moguntinae scientiarum utilium quae Erfurt est. - v.1 (1778-1779) 1780. Erfurt: Geo. Ad. Keyser, 1780. **SOURCES:** G439b, S2546b **SUBJECT:** Chemistry - Societies - Erfurt.

480   Commentationes medicae / Academia Electoralis Moguntinae scientiarum utilium quae Erfurt est. - v.1 (1778-1779) 1780. Erfurt: Geo. Ad. Keyser, 1780. **SOURCES:** G439b **SUBJECT:** Medicine - Societies - Erfurt.

481   Commentationes physicae / Academia Electoralis Moguntinae scientiarum utilium quae Erfurt est. - v.1 (1778-1779) 1780. Erfurt: Geo. Ad. Keyser, 1780. **SOURCES:** G439b, S2546c **SUBJECT:** Science - Societies - Erfurt.

482   Commentationes regiae societatis Gottingensis. - v.1-2, (1758-1768) 1763-1769. Bremen, 1763-1769. **CONTINUATION OF:** Commentarii Societatis regiae scientiarum gottingensis. A 2nd edition for the years 1758-1763 was published in Bremen. Kraus Reprint. **CONTINUED BY:** Novi commentarii societatis regiae scientiarum Gottingensis. **CONTINUED BY:** Commentationes societatis

regiae scientiarum Gottingensis. **CONTINUED BY:** Deutsche Schriften der königlichen Societät der Wissenschaften zu Göttingen. **INDEXES:** Index to the society's publications from 1751-1808 in v.16 of its Commentarii for 1804/1808. Also indexed in Vezenyi (Ref.1). **MICROFORMS:** IDC(8). **SOURCES:** G472a, S2434 **LOCATIONS:** BUC, ULS, ZDB **SUBJECT:** General - Societies - Göttingen.

483   Commentationes societatis regiae scientiarum Gottingensis. - v.1-16 (1778-1808) 1779-1808. Göttingen: Dieterich, 1779-1808. **CONTINUATION OF:** Commentationes regiae societatis Gottingensis. **CONTINUATION OF:** Commentarii Societatis regiae scientiarum gottingensis. **CONTINUATION OF:** Novi commentarii societatis regiae scientiarum Gottingensis. **EDITOR:** Christian Gottlob Heyne. **CONTINUED BY:** Deutsche Schriften der königlichen Societät der Wissenschaften zu Göttingen. **INDEXES:** Index to the society's publications from 1751-1808 in v.16 of its Commentarii for 1804/1808. Also indexed in Vezenyi (Ref.1). **MICROFORMS:** IDC. **SUBJECT:** General - Societies - Göttingen.

484   Commentationum medici / Societas regia scientiarum Gottingensis. - v.1, 1800. Göttingen: 1800. **EDITOR:** Heinrich August Wrisberg. **SOURCES:** G472f **SUBJECT:** General - Societies - Göttingen.

485   Commentationum recensitarum sylloge / Societas regia scientiarum Gottingensis. - v.1 (1756-1762) 1762, Sylloge altera, iv.1775, Sylloge commentationum pars 2, 1778. Göttingen: Vandenhoeck, 1762-1778. A 2nd ed. was published in 1784. **EDITOR:** Samuel Christian Hollmann. **SOURCES:** G472c, S2694d **SUBJECT:** General - Societies - Göttingen.

486   The Commercial and agricultural magazine. - v.7-14, 1803-1806. London: v.Griffiths, 1799-1802. **CONTINUATION OF:** Agricultural magazine, or Farmers' monthly journal. **MICROFORMS:** RP(1) (Goldsmiths'-Kress Library). **SOURCES:** B5878, OCLC **LOCATIONS:** BUC, CCF, ULS **SUBJECT:** Agriculture - Periodicals - England.

487   Commercium litterarium ad astronomiae incrementum. - v.1-3, 1731-1745. Nürnberg: 1731-1745. An edition in 1785 is cited (B1328). **EDITOR:** Michael Adelbulner. **SOURCES:** B1328, K3189, S1328 **LOCATIONS:** CCN **SUBJECT:** Astronomy - Periodicals - Germany.

488   Commercium litterarium ad rei medicae et scientiae naturalis incrementum institutum, quo quicquid novissime observatum, agitatum scriptum vel peractum est succincte dilucideque exponitur. v.1-v.15, 1731-1745. Nüremberg: Johann Ernst Adelbulner, 1731-1745. Weekly. **TRANSLATED AS:** Auswahl der medicinischen Aufsätze und Beobachtungen aus den Nürnbergischen gelehrten Unterhandlungen. **CONTINUATION OF:** Consultatio de universali commercio litterario ad rei medicae et scientia. Publisher varies. **EDITORS:** Christoph Jakob Trew, Johann Christoph Gotz and Johann Christoph Homann (Ref.46). **SOURCES:** B1340a, G628, K3521, OCLC, S3145 **LOCATIONS:** BUC,

CCF, CCN, ULS, ZDB   SUBJECT: Science - Periodicals - Germany.
SUBJECT: Medicine - Periodicals - Germany.

489   Compendieuse physikalische Bibliothek. - v.1, 1724. Leipzig: Johann Christian Martini, 1724. Revised and edited by Abraham Gotthelf Kästner. Issued in Leipzig: Johann Wendlein, 1754. EDITOR: Julius Bernhard von Rohr. SOURCES: K3184   LOCATIONS: BL, ZDB   SUBJECT: Science - Periodicals - Germany.

490   Compte-rendus annuels et sommaires des travaux / Société royale académique. Cherbourg. - SOURCES: S1127   SUBJECT: General - Societies - Cherbourg.

491   Comptes rendus de la Société royale d'agriculture de la généralité de Paris. - May 30, 1788-Sept.30, 1793.   Paris: 1788-1793.   LOCATIONS: BUC SUBJECT: Agriculture - Societies - Paris.

492   Conférence sur les sciences et les arts, presentée à Monseigneur le Dauphin. - No. 1-14 (July 1672-1674), no.15 (1683). Paris: 1672-1683. CONTINUATION OF: Recueil des mémoires et conférences qui ont esté présentées à Monseigneur le Dauphin. EDITOR: Jean Baptiste Denis. LOCATIONS: BUC, CCF, CCN, DSG   SUBJECT: Science - Periodicals - France.

493   Conjuror's magazine and physiognomical mirror including a superb edition of Lavater's Essay in physiognomy. v.1-v.2, 1792-1793. London: W. Locke, 1792-1793.   CONTINUED BY: Astrologer's magazine and philosophical miscellany.   SOURCES: GXX   LOCATIONS: BUC, ULS   SUBJECT: Alchemy - Periodicals - England.   SUBJECT: Astrology - Periodicals - England.

494   Connoissance des temps, ou, calendrier et ephémerides du lever et coucher du soleil, de la lune et des autres planets. - v.1-216, 1679-1894. Paris: 1678-1894. Title varies frequently (B1337).   It did not come under the purview of the Académie des sciences until 1702, and in 1766 expanded its scope to include short astronomical articles (Ref.19,p.63).   EDITORS: 1679-1684, Joachim Dalencé; 1684-1702, Jean Le Févre; 1702-1729, J. Lieutard; 1730-1733, Louis Godin; 1735-1759, Giacomo Filipppo Maraldi; 1760-1775, Joseph Jérome Lefrançois de Lalande; 1776-1787, Edmé Sébastien Jeurat; 1788-1794, Pierre François André Méchain (Ref.22:0351).   SOURCES: B1337, S1368 LOCATIONS: BUC, CCF, CCN   SUBJECT: Astronomy - Societies - Paris.

495   Le Conservateur de la santé, journal d'hygiene et de prophlylaxie. - v.1-5, 1799-1804. Paris: 1799-1804. EDITORS: Francois Philippe Bellay and P. Brion. SOURCES: G206   LOCATIONS: CCF   SUBJECT: Medicine - Periodicals - France - Popular.

496   Consultatio de universali commercio litterario ad rei medicae et scientiae naturalis. - No.1-2, 1730. Nürnberg: 1739. CONTINUATION OF: Commercium litterarium ad rei medicae et scientiae naturalis incrementum institutum.

SOURCES: B1340, S3155   **SUBJECT:** Science - Periodicals - Germany.
SUBJECT: Medicine - Periodicals - Germany.

497   Continuacíon del memorial literario y curiosos de la corte de Madrid. - v.1-18,
        July 1793-Dec. 1797. Madrid: 1793-1797. **CONTINUATION OF:** Memorial
        literario y curioso de la corte de Madrid. **LOCATIONS:** BUC, CCF, ULS
        SUBJECT: General - Periodicals - Spain.

498   Contribution to physical and medical knowledge, principally from the west of
        England. - v.1, 1799. Bristol: Biggs & Cottle, 1799. **EDITOR:** Thomas
        Beddoes. **SOURCES:** G346, OCLC   **LOCATIONS:** DSG   **SUBJECT:**
        Medicine - Periodicals - England.

499   Conversations académique tirée de l'Académie de Monsieur l'abbé Bourdelot. -
        Paris: Claude Barbin, 1672-1674. Imprint varies. Published under the title:
        Conférences de l'Académie de Monsieur l'abbé Boudelot, 1672-1673. Another
        edition in 1675 (Ref.67). **OTHER TITLE:** Conference de l'Académie de
        Monsieur l'abbé Bourdelot. **EDITOR:** Pierre Le Gallois. **SOURCES:** OCLC
        SUBJECT: Science - Societies - Paris.

500   Corps d'observations de la Société d'agriculture, du commerce et de arts, établi
        par les États de Bretagne. - v.1-2, 1757-1760. Rennes: J. Vater, 1757-1760.
        There was another edition of 1757/58 in 1761 (BUC), and one covering the
        years 1759/60 (Paris:Brunet,1771)(LC). **SOURCES:** OCLC **LOCATIONS:**
        BUL, ULS   **SUBJECT:** Agriculture - Societies - Rennes.   **SUBJECT:**
        Commerce - Societies - Rennes.

501   Correio mercantil e economica de Portugal. - Lisbon: 1790-1807. Attributed to
        Antonio Manuel Polycarpo da Silva (Ref.34, p.273). **MICROFORMS:** RP(1)
        (Goldsmith-Kress Library of Economic Literature). **LOCATIONS:** ULS
        SUBJECT: Economics - Periodicals - Portugal.

502   Cosmogonia, oder, neueste und älteste Naturgeschichte in brüderllichen
        Anschreiben. - Quartal 1-2, 1765. Amsterdam: Pesenecker, 1763. **EDITOR:**
        Christian Gottlieb König. **SOURCES:** K3235   **SUBJECT:** Science -
        Periodicals - Netherlands.

503   Crell's chemical journal, giving an account of the latest discoveries in chemistry
        with extracts from various foreign transactions; translated from the German,
        with occasional additions. - v.1-v.3, 1791-1793.   London: 1791-1793.
        EDITOR: Lorenz Florenz Friedrich von Crell.   **TRANSLATION OF:**
        Chemische journal für die Freunde der Naturlehre. **INDEXES:** Indexed in
        Geuss (Ref.53,v.4).   **SOURCES:** B1376, G625a, OCLC, S257
        LOCATIONS: BUC, ULS **SUBJECT:** Chemistry - Periodicals - Germany -
        Translations.

504   The Critical review, or, Annals of literature. - 1st ser. v.1-70, 1756-1790; 2nd
        ser. v.1-39, 1791-1803; 3rd ser. v.1-14, 1804-1811; 4th ser. v.1-6, 1812-1814;
        5th ser. v.1-5, 1815-1817. London: W. Simpkin and R. Marshall, 1756-1817.

Imprint varies. **EDITORS:** Tobias George Smollett and others. **MICROFORMS:** UM(1),NPL(1), RM(3). **SOURCES:** OCLC **LOCATIONS:** BUC, CCF, ULS, ZDB **SUBJECT:** General - Periodicals - England - Reviews.

505 Curieuser und nutzbarer Anmerkungen von Natur- und Kunstgeschichte. - No. 1-4, 1726-1729. Leipzig: 1726-1729. **CONTINUATION OF:** Sammlung von Natur- und Medecin. Issued a supplement to Sammlung von Natur- und Medecin. **MICROFORMS:** RP(1). **LOCATIONS:** BUC **SUBJECT:** Natural history - Periodicals - Germany. **SUBJECT:** Medicine - Periodicals - Germany.

506 Curious remarks and observations in physics, anatomy, chirurgery, botany and medicine, extracted from the history and memoirs of the Royal Academy of Sciences at Paris. - v.1-2, 1753-1754. London: C. Davis and P. Vaillant, 1753-1754. Extracted and translated by Peter Templeton. **TRANSLATION OF:** Histoire de l'Académie royale des sciences, Paris. **LOCATIONS:** BUC **SUBJECT:** Science - Societies - Paris - Translations.

507 Daedalus hyperboreus, eller nagra mathematiska och physicaliska försök och anmerckingar. - Nos. 1-6, 1716-1718. Stockholm: Johann Heinrich Werner, 1716-1718. Published for the Collegium curiosorum, which later became the Regia Societas Scientiarum Upsaliensis. **EDITOR:** Emanuel Swedenborg. **SOURCES:** B1398, G629, OCLC **LOCATIONS:** BUC, ULS **SUBJECT:** Science - Societies - Upsala.

508 Danmark og Norges oeconomiske Magasin. - v.1-8, 1754-1764. Copenhagen: 1757-1764. **SOURCES:** OCLC **LOCATIONS:** BUC, CCF, ZDB **SUBJECT:** Economics - Periodicals - Denmark.

509 De re rustica, or, The repository for select papers on agriculture, arts, and manufactures. - v.1-2 (nos.1-79) 1768-1770. London: R and L. Davis, 1768-1770. Included communications from the Royal Society of Arts. **MICROFORMS:** RP(1) (Goldsmiths' Kress Library). **SOURCES:** OCLC **LOCATIONS:** ULS **SUBJECT:** Agriculture - Periodicals - England.

510 La Décade philosophique, littéraire et politique. - v.1-42, 1792-1804. Paris: J.B. Say, 1792-1804. Three times a month. Subtitle varies. **CONTINUED BY:** Revue ou décade philosophique. **MICROFORMS:** L'Association pour le conservation et reproduction photographique de la press, Paris, 1971. **SOURCES:** G630, OCLC, S1374 **LOCATIONS:** BUC, CCF, ULS, ZDB **SUBJECT:** General - Periodicals - France.

511 Déliberations et mémoires de la Société de la Géneralité de Rouen. - v.1-3, 1763-1787. Rouen: Richard Lallemant, 1763-1787. Microforms RP(1) (Goldsmiths' Kress Library). **LOCATIONS:** BUC, CCF, ULS **SUBJECT:** Agriculture - Societies - Rouen.

512   Delicae medicae et chirurgicae, oder, Curiöse Anmerkungen, darinnen sich
      diejenigen, welche Medizin und Chirurgie lieben, nützlich ergötzen können. -
      v.1-2, 1703-1705. Leipzig: Christoph Hülze, 1703-1705. Supplements to pts.
      5-10 issued under the title: Anhang chymischer Ergötzligkeiten, by Aletophilus
      Chrysander.   SOURCES: GIII:1, K3515, OCLC   LOCATIONS: BUC
      SUBJECT: Medicine - Periodicals - Germany.   SUBJECT: Surgery -
      Periodicals - Germany. SUBJECT: Chemistry - Periodicals - Germany.

513   Delights for the ingenious, or, a monthly enterainment for the curious of both
      sexes. - Nos.1-8, 1711. London: 1711. The first 6 nos. were monthly.
      Includes mathematical problems. EDITOR: John Tipper. SOURCES: B8517
      LOCATIONS: BUC   SUBJECT: Mathematics - Periodicals - England -
      Popular.

514   Dell' Arte ostetricia, fogli periodici con rami colorati. - Bologna: San Tommaso
      d'Aqunio, 1788.   SOURCES: G261   LOCATIONS: BUC   SUBJECT:
      Obstetrics - Periodicals - Italy.

515   Denkwürdigkeiten für die Heilkunde und Geburtshülfe, aus den Tagebüchern der
      Kgl. practischen Anstalten zur Erlernung dieser Wissenschaften in Göttingen.
      - v.1-2, 1794-1795.   Göttingen: Vandenhoek und Ruprecht, 1794-1795.
      EDITOR: Friedrich Benjamin Osiander.   CONTINUED BY: Neue
      Denkwürdigkeiten für Aerzte und Geburtshelfer.   SOURCES: G123
      LOCATIONS: CCF, ULS, ZDB   SUBJECT: Medicine - Periodicals -
      Germany. SUBJECT: Obstetrics - Periodicals - Germany.

516   Derevensko zerkale / Ekonomecheskago obtchestva, St. Petersburg. - v.1-3,
      1798-1799. St. Petersburg: 1798-1799.   LOCATIONS: NUC   SUBJECT:
      Economics - Societies - St. Petersburg.

517   Description des arts et de métiers, approuvées par messieurs de l'Académie royale
      des science, Paris. - v.1-45, 1761-1788. Paris: Dessaint et Saillant, 1761-1786.
      TRANSLATED AS: Schauplatz der Künste und Handwerke.   There was
      another edition in quarto in 19 v.   Neuchâtel, Société typographique,
      1771-1783; v.20 printed in Paris. Contains 80 parts, some fewer than 20 pages,
      others of several volumes. INDEXES: Description des arts; a reel index and
      guide to the microform edition, by Edward A. Reno and Betty H. Erb.
      Wooodbridge, Conn., Research Publications, 1971. MICROFORMS: RM(3),
      RP(1) (Goldsmiths' Kress Library).   SOURCES: G17j, OCLC
      LOCATIONS: BL,ULS SUBJECT: Industrial arts - Societies - Paris.

518   Deutsche acta eruditorum, oder, Geschichte der Gelehrten, welche den
      gegenwärtigen Zustand der Litteratur in Europa begreiffen. - v.1-20,
      1712-1739. Leipzig: Gleditsch, 1712-1739.   EDITORS: Justus Gotthard
      Rabener, Christian Schöttgen and Johann Georg Walch. CONTINUED BY:
      Zuverlässigen Nachrichten der Litteratur in Europa, Veränderung und
      Wachstum der Wissenschaften. MICROFORMS: GO(8), RP(1). SOURCES:
      B1418, G632, K27, OCLC, S2915 LOCATIONS: BUC, CCF, CCN, ULS,
      ZDB   SUBJECT: General - Periodicals - Germany - Reviews.

519 Deutsche Gesundheits-zeitung. - Bd.1-6, 1786-1790. Hamburg und Altona: Eckdorf, 1786-1790. **CONTINUATION OF:** Aerzte ein medicinisches Wochenblatt. **EDITOR:** Georg Levison. **SOURCES:** GIII:72, K3627 **LOCATIONS:** ZDB **SUBJECT:** Medicine - Periodicals - Germany - Popular.

520 Der deutsche Obstgärtner, oder, gemeinschaftliches Magazin des Obstbaues in Deutschland. - v.1-22 (12 issues each) 1794-1804. Weimar: Industrie-Comptoir, 1794-1804. Subtitle varies. **EDITOR:** Johann Volkmar Sickler. **INDEXES:** Index to 1794 in v.12, 1794-1804 in v.22. **SOURCES:** K2999 **LOCATIONS:** BUC, CCF **SUBJECT:** Agriculture - Periodicals - Germany.

521 Deutsche Schriften der königlichen Societät der Wissenschaften zu Göttingen. - Bd.1, 1771. Göttingen: Dieterich, 1771. **CONTINUATION OF:** Commentationes regiae societatis Gottingensis. **CONTINUATION OF:** Commentarii Societatis regiae scientiarum gottingensis. **CONTINUATION OF:** Novi commentarii societatis regiae scientiarum Gottingensis. **CONTINUATION OF:** Commentationes societatis regiae scientiarum Gottingensis. Derived from a backlog of contributions particularly from nonmembers. **INDEXES:** Index to the society's publications from 1751-1808 in v.16 of its Commentarii for 1804/1808. Also indexed in Vezenyi (Ref.1). **SOURCES:** G472e, K284, S2695 **LOCATIONS:** BUC, CCN, ULS **SUBJECT:** General - Societies - Göttingen.

522 Deutsche Zeitung der Industrie und Spekulation für die k.k. Erbländer. - v.1-2, 1797. Prague: Schönfeld, 1797. **SOURCES:** K2780 **SUBJECT:** Manufactures - Periodicals - Czechoslovakia. **SUBJECT:** Economics - Periodicals - Czechoslovakia.

523 Deutsches Museum. - v.1-26, 1776-1788. Leipzig: Weygand, 1776-1788. Monthly. **EDITORS:** Heinrich Christian Boie and Christian Wilhelm Dohm. **CONTINUED BY:** Neues deutsches Museum. **INDEXES:** Indexed in Allgemeines Sachregister (q.v.). **MICROFORMS:** GO(9),RM(3). **SOURCES:** B1467, K4495, OCLC, S2920 **LOCATIONS:** BUC, CCF, CCN, ULS, ZDB **SUBJECT:** General - Periodicals - Germany - Reviews.

524 Deutsches theatrum chemicum. - v.1-3, 1728-1732. Nürnberg: Adam Jonathan Felssecker, 1728-1732. A collection of 53 monographs with separate title pages issued in volumes at two year intervals. Reprinted: New York, G.Olms, 1976. **EDITOR:** Friedrich Roth-Scholtz. **SOURCES:** K3186, OCLC **LOCATIONS:** BL **SUBJECT:** Chemistry - Periodicals - Germany. **SUBJECT:** Alchemy - Periodicals - Germany.

525 Deutschlands Flora oder botanisches Taschenbuch. - v.1-4, 1791-1804. Erlangen: 1791-1804. **EDITOR:** Georg Franz Hoffmann. **SOURCES:** B1470, G561, OCLC, S2565 **LOCATIONS:** BUC, ZDB **SUBJECT:** Botany - Periodicals - Germany.

526   Diana, oder Gesellschaftsschrift zur Erweiterung der Natur- Forst- und Jagdkunde.
      - v.1-4, 1800-1816. Gotha: Ettinger, 1800-1816. v.1-3, 1800-1805, v.4
      Kassel: Krieger, 1816. Subtitle and imprint vary. Official organ of: Societät
      der Forst-und Jagdkunde. EDITOR: Johann Mathaus Bechstein. SOURCES:
      K3030, OCLC LOCATIONS: CCF, ULS, ZDB   SUBJECT: Natural
      history - Periodicals - Germany.   SUBJECT: Forestry management -
      Periodicals - Germany. SUBJECT: Hunting - Periodicals - Germany.

527   Diarian miscellany, consisting of all the useful and entertaining parts, both
      mathematical and poetical, extracted from the Ladies' diary. - v.1-5, 1771-1775.
      London: G. Robinson and R. Baldwin, 1771-1775. CONTINUATION OF:
      Ladies' Diary. The first three volumes are mathematical. Vol. 6 published as
      Miscellanea mathematica (Ref.12,1.1).   OTHER TITLE: Miscellanea
      mathematica. EDITOR: Charles Hutton. LOCATIONS: ULS SUBJECT:
      Mathematics - Periodicals - England - Popular.

528   Diarian repository, or, Mathematical register containing a complete collection of
      all the mathematical questions which have been published in the Ladies' Diary
      from the commencement of that work in 1704 to the year 1760. - No.1-30,
      1771-1774. London: B. Robinson, 1771-1774. CONTINUATION OF:
      Ladies' diary. EDITOR: Samuel Clark (Ref.12,p.382). LOCATIONS: ULS
      SUBJECT: Mathematics - Periodicals - England - Popular.

529   Diario de los nuevas descumbrimentos de todas las ciencias. - Madrid: 1792-1793.
      (Ref.68,p.16). TRANSLATION OF: Médecine eclairée par les sciences
      physique. SOURCES: (Ref.68,p.16).   SUBJECT: Science - Periodicals -
      France - Translations.

530   Diario economico di agricoltura, manufatture i commercio. - Rome: 1776.
      Weekly. EDITOR: Arcangelo Casaletti. SOURCES: (Ref.138). SUBJECT:
      Agriculture - Periodicals - Italy. SUBJECT: Economics - Periodicals - Italy.

531   Diario filosófico, médica, cirurgía, colección de selectas observaciónes y curiosos
      fragmentos sobre la historia natural, física y medicina. - Nos.1-7, 1757
      (Ref.62,p.127). Madrid: Antonio Perez de Soto, 1757.   EDITOR: Juan
      Galisteo y Xiorro. SOURCES: B1473, OCLC, S1744 SUBJECT: Science
      - Periodicals - Spain. SUBJECT: Medicine - Periodicals - Spain.

532   Diario universal de medicina, cirurgia e pharmacia. - Madrid: Francisco Luiz
      Ameno, 1764-1772. EDITOR: Manoel Gomes de Lima.   SOURCES:
      (Ref.34,p.272). SUBJECT: Medicine - Periodicals - Portugal.

533   Diätetische Wochenblatt für alle Stände. - v.1-3, 1781-1783. Rostock: Müller,
      1781-1783. Weekly. EDITOR: Peter Benedict Christian Graumann.
      SOURCES: K3593, OCLC LOCATIONS: BUC, CCF, ULS SUBJECT:
      Nutrition - Periodicals - Germany - Popular.   SUBJECT: Medicine -
      Periodicals - Germany - Popular.

534 Dictionnaire raisonné universelle d'histoire naturelle. - v.1-6, 1765. Paris: Didot, 1765. **TRANSLATED AS:** Allgemeine Geschichte der Natur. An augmented and revised edition in 12 v. was published in Paris in 1768. Other editions were published in 1768, 1769, 1775, 1776 and 1800. **EDITOR:** Jacques Christophe Valmont de Bomare. The Allgemeine Geschichte der Natur, edited by Friedrich Heinrich Wilhelm Martini was derived from the Dictionnaire. A supplement to the first edition was published in Paris in 1768, and to the third in Laussanne in 1778. **SOURCES:** G532 **SUBJECT:** Natural history - Encyclopedias - France.

535 O Discipulo instruido pelos mestres mais sábios dos segredos naturaes des sienciàs. - Lisbon: Francisco Borges de Sousa, 1759-1760. **OTHER TITLE:** Semana proveitosas ao vivente racional. **EDITOR:** José Angelo de Moraes. **SOURCES:** (Ref.34,no.16). **SUBJECT:** Science - Periodicals - Portugal.

536 Discours prononcé dans l'assemblée publique de l'Académie royale de sciences et belles-lettres, Berlin. - v.1, 1787. Berlin: 1787. **INDEXES:** Cumulative index (Ref.126). **LOCATIONS:** ZDB **SUBJECT:** General - Societies - Berlin.

537 Discoursos mercuriales económico-politicos. - Covered commerce, trade agriculture and social problems. **EDITOR:** Juan Enrique de Graef. **SOURCES:** (Ref.62,p.127). **SUBJECT:** Economics - Periodicals - Spain.

538 Discurscuriöser Sachen insonderheit hermetischer, philosophischer, physikalischer, medicinischer und anderer Wissenschaften. - Leipzig: Johann Heinichen, 1708. Issued from Jan. to April, 1708. **EDITOR:** Johann Gottfried Meerheim. **SOURCES:** K3179, OCLC **SUBJECT:** Alchemy - Periodicals - Germany. **SUBJECT:** Science - Periodicals - Germany.

539 Disputationes ad morborum historiam et curationem facientes. - v.1-v.7, 1757-1760. Lausanne: Sigismundi d'Arnay, 1757-1760. **TRANSLATED AS:** Beiträge zur Beförderung der Geschichte und Heilung der Krankheiten. **TRANSLATED AS:** Sammlung academischer Streitschriften die Geschichte und Heilung der Krankheiten betreffend. Some volumes have the imprint: Venice: Baglioni. **EDITOR:** Albrecht von Haller. **SOURCES:** G353 OCLC **LOCATIONS:** NUC **SUBJECT:** Medicine - Periodicals - Switzerland - Dissertations.

540 Disputationes chirurgicae selectae, collegit et prefatus Alb. Haller. Bd.1-Bd.5, 1755-1766. Lausanne: Bosquet, 1755-1766. **TRANSLATED AS:** Collection de thèses médico-chirurgicales. **TRANSLATED AS:** Auserlesene chirurgische Disputationen. **EDITOR:** Albrecht von Haller. **SOURCES:** G352 **LOCATIONS:** BL, NUC **SUBJECT:** Surgery - Periodicals - Switzerland - Dissertations.

541 Dissertations pour le prix / Akademie der Wissenschaften, Berlin. - v.1-4, 1742-1760. Berlin: Haude und Spener, 1742-1760. **OTHER TITLE:** Dissertationum occasione premi, Academia regia scientiarum Borussica.

INDEXES: Cumulative index (Ref.126). **LOCATIONS:** BL **SUBJECT:** General - Societies - Berlin - Prize essays.

542 Dissertationum experimentalium ex commentarii / Academia scientiarum imperialis Petropolitana. - v.1, 1762. St. Petersburg: 1762. **LOCATIONS:** BUC **SUBJECT:** Science - Societies - St. Petersburg.

543 Dissertazioni medico-chirurgico-practiche. - v.1, 1790. Vienna: 1790. Translated by Luig Careno. **TRANSLATION OF:** Acta Academiae Caesareae regia Josephinae medico-chirurgicae Vindobonensis. **SOURCES:** G75b **LOCATIONS:** ULS **SUBJECT:** Medicine - Societies - Vienna - Translations. **SUBJECT:** Surgery - Societies - Vienna - Translations.

544 Dresdnisches Magazin, oder, Ausarbeitungen und Nachrichten zum Behufe der Naturlehre, der Arzneykunst, der Sitten und der schönen Wissenschaften. - Bd.1-2 (each 8 issues) 1759-1765. Dresden: Michael Gröll, 1759-1765. **SOURCES:** B1486, K225, S2505 **LOCATIONS:** BUC, CCN, ULS, ZDB **SUBJECT:** Science - Periodicals - Germany.

545 Early proceedings of the American philosophical society from the manuscript minutes of the meetings, 1744-1821. - Philadelphia: 1884. (Ref.69). **LOCATIONS:** BUC, ULS **SUBJECT:** Science - Societies - Philadelphia.

546 Efemerides barométrico-medicas Matritenses / Academia médico-matritenses, Madrid. - Madrid: Imprenta real, 1737-1746. **EDITORS:** Francisco Fernandez Navarrete, Alejandro Martinez Argando Alna and José Arcadio de Ortega. **SOURCES:** GIII:8 **LOCATIONS:** BL **SUBJECT:** Medicine - Societies - Madrid.

547 Effemeridi astronomiche calcolate pel meridiano di Milano. - Milan: Osservatorio di Brera, 1774-1873. **OTHER TITLE:** Ephemeridii astronomicae ad meridiana mediolanensem. **EDITOR:** Giovanni Angelo de Cesaris. **SOURCES:** B1531, OCLC, S1889 **LOCATIONS:** BUC, CCF **SUBJECT:** Astronomy - Periodicals - Italy.

548 Effemeridi letterarie di Roma. - v.1-35, 1772-1806, ns. v.1-13, 1820-1823. Rome: Homer, 1772-1823. **OTHER TITLE:** Efemeridi letterarie di Roma. **EDITOR:** Giovanni Ludovico Bianconi (Ref.70). **LOCATIONS:** BUC, CCF, ULS, ZDB **SUBJECT:** General - Periodicals - Italy.

549 Eisenhütten Magazin, darin Alles was zum Eisenhütten gehöret. - v.1-2, 1791-1793. Wernigerode: 1791-1793. **EDITORS:** Johann Friedrich Tölle and L.E.S. Gärtner. **SOURCES:** B1543 **LOCATIONS:** BUC **SUBJECT:** Mineral industries - Periodicals - Germany.

550 Der Eklektiker, ein medizinisches Wochenblatt für unmedizinische Familienväter. - v.1, 1771. Hamburg: 1771. **SOURCES:** K3558 **SUBJECT:** Medicine - Periodicals - Germany - Popular.

551 Ensayo de la Sociedad bascongada de los amigos del pais. - v.1, 1766-1768.
Vitoria: 1766-1768. (Ref.6,p.290) SOURCES: S1779 LOCATIONS: BUC,
ULS SUBJECT: Economics - Societies - Vitoria.

552 Die Entdeckungen der neuesten Zeit in der Arzneygelahrtheit. - v.1 (1770-71),
1778, v.2 (1772-73) 1782), v.3 (1774-76) 1786, v.4 (1774-1779) 1788.
Nördlingen: Carl Gottlob Beck, 1778-1788. EDITOR: Johann Augustin Philipp
Gesner. SOURCES: G278, K3578, OCLC LOCATIONS: BUC, ZDB
SUBJECT: Medicine - Periodicals - Germany.

553 Entomologisches Taschenbuch, für die Anfänger und Liebhaber dieser
Wissenschaft. - v.1-2, 1796-1797. Regensburg: Montag und Weiss, 1796-1797.
EDITOR: David Heinrich Hoppe. SOURCES: B1590, G579, K3366, S3194
LOCATIONS: BUC, ULS, ZDB SUBJECT: Entomology - Periodicals -
Germany.

554 Ephemeriden der Handlung, oder Beiträge und Versuche für Kaufleute. - Lübeck:
Donatius, 1784. EDITOR: Johann Christian Schedel. CONTINUED BY:
Allgemeines Journal für die Handlung. SOURCES: K2736 SUBJECT:
Commerce - Periodicals - Germany.

555 Ephemeriden für die Naturkunde, Oekonomie, Handlung, Künste und Gewerbe.
- No.1-4, 1792-1796. Leipzig: 1792-1796. Imprint and title vary. OTHER
TITLE: Zeitschrift für Nachdenkende; Zeitschrift für die Naturkunde;
Ephemeriden für Kunste und Gewerbe. EDITOR: Johann Christian Schedel.
SOURCES: B4904, G754, 2988, K3004, S3044 SUBJECT: Natural history
- Periodicals - Germany. SUBJECT: Commerce - Periodicals - Germany.

556 Ephemerides astronomicae ad meridianum vindobonensem. - v.1-50, 1756-1895.
Vienna: 1756-1805. EDITORS: Maximilian Hell (1720-1792), Franz von Paul
Triesnecker and Johann Tobias Burg (1793-1805). SOURCES: B1598, S3583
LOCATIONS: BUC, CCF, ZDB SUBJECT: Astronomy - Periodicals -
Austria.

557 Ephemerides des mouvements célestes. - Paris: 1716-1800. Publisher varies.
EDITORS: Philippe Desplaces 1715-43 (1716-44), Nicolas Louis de Le Caille
1744-1763 (1745-75), Joseph Jérome Le François Lalande 1774-1792
(1775-1800). SOURCES: B1604, S1381 LOCATIONS: BUC, CCF, ZDB
SUBJECT: Astronomy - Periodicals - France.

558 Ephemerides eruditorum. - v.1-4, 1667-1671. Leipzig: Schürer und Fritzsch,
1667-1671. TRANSLATION OF: Journal des sçavans. LOCATIONS: ZDB
SUBJECT: General - Periodicals - France - Translations.

559 Ephemerides historiae et observationes / Societas meteorologica
Theodoro-Palatina. - v.1-12 (1781-1792) 1783-1795 (Ref.71). Mannheim:
1783-1794. SOURCES: G507, S3079 LOCATIONS: CCF, ZDB
SUBJECT: Meteorology - Societies - Mannheim. SUBJECT: Medicine -
Societies - Mannheim.

560   Ephemerides meteorologico-medicae annorum 1780-93. - v.1-v.5, 1794. Vienna: Alb. Ant. Patzowsky, 1794. **TRANSLATED AS:** Medicinische Ephemeriden. A year by year account of the prevailing weather and accompanying morbidity published in 1794. Not a periodical. **EDITOR:** Samuel Benkö. **SOURCES:** B1602, G280, OCLC, S3584 **LOCATIONS:** BUC **SUBJECT:** Meteorology - Periodicals - Germany. **SUBJECT:** Medicine - Periodicals - Germany.

561   Ephemerides nauticas, ou, diario astronomica / Academia real das sciencias de Lisboa. - v.1-10(1789-1790)1788-1796. Lisbon: 1789-1796. Continued until 1860 (Ref.34:31). Edited 1789-1791 by Custodio Gomes Villas-Boas, F. A. Ciera and F. de Borja Gaçao Stockler; 1792-1795 by Villa-Boas alone, and 1796-1798 by José Maria Dantas Pereira. **SOURCES:** B1603, S1720d **LOCATIONS:** CCF, ULS **SUBJECT:** Naval art and science - Societies - Lisbon. **SUBJECT:** Astronomy - Societies - Lisbon.

562   Ephemerides physico-astronomicae. - Salzburg: F. Duylle, 1789. **EDITOR:** Dominikus Beck. **SOURCES:** K3330 **LOCATIONS:** ZDB **SUBJECT:** Science - Periodicals - Austria.

563   Ephémérides pour servir à l'histoire de toutes les parties de guérir. - v.1, 1790. Paris: 1790. **EDITORS:** Pierre Lassus and Philippe Jean Pelletan. **SOURCES:** G87 **SUBJECT:** Medicine - Periodicals - France.

564   Ephemerides sive observationum medico-physicarum / Academiae Caesareae Leopoldinae naturae curiosoum. - v.1-5, centuria I-X, 1712-1722. Frankfurt; Leipzig: Christopher Rigel, 1712-1722. **CONTINUATION OF:** Miscellanea curiosa Academia naturae curiosorum. Publisher varies. **OTHER TITLE:** Observationum medico-physicarum; Acta Wratislavensis. **CONTINUED BY:** Acta physico-medica, Academia Caesarea Leopoldina- Carolina naturae curiosorum. **CONTINUED BY:** Nova acta physico-medica academiae Caesareae Leopoldinae-Carolinae naturae curiosorum. **INDEXES:** Index universalis et absolutissimus rerum memorabilium ac notabilium medico-physicarum quae in Decuriis 3 ac Centuriis 10 ab anno 1670 usque ad anum 1722 extant. Nuremberg, 1739. **MICROFORMS:** RM(3), IDC(8). **SOURCES:** B3051a, G7b, K3180, OCLC, S2580a **LOCATIONS:** BUC, CCF, ULS, ZDB **SUBJECT:** Science - Societies - Halle. **SUBJECT:** Medicine - Societies - Halle.

565   Erfurtische gelehrte Nachrichten. - v.1-14, 1754-1768. Erfurt: Johann Andreas Görling, 1754-1768. Imprint varies. **EDITOR:** Johann Wilhelm Baumer. **CONTINUED BY:** Erfurtische gelehrte Zeitungen. **CONTINUED BY:** Erfurtische gelehrte Zeitung. **CONTINUED BY:** Nachrichten von gelehrten Sachen von dem Akademie zu Erfurt. **SOURCES:** B1623, K199, S2552 **LOCATIONS:** CCF, ULS, ZDB **SUBJECT:** General - Periodicals - Germany - Reviews.

566   Erfurtische gelehrte Zeitung. - v.1-17, 1780-1796. Erfurt: Reval, 1780-1796. **CONTINUATION OF:** Erfurtische gelehrte Nachrichten. **CONTINUATION OF:** Erfurtische gelehrte Zeitungen. Imprint varies. **EDITOR:** Justus

Friedrich Froriep. **CONTINUED BY:** Nachrichten von gelehrten Sachen von dem Akademie zu Erfurt. **SOURCES:** B1623a, K348, S2552,2 **LOCATIONS:** CCF, ULS, ZDB **SUBJECT:** General - Periodicals - Germany - Reviews.

567 Erfurtische gelehrte Zeitungen. - v.1-11, 1769-1779. Erfurt: 1769-1779. **CONTINUATION OF:** Erfurtische gelehrte Nachrichten. **EDITOR:** J.D. Reidel. **CONTINUED BY:** Erfurtische gelehrte Zeitung. **CONTINUED BY:** Nachrichten von gelehrten Sachen von dem Akademie zu Erfurt. **SOURCES:** K272, S2552 **LOCATIONS:** CCF, ZDB **SUBJECT:** General - Periodicals - Germany - Reviews.

568 Erläuterung der medicinischen und chirurgischen Praxis. - v.1, 1795. Leignitz; Leipzig: Xavier Siegert, 1795. Translations of v.5 of Medical facts and observations. **SOURCES:** G102a **SUBJECT:** Medicine - Periodicals - England - Translations.

569 Esculapius, door, denwelken aan't publycq vorgestelt wordt de konst om de gesondheid te onderhouden al derhande zoort van ziekten te genezen door verandering van lucht door maatregels omtrent de voeding. - May 3, 1723-Dec.6, 1723. Amsterdam: 1723. Weekly. **EDITOR:** Willem van Ranouw. **SOURCES:** GIII5 **LOCATIONS:** CCN **SUBJECT:** Medicine - Periodicals - Netherlands.

570 Esprit des journaux français et étrangers. - v.1-32, Jul. 1772-Mar. 1803; v.1-13, Sept.1804-1814 and Apr. 1817-Apr. 1818. Liége; Brussels: Tutot, 1772-1818. 6 vols. a year. Imprint and subtitle vary. **EDITORS:** J.L. Coster, L.F. Lignac and Abbé Curtin (Ref.86,p.51). **INDEXES:** There are indexes to 1772-1784 and 1803-1811. **SOURCES:** G639 **LOCATIONS:** BUC, CCF, CCN, ULS, ZDB **SUBJECT:** General - Periodicals - Belgium - Reviews.

571 Essai de médecine theorique et pratique. - Geneva; Lyon: 1784. **EDITORS:** Pierre Brion and d'Yvoiry. **SOURCES:** GIII:59 **LOCATIONS:** DSG **SUBJECT:** Medicine - Periodicals - Switzerland.

572 Essais de médecine et d'histoire naturelle. - Paris: Carpentras, 1797-1798. **EDITORS:** Waton and Guérin. **CONTINUED BY:** Observations médicale. **SOURCES:** G138 **LOCATIONS:** CCF **SUBJECT:** Medicine - Periodicals - France.

573 Essais et observations de médecine, de la société de Edinburgh. - v.1-7, 1740-1747. Paris: Guerin, 1740-1747. A translation by Pierre Demours. **TRANSLATION OF:** Medical essays and observations, Medical Society in Edinburgh. **INDEXES:** Index to v.1-7 in v.7. **SOURCES:** GIII7a, OCLC, NUC **LOCATIONS:** CCF, CCN, ULS **SUBJECT:** Medicine - Societies - Scotland - Translations.

574 Essais et observations physique et littéraire de la Société d'Edinbourg. - v.1, 1759. Paris: Guerin, 1759. Imprint varies. Translated by Pierre Demours.

TRANSLATION OF: Essays and observations, physical and literary, Medical Society in Edinburgh. **SOURCES:** B1635, G10b,d S1384 **LOCATIONS:** CCF, ULS, ZDB   **SUBJECT:** Medicine - Societies - Edinburgh - Translations.

575   Essayes of natural experiments made in the Academie del Cimento. v.1, 1684. London; Benjamin Allen, 1684. Translated by Richard Waller. Reprinted in 1954 by Johnson Reprint Corp. with an introduction by A. Rupert Hall. **TRANSLATION OF:** Saggi di naturali esperienze fatte nell Accademia del Cimento. **MICROFORMS:** RM(3).   **SOURCES:** OCLC   **SUBJECT:** Science - Societies - Florence - Translations.

576   Essays and observations, physical and literary / Philosophical Society in Edinburgh. - v.1-3, 1754-1765.   Edinburgh: G. Hamilton; J. Balfour, 1754-1765.   **TRANSLATED AS:** Essais et observations de médecine de la Société d'Edinbourg. **CONTINUATION OF:** Medical essays and observations, revised and published by a Society in Edinburgh. **TRANSLATED AS:** Neue Versuche und Bemerkungen aus der Arzneykunst. 2d edition, 3 v.1770-1771. Johnson Reprint. **CONTINUED BY:** Medical and philosophical commentaries by a Society in Edinburgh.   **CONTINUED BY:** Annals of medicine. **CONTINUED BY:** Edinburgh medical and surgical journal. **INDEXES:** Index to v.1-3 in The Edinburgh medical and surgical journal, v.20, 1824. **MICROFORMS:** RM(3).   **SOURCES:** B1636, G10, OCLC, S117 **LOCATIONS:** BUC, CCF, CCN, ULS, ZDB   **SUBJECT:** Science - Societies - Edinburgh.   **SUBJECT:** Medicine - Societies - Edinburgh.

577   État de médecine, chirurgie et pharmacie en Europe. - v.1-2, 1776-1777. Paris: P.F. Didot, 1776-1777. (Ref.35:564). **EDITORS:** Guillaume-René LeFebure de Saint-Ildephont and De Horne.   **SOURCES:** G38   **LOCATIONS:** BN, BUC, NUC   **SUBJECT:** Medicine - Periodicals - France.

578   Etwas für Alle, oder, neue Stuttgardter Realzeitung. - Stuttgart: Johann Christian Erhard, 1765-1766.   **CONTINUATION OF:** Physicalisch-ökonomische Wochenschrift.   **EDITOR:** Johann Christian Erhard. **CONTINUED BY:** Stuttgarder allgemeine Magazin.   **SOURCES:** B1644, K5297, S3249 **LOCATIONS:** ZDB   **SUBJECT:** Economics - Periodicals - Germany.

579   L'Europe sçavante. - v.1-12 in 2, 1718-1720. La Haye: Alexander de Rogissart, 1718-1729. Reprinted by Slatkine Reprints, 1969. **OTHER TITLE:** Europe savante. **EDITORS:** Thémisuel de Saint-Hyacinthe, Jean Lévesque de Burigny, Louis-Jean Lévesque de Pouilly, Gérard Lévesque de Champeaux, and Juste van Effen (Ref.56).   **SOURCES:** B1645, G640, OCLC, S822   **LOCATIONS:** BUC, CCF, CCN, ULS, ZDB   **SUBJECT:** General - Periodicals - Netherlands.

580   Experiments in agriculture / Dublin Society. - (1765-1770) 1766-1771. Dublin: Powell, 1766-1771.   **SOURCES:**(Ref.106,no.38).   **SUBJECT:** Agriculture - Societies - Dublin.

581 Extracto de los actas de la sociedad ecónomicas de los amigos del país de Valencia. - Valencia: 1785-1792. **LOCATIONS:** BUC **SUBJECT:** Economics - Societies - Valencia.

582 Extractos de las juntas generales celebradas de la sociedad vascongada de los amigos del pais / Madrid. - v.1-13, 1771-1784. Madrid: 1771-1784. **SOURCES:** S1762 **LOCATIONS:** BUC **SUBJECT:** Economics - Societies - Madrid.

583 Extraits des observations astronomique et physiques fait à l'Observatoire royale, Paris. - Paris: 1785-1791. **LOCATIONS:** BUC, CCF **SUBJECT:** Astronomy - Societies - Paris.

584 Ezhemiesiachnyia sochinenïia uchenykh dielaks / Akademiia nauk, St. Petersburg. - v.1-20, 1755-1764. St. Petersburg: 1755-1764. **CONTINUATION OF:** Kratkoe opisaïe kommentariev, Academia scientiarum imperiales Petropolitana. **CONTINUATION OF:** Soderzhanïe uchenykh razsuzhdenïi, Academia scientiarum imperiales Petropolitana. **LOCATIONS:** BUC, ULS **SUBJECT:** General - Societies - St. Petersburg.

585 The Farmers' magazine, a periodical work devoted to agriculture and rural affairs. - v.1-26 (nos.1-104) 1800-1825. Edinburgh: James Symington, 1800-1825. Quarterly.A sixth edition of v.1 appeared in 1806. Imprint varies. **EDITOR:** Robert Brown. (Ref.72. II:9). **INDEXES:** There is an index for v.1-24. **MICROFORMS:** UM(1), RP(1) (Goldsmiths' Kress Library). **SOURCES:** B6205, OCLC **LOCATIONS:** BUC, CCF, CCN, ULS **SUBJECT:** Agriculture - Periodicals - Scotland.

586 Farmers' magazine and useful family companion. - v.1.-5, 1776-1780. London: C. Dilley; Bath: Crutwell, 1776-1780. Agricola Sylvan is said to be a pseudonym for Henry Home. (Ref.73). **EDITOR:** Agricola Sylvan. **SOURCES:** B6206, OCLC **LOCATIONS:** BUC, CCF, ULS **SUBJECT:** Agriculture - Periodicals - England.

587 Fasti consolari dell' Accademia fiorentina / Accademia del Cimento. v.1, 1717. Florence: G.G.Tartini, 1717. **EDITOR:** Salvino Salvini. **SOURCES:** S1821b **LOCATIONS:** NUC **SUBJECT:** Science - Societies - Florence.

588 Feldbau- Haus- und Landwirthschafts- Wochenblatt für Güterbesitzer, Wirthschaftsämter, etc. - v.1-3, 1791. Vienna: 1791. **SOURCES:** K2978 **SUBJECT:** Home economics - Periodicals - Austria. **SUBJECT:** Agriculture - Periodicals - Austria.

589 Feuille de santé. - Paris: 1791. **EDITOR:** Chambon. **SOURCES:** G190 **SUBJECT:** Medicine - Periodicals - France.

590 Feuille du cultivateur. - T.1-8, 6 Oct. 1790- 13 Sept.1798. Paris: 1790-1798. **CONTINUATION OF:** Journal de commerce et d'agriculture. **CONTINUATION OF:** Journal d'agriculture, du commerce et des finances.

CONTINUATION OF: Journal d'agriculture, du commerce, des arts et des finances. Appeared as supplement to: Journal générale de France, 15 Jan. 1788- 5 May 1790, and as Feuille d'agriculture et d'économie rurale, 12 May-29 Sept.(nos. 19-39)1790. There were new editions in 1795 and 1802. OTHER TITLE: Feuille d'agriculture et d'économie rurale; Journal générale de France, partie d'agriculture. EDITORS: Jean Baptiste Dubois de Jancigny, P.M.A.Brussonet, Armand Bernardin Lefevre, and Antoine Augustin Parmentier.   SOURCES: B6224   LOCATIONS: BUC, CCF, ULS SUBJECT: Agriculture - Periodicals - France.

591   Feuilles hebdomadaire sur la médecine, la chirurgie, la pharmacie et les sciences qui y ont rapport. - v.1, 1791. Montpellier: Jean Martel, 1791. Weekly. CONTINUED BY: Journal d'instruction sur toutes les parties de l'art de guérir. SOURCES: G96   LOCATIONS: CCF   SUBJECT: Medicine - Periodicals - France.

592   Flora batava, of, afbeelding en beschrijving van Nederlandsche gewassen. - v.1-28, 1800-1934. Amsterdam: 1800-1934.   LOCATIONS: BUC, CCN, ZDB   SUBJECT: Botany - Periodicals - Netherlands.

593   Flora, oder Nachrichten von merkwürdigen Blumen. - Heft 1-6, 1788-1790. Stuttgart: Johann Benedict Metzler, 1788-1790. This title may be considered a book-in-parts.   SOURCES: B1707, K3325   LOCATIONS: BN, ZDB SUBJECT: Botany - Periodicals - Germany.

594   Foreign medical review. - v.1 (part 1-4) 1779-1780. London: J. Murray, 1779-1780. May be associated with The London medical journal which followed (Ref.74). CONTINUED BY: London medical journal.   SOURCES: G40 LOCATIONS: BUC, ULS   SUBJECT: Medicine - Periodicals - England - Reviews.

595   Forst Archiv zur Erweiterung der Forst und Jagd Wissenschaft und der Forst und Jagd Litteratur. - v.1-17, 1788-1796. Ulm: Stettin, 1788-1796. EDITOR: Wilhelm Gottfried Moser.   CONTINUED BY: Neues Forst Archiv. CONTINUED BY: Annalen der Forst und Jagdwissenschaft.   SOURCES: B6270, K2957,OCLC   LOCATIONS: BUC, CCF, CCN, ULS, ZDB SUBJECT: Forestry management - Periodicals - Germany.

596   Forstjournal. - v.1-2, 1797-1801. Leipzig: Weinbrack, 1797-1801. Imprint varies.   EDITOR: Friedrich Casimir Medius.   SOURCES: K3013 LOCATIONS: ULS, ZDB   SUBJECT: Forestry management - Periodicals - Germany.

597   Forst und Jagd-kalender. - v.1-10, 1794-1803. Leipzig: Kuchler, 1794-1803. LOCATIONS: ZDB   SUBJECT: Forestry management - Periodicals - Germany.   SUBJECT: Hunting - Periodicals - Germany.

598   Forst und Jagd-taschenbuch. - v.1, 1794. Leipzig: Graff, 1794. **LOCATIONS:**
       BUC, ZDB   **SUBJECT:** Forestry management - Periodicals - Germany.
       SUBJECT: Hunting - Periodicals - Germany.

599   Forst und Jagdbibliothek. - St.1-3, 1788-1789. Stuttgart: Johann Benedict Meyer,
       1788-1789. **CONTINUATION OF:** Allgemeines ökonomisches Forstmagazin.
       EDITOR: Johann Friedrich Stahl. **SOURCES:** B6271,K2958 **LOCATIONS:**
       BUC,ZDB   **SUBJECT:** Forestry management - Periodicals - Germany.

600   Der Förster, oder, ein Beitrag zum Forstwesen. - v.1-2 (each of 3 issues)
       1797-1803. Nurnberg: Stein, 1797-1803.   **EDITOR:** Franz Xaver Georg
       Heidenberg.   **SOURCES:** K3016a   **LOCATIONS:** ZDB   **SUBJECT:**
       Forestry management - Periodicals - Germany.

601   Fortgesetzte Auszüge aus den besten alten medicinischen Probeschriften. Bd.1,
       1783. Offenbach am Main: U. Weiss und C. L. Brede, 1783.   Not identical
       with Theil 3 of the Auszüge der besten medicinischen Probeschriften des 16ten
       und 17ten Jahrhunderts (Ref.21: 23:1240). **SOURCES:** G281   **SUBJECT:**
       Medicine - Periodicals - Germany - Dissertations.

602   Fortgesetzte Beiträge zur Naturkunde. - St.7-12, 1765. Berlin: Verlag der
       Realschule, 1765.   **CONTINUATION OF:** Physicalische Briefe.
       CONTINUATION OF: Monatliche Beiträge zur Naturkunde.   **EDITOR:**
       Johann Daniel Denso.   **CONTINUED BY:** Neue Monatliche Beiträge zur
       Naturkunde. **MICROFORMS:** RM(3). **SOURCES:** B3646b, K3236, OCLC
       LOCATIONS: BUC, ZDB   **SUBJECT:** Natural history - Periodicals -
       Germany.

603   Fortgesetzte Bemühungen zum Vortheil der Naturkunde und des gesellschaftlichen
       Lebens der Menschen. - St.1-4, 1759-1761.   Berlin; Stettin: 1759-1761.
       CONTINUATION OF: Neue Wahrheiten zum Vortheil der Naturkunde.
       EDITOR: Johann Heinrich Gottlob von Justi.   **SOURCES:** K2872
       LOCATIONS: BN, ZDB   **SUBJECT:** Natural history - Periodicals -
       Germany. **SUBJECT:** Economics - Periodicals - Germany.

604   Fortgesetzte Nachrichten von dem Zustande der Wissenschaften und Künste in den
       Kgl. dänischen Reichen und Ländern. - v.1-4 (6 St. each) 1758-1768.
       Copenhagen; Leipzig: Friedrich Christian Pelt, 1758-1768. **CONTINUATION
       OF:** Nachrichten von dem Zustande der Wissenschaften und Künste in den Kgl.
       dänischen Reichen und Ländern.   **EDITOR:** Anton Friedrich Büsching.
       MICROFORMS: RM(3).   **SOURCES:** B3217a,d K216, OCLC, S628
       LOCATIONS: BUC, ULS, ZDB   **SUBJECT:** General - Periodicals -
       Denmark - Reviews.

605   Fragmente, Nachrichten und Abhandlungen zur Beförderung der Finanz-, Polizei-,
       Ökonomie und Naturkunde. - Heft 1-3, 1788-1791. Berlin: Friedrich Mauer,
       1788-1791.   **SOURCES:** K2755   **SUBJECT:** Economics - Periodicals -
       Germany.

606   Fragmente zur Arnei- und Naturkunde und Geschichte. - v.1-2, 1780-1782.
      Frankfurt; Leipzig: Johann Georg Fleischer, 1780-1782. OTHER TITLE:
      Fragmente zur Arzneikunde und Naturgeschichte. EDITOR: Johann Philipp
      Berchelmann. SOURCES: B1771, G360, K3583, S2588 LOCATIONS:
      ZDB SUBJECT: Medicine - Periodicals - Germany. SUBJECT: Natural
      history - Periodicals - Germany.

607   Fragmente zur mineralogischen und botanisches Geschichte Steymarks und
      Kärnthens. - v.1, 1783. Klagenfurth: 1783. LOCATIONS: BUC, ZDB
      SUBJECT: Mineralogy - Periodicals - Germany. SUBJECT: Botany -
      Periodicals - Germany.

608   Franckfurtische gelehrte Zeitungen. - v.1-36, 1736-1771. Frankfurt: Samuel
      Tobias Hocker, 1736-1771. EDITOR: Samuel Tobias Hocker. CONTINUED
      BY: Frankfurter gelehrte Anzeigen. SOURCES: B1779, K109
      LOCATIONS: CCF, ULS, ZDB SUBJECT: General - Periodicals - Germany
      - Reviews.

609   Frankfurter Beiträge zur Ausbreitung nützlicher Künste und Wissenschaften. -
      v.1-3, 1780-1781. Offenbach: 1780-1781. EDITORS: Heinrich Wilhelm
      Seyfried and Philippe Jacques Rühl. SOURCES: B1776, K5541, OCLC,
      S3174 LOCATIONS: CCF, ULS SUBJECT: General - Periodicals -
      Germany.

610   Frankfurter gelehrte Anzeigen. - v.1-19, 1772-1790. Frankfurt: Eichenberg,
      1772-1790. CONTINUATION OF: Franckfurtische gelehrte Zeitungen.
      EDITORS: Johann Heinrich Merck, Johann Georg Schlosser and Karl Friedrich
      Bahrdt. MICROFORMS: GO(9). SOURCES: B1777, K288, S2589
      LOCATIONS: CCF, ULS, ZDB SUBJECT: General - Periodicals - Germany
      - Reviews.

611   Frankfurter medicinische Annalen für Aerzte, Wundärzte, Apotheker und
      denkende Leser aus allen Ständen. - v.1, 1789. Frankfurt: Jäger, 1789.
      CONTINUATION OF: Medicinisches Wochenblatt für Aerzte, Wundärzte und
      Apotheker. CONTINUATION OF: Neues medicinisches Wochenblatt für
      Aerzte, Wundärzte und Apotheker. EDITORS: Johann Valentin Müller and
      Georg Friedrich Hoffmann. CONTINUED BY: Medicinisches Wochenblatt
      oder forgesetzte medicinische Annalen. CONTINUED BY: Medicinischer
      Rathgeber für Aerzte und Wundärzte. CONTINUED BY: Medicinisches
      Repertorium für Gegenstände aus allen Fächern der Arzneiwissenschaft.
      SOURCES: G81, K3649, OCLC LOCATIONS: ULS, ZDB SUBJECT:
      Medicine - Periodicals - Germany. SUBJECT: Surgery - Periodicals -
      Germany. SUBJECT: Pharmacy - Periodicals - Germany.

612   Fränkische Acta erudita et curiosa. - v.1-4 (nos. 1-24) 1726-1732. Nürnberg:
      Endter & Engelbrecht, 1726-1732. CONTINUATION OF: Nova litteraria
      circuli Franconici. EDITOR: E. Fr. Just Heinrich. CONTINUED BY:
      Nützliche und auserlesene Arbeiten der Gelehrten im Reich. INDEXES: Index
      nos.1-12 in 12, 13-24 in 24. SOURCES: B1775, G641, K75, S3157

LOCATIONS: BUC, CCF, CCN, ULS, ZDB   SUBJECT: General - Periodicals - Germany - Reviews.

613   Fränkische ökonomisch-landwirthschaftliche Mannichfaltigkeiten. - v.1-3, 1777-1788. Schwabach: Weber, 1777-1788. SOURCES: K2918 LOCATIONS: CCF   SUBJECT: Agriculture - Periodicals - Germany. SUBJECT: Economics - Periodicals - Germany.

614   Fränkische Sammlungen von Anmerkungen aus der Naturlehre, Arzneygelahrtheit, Ökonomie und den damit verwandten Wissenschaften. - v.1-8 (nos.1-48) 1756-1768. Nürnberg: Monath, 1756-1768. EDITOR: Heinrich Friedrich Delius. SOURCES: B1781, G642, K3223, S3159 LOCATIONS: BUC, CCF, ULS, ZDB   SUBJECT: Natural history - Periodicals - Germany. SUBJECT: Medicine - Periodicals - Germany. SUBJECT: Economics - Periodicals - Germany.

615   Fränkish-arzneikundliche Annalen, grössentheils aus den Tagebüchern des Bamberger Krankenhauses gezogen. - Heft 1-4, 1792. Bamberg: 1792. EDITOR: Adalbert Friedrich Marcus. SOURCES: K3674 SUBJECT: Medicine - Periodicals - Germany.

616   Französische medicinische Literatur, oder, Anzeigen und Auszüge aus den neuesten Französischen Werken über Physik, Medicin und Oeconomie. - St.1-2, 1790. Heidelberg: Pfähler, 1790. EDITOR: Heinrich Tabor. SOURCES: G282, K3664 SUBJECT: Medicine - Periodicals - France - Translations. SUBJECT: Science - Periodicals - France - Translations.

617   Freemason's magazine, or, general and complete library. - London: 1793-1798. v.1-11, June 1793-1798. The single no. for 1797 is entitled: Scientific magazine. OTHER TITLE: Scientific magazine and freemason's repository; Freemason's repository. MICROFORMS: RP(1). SOURCES: OCLC LOCATIONS: BUC, ULS   SUBJECT: Science - Periodicals - England - Popular.

618   Freiberger gemeinnützige Nachrichten für das chursächische Erzgebürge. - Freiberg: 1800-1812. SOURCES: K5978 LOCATIONS: BUC, CCF SUBJECT: Mineral industries - Periodicals - Germany.

619   Freimüthige Nachrichten von neuen Büchern und anderen zur Gelahrtheit gehörigen Sachen. - v.1-20, 1744-1763. Zürich: Heidegger, 1744-1763. EDITORS: S. Wolf and Heidegger (Ref.123, p.40). CONTINUED BY: Wöchentliche Anzeigen zum Vortheil der Liebhaber der Wissenschaften und Künste. SOURCES: K140 LOCATIONS: BUC, CCF, ULS SUBJECT: General - Periodicals - Switzerland - Reviews.

620   Früchte der einsamkeit in unterschiedlichen physikalischen, medicinischen und chirurgischen Zeitverkürzungen. - St.1-6, 1775-1776. Breslau: Wilhelm Gottlieb Korn, 1775-1776. EDITOR: Christian Ehrenfried Rückert.

SOURCES: K3261, OCLC   **SUBJECT:** Science - Periodicals - Germany - Reviews.   **SUBJECT:** Medicine - Periodicals - Germany - Reviews.

621   Für die Litteratur und Kenntniss der Naturgeschichte, sonder der Conchylien und der Steine. - v.1-2, 1782. Weimar: C.L. Hoffmann, 1782. **CONTINUATION OF:** Journal für die Liebhaber des Steinreichs. **EDITOR:** Johann Samuel Schröter. **CONTINUED BY:** Neue Litteratur und Beiträge zur Kenntniss der Naturgeschichte. **SOURCES:** B249a, B6306, G534, K3287, OCLC, S3305 **LOCATIONS:** BUC, CCF, CCN, ULS   **SUBJECT:** Natural history - Periodicals - Germany.   **SUBJECT:** Mollusks - Periodicals - Germany. **SUBJECT:** Mineralogy - Periodicals - Germany.

622   Galleria di Minerva, overo, Notizie universali di quanto è stato scritto da letterati di Europa. - v.1-7, 1698-1717. Venice: 1696-1717. **MICROFORMS:** RP(1). **LOCATIONS:** BUC, ULS, ZDB   **SUBJECT:** General - Periodicals - Italy - Reviews.

623   Gartenkalender. - v.1-v.7, 1782-1789.   Kiel: Hirschfeld; Braunschweig: Schulbuchhandlung, 1782-1789. **CONTINUATION OF:** Taschenbuch für Gartenfreunde. **EDITOR:** Christian Cay Lorenz Hirschfeld. **CONTINUED BY:** Kleine Gartenbibliothek.   **SOURCES:** B1821, K2936, OCLC **LOCATIONS:** BN, ZDB   **SUBJECT:** Gardening - Periodicals - Germany.

624   Gavnlig og underholdende laesning i naturvidenskaben. - v.1-3, 1793-1803. Copenhagen: 1793-1803. **EDITOR:** Odin Henrik Wolf. **SOURCES:** B1823, G643, S620   **SUBJECT:** Natural history - Periodicals - Denmark.

625   Gazette d'agriculture, commerce, arts et finances. - Paris: 1769-1783. **CONTINUATION OF:** Gazette du commerce, de l'agriculture et des finances. **EDITOR:** P.J.A. Roubard. **SOURCES:** OCLC **LOCATIONS:** CCF, ULS **SUBJECT:** Agriculture - Periodicals - France.   **SUBJECT:** Commerce - Periodicals - France.   **SUBJECT:** Economics - Periodicals - France.

626   Gazette de médecine. - T.1-4, No.52, T.5, no.1, 1761-1763 (Ref.10). Paris: Grangé, 1761-1763. A pirated edition appeared three weeks after the first issue. **OTHER TITLE:** Gazette d'Epidaure. **EDITOR:** Jacques Barbeu-Dubourg. **SOURCES:** G14   **LOCATIONS:** CCF, ULS   **SUBJECT:** Medicine - Periodicals - France.

627   Gazette de médecine et d'hippiatrique. - No.1-3, 1777-1778. Cap-Hatien: 1777-1778. **OTHER TITLE:** Gazette de médecine pour les colonies. **EDITOR:** Duchemin L'Étang. **SOURCES:** Ref.129,p.343 **LOCATIONS:** CCF   **SUBJECT:** Medicine - Periodicals - Haiti.

628   Gazette de santé, contenant les nouvelles descouvertes sur le moyens de bien porter, et de querir quand on est malade. - v.1-56, 1773-1829. Paris: Ruault, 1773-1829. Every ten days. Publisher and subtitle vary. Edited by a "Société de Médecins", Année 4-25. **EDITORS:** Jean Jacques de Gardane, (Année 1-3) 1773-1775; Jean Jacques Paulet (Année 4-25) 1776-1799; Philippe Pinel,

1784-1789 (Ref.11); Marie de Saint Ursin, (Année 26-35) 1800-1809. **CONTINUED BY:** Gazette médicale de Paris. Supplements accompany some issues. **SOURCES:** G32 **LOCATIONS:** BUC, CCF, ULS **SUBJECT:** Medicine - Periodicals - France.

629   Gazette de Santé, oder, gemeinnützliches medizinisches Magazin für Leser aus allen Ständen. - Jahrgang 1-4, 1782-1786.Zürich: J.C. Fuessli, 1782-1786. Six issues a year. **OTHER TITLE:** Gemeinnütziges medicinisches Magazin. **EDITOR:** Johann Heinrich Rahn. **CONTINUED BY:** Archiv gemeinnützger physischer und medicinischer Kenntnisse. **CONTINUED BY:** Gemeinnütziges Wochenblatt physischen und medizinischen Kentnisse. **CONTINUED BY:** Magazin für gemeinnnützige Arzneikunde und medicinische Polizei. **SOURCES:** G51, K3596, OCLC **LOCATIONS:** BUC, ULS **SUBJECT:** Medicine - Periodicals - Switzerland.

630   Gazette du commerce, de l'agriculture et des finances. - Paris: Knapen, 1763-1768. From April 1, 1763 to May 28,1765 as Gazette du commerce. **OTHER TITLE:** Gazette du commerce. **CONTINUED BY:** Gazette d'agriculture, commerce, arts et finances. Includes supplements in 1766-1767. **MICROFORMS:** RP(1). (Goldsmith Kress library), ACR(1). **LOCATIONS:** BUC, CCF, ULS **SUBJECT:** Agriculture - Periodicals - France. **SUBJECT:** Commerce - Periodicals - France.

631   Gazette salutaire, composée de tout ce qui contiennent d'interessant pour l'humanité, livres nouveaux, les journaux et autres écrits publiés concernant la médecine, la chirurgie, la botanique, etc. - v.1-32, Apr. 1761-Nov. 1793. Bouillon: 1761-1793. Weekly. **CONTINUATION OF:** Gazette de médecine. Subtitle varies. **OTHER TITLE:** Gazette d'Epidaure. **EDITOR:** Friedrich Emmanuel Grünwald. **SOURCES:** GIII:14, OCLC **LOCATIONS:** BUC, CCF, CCN, ULS **SUBJECT:** Medicine - Periodicals - France.

632   Gazety slaskie, dla ludu pospolitego. - Kartka IX, 1790. Warsaw: 1790. **SOURCES:** G88 **SUBJECT:** Medicine - Periodicals - Poland.

633   Geist der neuesten medicinischen Litteratur in Frankreich zum Behuf deutsche Aerzte, in Auszügen aus den neuesten Originalwerken. - v.1 (nos.1-2) 1798-1799. Breslau: 1798-1799. **EDITOR:** August Theodor Zanth (formerly Abraham Zadig) (Ref.21 v.24:106). **SOURCES:** G284 **SUBJECT:** Medicine - Periodicals - Poland - Reviews.

634   Geist und Kritik der medicinischen und chirurgischen Zeitschriften Deutschlands für Aerzte und Wundärzte. - v.1-18 (9 parts) 1798-1806. Leipzig; Breslau: 1798-1806. Imprint varies. **EDITOR:** Johann Joseph Kausch. **SOURCES:** G283, K3720 **LOCATIONS:** BUC, CCF, ULS, ZDB **SUBJECT:** Medicine - Periodicals - Germany - Reviews.

635   Gelehrte Ergötzlichkeiten und Nachrichten. - v.1-2 (6 issues each) 1774. Stuttgart: Cotta, 1774. **EDITOR:** Balthasar Haug. **CONTINUED BY:** Schwäbisches Magazin von gelehrten Sachen. **CONTINUED BY:** Zustand der

Wissenschaften und Künste in Schwaben.   **SOURCES:** G647,K301
LOCATIONS: BUC, CCF, ULS, ZDB   SUBJECT: General - Periodicals -
Germany - Reviews.

636   Gemeinnützige Arbeiten der Churfürstliche Bienengesellschaft in der Oberlausitz,
die Physik und oeconomie der Bienen betreffend, nebst andern dahin
einschlagenden natürlichen Dingen. - Bd.1-2, 1773-1776. Leipzig; Berlin:
1773-1776.   **CONTINUATION OF:** Abhandlungen und Erfahrungen der
physikalisch- oeconomischen Bienengesellschaft in Oberlausitz zur Aufnahme
der Bienenzucht in Sachsen.   **EDITOR:** Adam Gottlob Schirach.   **SOURCES:**
K2909, S2382   **LOCATIONS:** ZDB   **SUBJECT:** Bee culture - Societies -
Oberlausitz.

637   Der Gemeinnützige mathematische Liebhaber, oder, Aufgaben aus der Arithmetic,
Geometrie, Trigonometrie, Astronomie, Geographie, Mechanic, Hydrostatic,
Navigation und Algebra, mit ihren gründlichen Auflösungen, zur Uebung und
Beförderung der mathematischen Wissenschaften. - Hamburg: 1767-1769.
EDITOR: Johann Reimer.   **SOURCES:** K3241   **LOCATIONS:** CCN, ULS
SUBJECT: Mathematics - Periodicals - Germany.

638   Gemeinnützige medicinische Erfahrung, ein Magazin practischer Kenntnisse für
angehende Ärzte. - Leipzig: 1794.   **SOURCES:** G285, OCLC
LOCATIONS: ULS   **SUBJECT:** Medicine - Periodicals - Germany.

639   Gemeinnützige Nachrichten und Bemerkungen besonders / Oeconomische
Gesellschaft des Kantons Bern. - Bd.1, 1798. Bern, 1798. **SOURCES:** S2135
LOCATIONS: ULS **SUBJECT:** Agriculture - Societies - Bern. **SUBJECT:**
Natural history - Societies - Bern.

640   Gemeinnützige Unterhaltungen aus der Arzneykunst, Naturgeschichte und
Ökonomie. - v.1-2, 1790-1791. Kiel: 1790-1791. Weekly. **EDITOR:** Johann
Georg Reyher.   **SOURCES:** B1865, G286, K3342, S2850   **SUBJECT:**
Medicine - Periodicals - Germany.   **SUBJECT:** Natural history - Periodicals
- Germany.   **SUBJECT:** Economics - Periodicals - Germany.

641   Gemeinnütziges Fränkisches Magazin, oder, Sammlung merkwürdiger nützlicher
Grundsätze und Erfahrungen aus der Naturlehre, Naturgeschichte, Arzneikunde,
Moral, Landwirthschaft. - St.1-4, 1770-1780. Nürnberg: Zeh, 1779-1780.
EDITOR: Johann Christoph Heppe.   **SOURCES:** B1870, G648, K5530, S3160
LOCATIONS: CCF, ULS, ZDB   **SUBJECT:** Science - Periodicals - Germany
- Popular.

642   Gemeinnütziges Journal über die Bäder und Gesundbrunnen Deutschlands. -
St.1-2, 1800-1802. Marburg: Krieger, 1800-1802.   **EDITOR:** Christoph
Heinrich Fenner von Fennberg.   **SOURCES:** K3736   **SUBJECT:** Balneology
- Periodicals - Germany.   **SUBJECT:** Medicine - Periodicals - Germany.

643   Gemeinnütziges Magazin der Natur und Kunst, oder, Abhandlungen zur
Beförderung der Naturkunde, der Künste, Manufacturen und Fabriken. - Pt.1-3,

1763-1767. Berlin: 1763-1767. SOURCES: B1871, S2320 SUBJECT: Natural history - Periodicals - Germany. SUBJECT: Economics - Periodicals - Germany.

644 Gemeinnütziges Natur-und Kunstmagazin, oder, Abhandlungen zur Beförderung der Naturkunde. - v.1 (St.1-6)-2(St.1-3) 1755-1764. Berlin; Jena: Hartung, 1755-1764. Bd.1, St.2-6 as Natur- und Kunst-Kabinet. Imprint varies. OTHER TITLE: Natur-und Kunst-Kabinet. SOURCES: B3229, G535, K3220, OCLC, S2831 LOCATIONS: BUC, CCN SUBJECT: Natural history - Periodicals - Germany. SUBJECT: Economics - Periodicals - Germany.

645 Gemeinnützliches Wochenblatt physischer und medizinischer Kenntnisse. Bd.1-Bd.2, 1792. Zürich: Orell, Gessner und Rüssli, 1792. CONTINUATION OF: Gazette de Santé, oder, gemeinnütziges medizinisches Magazin für Leser aus allen Ständen. CONTINUATION OF: Archiv gemeinnützger physischer und medicinischer Kenntnisse. EDITOR: Johann Heinrich Rahn. CONTINUED BY: Magazin für gemeinnnützige Arzneikunde und medicinische Polizei. SOURCES: G104, K3673 LOCATIONS: ULS SUBJECT: Medicine - Periodicals - Switzerland.

646 Gemeinnützlicher Vorrath auserlesener Aufsätze zur Beförderung der Haushaltungswissenschaft, Künste, Manufakturen und Fabriken, wie auch der Arzneygelahrtheit und Naturkunde. - v.1-3, 1767-1768. Leipzig: Hilscher, 1767-1768. EDITOR: Johann Georg Krünitz. SOURCES: B1869, K2889, S2933 SUBJECT: Economics - Periodicals - Germany. SUBJECT: Medicine - Periodicals - Germany.

647 Genees-en heelkundige proeven en annmerkingen, verat in de Philosophical transactions. - v.1, 1745. n.p.: 1745. Includes translations from v.1-9, 1665-1674 of the Philosophical transactions, some taken from the abridgement by Samuel Mihles. TRANSLATION OF: Philosophical transactions. LOCATIONS: NUC SUBJECT: Science - Societies - London - Translations.

648 Genees-heel-arzteny-en vroedkundig magazyn. - v.1 (St.1-4) 1782-1784. Rotterdam: 1782-1784. EDITOR: Martinus Pruys en Lambertus Nolst. SOURCES: GIII:48 LOCATIONS: BUC, CCN SUBJECT: Medicine - Periodicals - Netherlands.

649 Genees-heel-en naturkundige verlustigingen getrokken uit verscheide werken, over deeze wetenschappen in Europe. - St.1-2, 1766. Amsterdam: 1766. SOURCES: G287, S732 LOCATIONS: CCN SUBJECT: Medicine - Periodicals - Netherlands - Reviews. SUBJECT: Natural history - Periodicals - Netherlands - Reviews.

650 Genees-heel-ontleed-en natuurkundige courant of bericht van alle nuttige wetenschappen. - No. 1-53, 1 Feb.1, 1765-Jan.30, 1767 (Ref.75, no.9). Amsterdam: Eleveld, 1765-1767. Weekly. OTHER TITLE: Geneeskundige courant of weeklyks bericht. SOURCES: GIII:19 LOCATIONS: CCN, ULS

SUBJECT: Medicine - Periodicals - Netherlands.   SUBJECT: Science - Periodicals - Netherlands.

651   Genees- natuur en huishoudkundig kabinet, of, uitgezogte verzameling van de voornaamste buitenlandsche academien en genootschappenen de beroemste in-en uit-landsche schryveren. - v.1-4, 1779-1791.   Leyden: J. van Tiffelen, 1779-1791.   EDITOR: Jacob Voegen van Engelen.   SOURCES: B1873, G650, OCLC, S870   LOCATIONS: BL, CCN   SUBJECT: Medicine - Periodicals - Netherlands - Reviews.   SUBJECT: Science - Periodicals - Netherlands - Reviews.

652   Genees- natuur- en huishoudkundige jaarboeken, of, verzameling van de nieuwste verhandelingen, ontdekkingen, uitbindingen en waarnemingen der eerste geleerden in Europa, door een Genootschap van genees en natuurkundigen. - v.1-6 in 13 issues. Dordrecht: A. Blusse en W. Holtrop, 1778-1782. CONTINUED BY: Nieuwe genees- natuur- en huishoudkundige jaarboeken. CONTINUED BY: Algemeene genees- natuur- en huishoudkundige jaarboeken. SOURCES: B1872, G649, S1872   LOCATIONS: BUC, CCN, ULS SUBJECT: Medicine - Societies - Dordrecht.   SUBJECT: Natural history - Societies - Dordrecht.   SUBJECT: Home economics - Societies - Dordrecht.

653   Het Geneeskundig journal van London. - Bruges: J. Bogaert, 1786. TRANSLATION OF: London medical journal.   SOURCES: (Ref.17,no.21) SUBJECT: Medicine - Periodicals - England - Translations.

654   Geneeskundig tijdschrift, of, versammeling van ontleed-heel-artzeny- en naturkundige waarnemingen en nieuwe entdekkingen. - v.1-4 (Nos.1-100) 1767-1770. Rotterdam: Abraham Bothall, 1768. Biweekly.   SOURCES: GIII:24 LOCATIONS: BUC, CCN, ULS SUBJECT: Medicine - Periodicals - Netherlands - Reviews.

655   Geneeskundige bibliothek. - St.1-2, 1785-1786. Utrecht: 1785-1786. SOURCES: G363 LOCATIONS: BUC, ULS   SUBJECT: Medicine - Periodicals - Netherlands - Reviews.

656   Geneeskundige proeven en aanmerkingen. - 4 v.in 1. 1739-1741. Amsterdam: Hartig, 1739-1741. TRANSLATION OF: Medical essays and observations, Philosophical society of Edinburgh. SOURCES: OCLC LOCATIONS: DSG SUBJECT: Medicine - Societies - Scotland - Translations.

657   Geneeskundige Verhandelingen, aan de koninglyke Sweedsche Academie / Kungliga Svenska Vetenskapsakademien. - v.1-4, 1775-1778.   Leiden: 1775-1778.   CONTINUATION OF: Handlingar Kongliga Svenska Vetenskapsakademien.   EDITOR: Jean Bernard Sandifort.   SOURCES: G477c, OCLC, S882 LOCATIONS: CCN SUBJECT: Medicine - Societies - Stockholm.

658   A General collection of discourses of the virtuosi of France, upon questions of all sorts of philosophy, and other natural knowledge. - v.1-2, 1664-1665. London:

Thomas Dring and John Starkey, 1664-1665. The 2nd vol. bore the title: Another collection of the philosophical conferences of the French virtuosi. Translated by George Havers and John Davies. **OTHER TITLE:** Another collection of the philosophical conferences of the French virtuosi. **TRANSLATION OF:** Centurie des questions traitées ès conférences du bureau d'addresse. **MICROFORMS:** UM(1). **SOURCES:** OCLC **SUBJECT:** General - Societies - Paris - Translations.

659   General history of discoveries and improvements. - v.1-4, 1726-1728. London: J. Roberts, 1726-1728. Another edition was published in 1727 as The history of principal discoveries and improvements in the several arts and sciences. **EDITOR:** Daniel Defoe. **MICROFORMS:** UM(1). **LOCATIONS:** BUC **SUBJECT:** Science - Periodicals - England.

660   General history of trade, and especially consider'd as it represents the British commerce. - No 1-4, 1713. London: J. Baker, 1713. **EDITOR:** Daniel Defoe. **MICROFORMS:** RP(1). (Goldsmiths' Kress Library). **SOURCES:** OCLC **LOCATIONS:** BUC, ULS **SUBJECT:** Economics - Periodicals - England.

661   The General magazine of arts and sciences, philosophical, philological, mathematical and mechanical. - v.1-14, 1755-1766. London: W. Owen, 1755-1766. Facsimile edition, Philadelphia, 1938. **EDITOR:** Benjamin Martin. **MICROFORMS:** RM(3). **SOURCES:** B1875, OCLC, S286 **LOCATIONS:** BUC, ULS **SUBJECT:** Science - Periodicals - England - Popular.

662   The Gentleman's companion, containing new enigmas, rebuses, charades, likewise some useful papers from the transactions of the Royal Society. - v.1-5 (nos.1-30) 1798-1827. London: W. Baynes and J. Wright, 1798-1827. Publisher varies. 2nd edition in 1809. **OTHER TITLE:** Companion to the Gentleman's diary; Gentleman's mathematical companion. **EDITOR:** William Davis. **INDEXES:** There is an index to the series in an appendix. **MICROFORMS:** RM(3). **SOURCES:** S288 **LOCATIONS:** BUC, ULS **SUBJECT:** Mathematics - Periodicals - England - Popular.

663   The Gentleman's diary, or, the mathematical repository, containing many useful and entertaining particulars, peculiarly adapted to the ingenious gentlemen engaged in the delightful study of the mathematics. - v.1-v.100, 1741-1840. London: Company of Stationers, 1741-1840. Annual. It was reprinted in 3 v., 1814-1817, covering the years 1741-1800 and including the supplements under the title: The Gentleman's Diary, or,Mathematical Repository, Davis's edition (Ref.12, p.383). Also issued in the British Palladium. **OTHER TITLE:** Mathematical repository. **EDITORS:** John Badder (1741-1756), George Ingman (1741-1756), Thomas Peat (1741-1780), Charles Wildbore (1781-1803), Olinthus Gilbert Gregory (1804-1819), Thomas Leybourn (1820-1840). Supplements for the years, 1741-1745 published under the title: A Collection of mathematical problems and aenigmas. **MICROFORMS:** Northeast Document Conservation Center. **SOURCES:** B1881, OCLC, S287

LOCATIONS: BUC, ULS, ZDB **SUBJECT:** Mathematics - Periodicals - England.

664   The Gentleman's magazine. - v.1-303, 1731-1907. London: Edward Cave, 1731-1907. Subtitle and imprint vary. There were reprints of many of the issues (Ref.75). **EDITORS:** Edward Cave, John Nichols and D. Henry. **INDEXES:** Johnson, Samuel. A general index to the first twenty volumes. London, E. Cave, 1753. Ayscouth, Samuel. A general index to the first fifty-six volumes, 1731-1786. London, John Nichols, 1789. John Nichols, General index to the Gentleman's magazine from the year 1787 to 1818. London, Nichols, 1821. **MICROFORMS:** PMC(1), UM(1), RP(1). **SOURCES:** G651   **LOCATIONS:** BUC, CCF, CCN, ULS, ZDB **SUBJECT:** General - Periodicals - England.

665   Geographische Ephemriden. - v.1-51, 1798-1816. Weimar: Industr. Cpt., 1798-1816. **OTHER TITLE:** Allgemeine geographische Ephemeriden. **EDITORS:** Franz Xaver von Zach and Friedrich Justin Bertuch. **CONTINUED BY:** Neue allgemeine geographische Ephemeriden. **SOURCES:** B1888, K1308   **LOCATIONS:** ULS, ZDB   **SUBJECT:** Geography - Periodicals - Germany.

666   Geographischer Buchersall, zum Nutzen und Vergnügen eroffnet. - v.1-3 (10 issues in each v.) (1764/66) 1766-(1775/1778) 1778. Chemnitz: Johann David Stoszele, 1766-1778. **EDITOR:** Johann Georg Hagers. **LOCATIONS:** ULS, ZDB   **SUBJECT:** Geography - Periodicals - Germany - Reviews.

667   Geographisches Magazin. - Bd.1-Bd.4(St.1-16), 1783-1785. Dessau und Leipzig: Buchhandlung der Gelehrten, 1783- 1785. **EDITOR:** Johann Ernst Fabri. **CONTINUED BY:** Neues geographisches Magazin. **CONTINUED BY:** Historische und geographische Monatsschrift. **CONTINUED BY:** Beiträge zur Geographie, Geschichte und Staatenkunde. **CONTINUED BY:** Magazin für die Geographie, Staatenkunde und Geschichte. **SOURCES:** B1892, K1142 **LOCATIONS:** ZDB   **SUBJECT:** Geography - Periodicals - Germany.

668   Geoponica, eine ökonomische Monatschrift für Kur-und Livland. - v.1-2, 1798-1799. Mitau: 1798-1799. Weekly. **EDITORS:** C. Georg and Johann Martin Franz August Zernevsky. **SOURCES:** K3018 **SUBJECT:** Economics - Periodicals - Yugoslavia.

669   Georgica Bavarica, oder, ökonomische Auszüge und gründliche Nachrichten wie sowohl adeliche Land- als gemeine Baueren-Güther verbessert. - v.1 (St.1-13) 1752. Munich: Johann Jacob Vötter, 1752. **SOURCES:** K2864 **SUBJECT:** Agriculture - Periodicals - Germany.

670   Georgical essays, in which the food of plants is particularly considered. - v.1-4, 1770-1772. London: T.Durham, 1770-1772. Reprinted in 1773, 1777, and 1803-4. **EDITOR:** Alexander Hunter. **SOURCES:** OCLC **LOCATIONS:** BUC **SUBJECT:** Agriculture - Periodicals - England.

671   Gerichtlich-medicinische Beobachtungen. - Bd.1-2, 1778-1780. Königsberg: Hartung, 1778-1780. Reprinted 1819 in Vienna as a supplement to Metzger's Kurzgefassten System der gerichtlichen Arzneiwissenschaft. **EDITOR:** Johann Daniel Metzger. **CONTINUED BY:** Neue gerichtlich-medicinische Beobachtungen. **SOURCES:** G364, K3377, OCLC **LOCATIONS:** ZDB **SUBJECT:** Medical jurisprudence - Periodicals - Germany.

672   Gesammelte Nachrichten der oeconomischen Gesellschaft in Francken. - v.1-3, 1765-1767. Anspach: Posch, 1765-1767. Weekly. **EDITOR:** Johann Christoph Hirsch. **CONTINUED BY:** Magazin der patriotisch-ökonomischen Gesellschaft in Franken. **SOURCES:** K2882 **LOCATIONS:** ZDB **SUBJECT:** Economics - Societies - Anspach.

673   Geschichte und Versuche einer chirurgischen Privatgesellschaft zu Copenhagen. - v.1, 1774. Copenhagen: Rothe, 1774. **EDITOR:** Johann Clemens Tode. **SOURCES:** G33 **SUBJECT:** Surgery - Societies - Copenhagen.

674   Gesellschaftliche Erzählungen für die Liebhaber der Naturlehre, Haushaltungs-wissenschaft, Arzneykunst und Sitten. - v.1-4 (St.1-104) 1753-1757. Hamburg: Grund, 1753-1757. **EDITOR:** Johann August Unzer. **CONTINUED BY:** Neue gesellschaftliche Erzählungen für die Liebhaber der Naturlehre, Haushaltungs-wissenschaft, Arzneykunst und Sitten. **SOURCES:** B1905, K3215, S2749 **LOCATIONS:** BUC, CCF, ZDB **SUBJECT:** Science - Periodicals - Germany.

675   Gesundheits-almanach, oder, medicinisches Taschenbuch. - v.1, 1796. Frankfurt: Andre, 1796. **EDITOR:** Johann Valentin Müller. **SOURCES:** G198 **SUBJECT:** Medicine - Periodicals - Germany - Popular.

676   Gesundheitsalmanach zum Gebrauch für die aufgeklärten Stände Deutschlands. - Leipzig: 1793. **EDITOR:** Johann Carl Friedrich Leune. **SOURCES:** G193 **SUBJECT:** Medicine - Periodicals - Germany - Popular.

677   Der Gesundheitstempel, eine diätetische Monatsschrift. - v.1 (pt.1-6), 1797-1799. Leipzig: Jacobäer, 1797-1799. **EDITOR:** Christian Friedrich Niceus. **SOURCES:** G203, K3707 **SUBJECT:** Nutrition - Periodicals - Germany - Popular. **SUBJECT:** Medicine - Periodicals - Germany - Popular.

678   Giornale aerostatico. - v.1, no.1-3, Jan.-Mar. 1784. Milan: Gaetano Motto, 1784. **LOCATIONS:** BUC, NUC **SUBJECT:** Balloon ascension - Periodicals - Italy.

679   Giornale astro-meteorologico. - 1-3, 1789-1791. Venice: 1789-1791. **SOURCES:** B1930, S2092 **LOCATIONS:** BUC **SUBJECT:** Astronomy - Periodicals - Italy. **SUBJECT:** Meteorology - Periodicals - Italy.

680   Giornale de' letterati oltramontain. - Nos. 1-144, 1722-1757. Venice: Luigi Pavino, 1722-1757. **SOURCES:** B1957, S1837 **LOCATIONS:** ULS **SUBJECT:** General - Periodicals - Italy - Reviews.

681   Giornale de' litterati. - v.1-14, 1668-1681. Rome: Nicolo Angelus Tinassi, 1668-1681. Publisher varies. (Benedetto Carrara after 1675.) Issued by Ciampini under the same title in 7 v., 1675-1668. Translation in Collection académique v.4 and 7. **EDITOR:** Franceso Nazari. **SOURCES:** B1955, GIII:6, OCLC, S2033 **LOCATIONS:** BUC, CCF, ULS **SUBJECT:** General - Periodicals - Italy - Reviews.

682   Giornale dei letterati del A. Fabroni. - v.1-102, 1771-1796. Pisa: A. Pizzorno, 1771-1796. Imprint varies. **EDITOR:** A. Fabroni. **CONTINUED BY:** Nuovo giornale dei letterati. **CONTINUED BY:** Giornale pisano di letteratura, scienze ed arti. **CONTINUED BY:** Giornale scientifico e letterario. **SOURCES:** B1954, OCLC, S2005 **LOCATIONS:** BUC, CCF, ULS **SUBJECT:** General - Periodicals - Italy - Reviews.

683   Giornale dei letterati del Pagliarini. - v.1-22, 1745-1766. Rome: 1745-1766. Monthly. **OTHER TITLE:** Novelle letterarie oltra montane; Notizie letterarie oltramontane. **EDITOR:** Pagliarini. **MICROFORMS:** DA(1). **SOURCES:** B1956 **LOCATIONS:** ULS **SUBJECT:** General - Periodicals - Italy - Reviews.

684   Giornale dei letterati (Firenze). - v.1-7, 1742-1757, ns. v.1-8, 1785-1788. Florence: P. Giovanneli, 1742-1785. Monthly. **SOURCES:** B1957, S1837 **LOCATIONS:** BUC, CCF, ULS **SUBJECT:** General - Periodicals - Italy - Reviews.

685   Giornale dei letterati (Parma). - v.1-12, 1668-1679, ser. 2, v.1-7, 1686-1697. Parma; Modena: Guiseppe Giovanelli, 1668-1697. Ser. 2 edited by Bernardino Bacchini. **SOURCES:** B1951, GI:2, S1989 **LOCATIONS:** BUC, CCF, ULS, ZDB **SUBJECT:** General - Periodicals - Italy - Reviews.

686   Giornale di chirurgia di Firenze. - v.1, no.1, 1788. Venice: 1788. **LOCATIONS:** ULS **SUBJECT:** Surgery - Periodicals - Italy.

687   Giornale di medicina. - v.1-13, 1762-1766. Venice: Bendetto Milocco, 1762-1776. Weekly. Suspended Sept.4, Jan. 5, 1774-1776. **OTHER TITLE:** Gazetta medica d'oltramonti; Gazetta medica di Parma. **EDITOR:** Pietro Orteschi and Jacopo Panzani. **CONTINUED BY:** Nuovo giornale di medicina. **SOURCES:** GIII:15, OCLC **LOCATIONS:** BUC, ULS **SUBJECT:** Medicine - Periodicals - Italy - Reviews.

688   Giornale d'Italie, spettante alle science naturali e principalmente all'agricoltura, alle arti ed al commercio. - v.1-12, 1765-1776. Venice: Levi-Malvano, 1765-1776. **CONTINUED BY:** Nuovo giornale d'Italie, spettante alle science naturali e principalmente all'agricoltura, alle arti ed al commercio. **SOURCES:** B1948, G653, S2094 **LOCATIONS:** BUC, CCF, ULS **SUBJECT:** Science - Periodicals - Italy.

689   Giornale enciclopedico di Liege. - Lucca: 1756-1783. Some original material added. **TRANSLATION OF:** Journal encyclopédique ou universel.

SOURCES: Ref.92,p.220. LOCATIONS: CCF SUBJECT: General - Periodicals - France - Translations.

690 Giornale enciclopedico d'Italia, o, sia Memorie scientifiche e letterarie raccolti da' Giornali di Bologna, di Vicenza, di Milano, de Bouillon, etc. - v.1-8, 1785-1789. Naples: Giuseppi Campo, 1785-1789. LOCATIONS: ULS SUBJECT: General - Periodicals - Italy - Reviews.

691 Giornale fiorentino di agricoltura, arti, commercio ed economica. - v.1-3, 1786-1788. MICROFORMS: RP(1) (Goldsmiths' Kress Library). SOURCES: (Ref.76). LOCATIONS: BUC SUBJECT: Economics - Periodicals - Italy. SUBJECT: Agriculture - Periodicals - Italy.

692 Giornale fisico-medico, ossia raccolta di osservazioni sopra la fisica, matematica, chimica, storia naturale, medicina, chiurgia, arti ed agricoltura. - v.1.-16, 1792-1795. Pavia: 1792-1795. CONTINUATION OF: Biblioteca fisica d'Europa. EDITOR: Luigi Gaspero Brugnatelli. CONTINUED BY: Avanzamenti della medicina e fisica. SOURCES: B1938, G652, OCLC LOCATIONS: BUC, CCF, ULS, ZDB SUBJECT: Science - Periodicals - Italy.

693 Giornale medico e letterario di Trieste. - v.1-4, 1790-1791. Trieste: Giovanni Tommaso Höchenberger, 1790-1791. EDITOR: Benedetto Frizzi. SOURCES: G90 LOCATIONS: BUC SUBJECT: Medicine - Periodicals - Italy.

694 Giornale per servire alla storia ragionata della medicina di questo seculo. - v.1-13, 1783-1800. Venice: Pasquale, 1783-1800. EDITOR: Francesco Aglietti. SOURCES: G56, OCLC LOCATIONS: BUC, CCF, ULS, ZDB SUBJECT: Medicine - Periodicals - Italy.

695 Giornale scientifico, letterario e delle arti di una società filosofica di Torino. - v.1-13, 1789-1790. Turin: 1789-1790. EDITORS: Giovanni Antonio Giobert and Carlo Guilo. SOURCES: B1943, S2067 LOCATIONS: CCF, ULS SUBJECT: Science - Periodicals - Italy.

696 Gnothi sauton, oder, Magazin der Erfahrungsseelenkunde als ein Lesebuch für Gelehrte und Ungelehrte. - v.1-10, 1783-1793. Berlin: Mylius, 1783-1793. Reprinted 1787-1788 (Berlin, Mylius, 1787-1788). Facsimile reprint (Lindau, Antiqua Verlag, 1978/9) (Ref.77). OTHER TITLE: Magazin der Erfahrungsseelenkunde. EDITORS: Carl Philipp Moritz, Friedrich Pockels and Salomon Maimon. CONTINUED BY: Psychologisches Magazin. CONTINUED BY: Zeitschrift für die spekulative Physik. INDEXES: General index in v.10. SOURCES: G241, K549, OCLC, S2324 LOCATIONS: BUC, CCF, CCN, ULS, ZDB SUBJECT: Psychology - Periodicals - Germany.

697 Gothaische gelehrte Zeitungen, ausländische Literatur. - v.1-32, 1774-1804. Gotha: Carl Wilhelm Ettinger, 1774-1804. EDITOR: Emil Christian Klüpfel.

SOURCES: B2025, G657, K3197, OCLC, S2755 LOCATIONS: BUC, CCF, CCN, ULS, ZDB SUBJECT: Science - Periodicals - Germany. SUBJECT: Natural history - Periodicals - Germany.

712 Hanauisches Magazin. - v.1-8, 1778-1785. Hanau: Waisenhaus, 1778-1785. Weekly. SOURCES: B2028, K336, S2774 LOCATIONS: BUC, CCF, ULS, ZDB SUBJECT: General - Periodicals - Germany.

713 Handbuch für Liebhaber der Natur und Öconomie. - v.1-4, 1788-1789. Grätz: Weigand, 1788-1789. EDITOR: Patriz Dengg. SOURCES: B2029, K2959, S3404 LOCATIONS: SUBJECT: Natural history - Periodicals - Austria. SUBJECT: Economics - Periodicals - Austria.

714 Handelingen der Nederlandsche Maatschappij vor nijverheid en handel. - v.1-100, 1778-1876. Haarlem: J. Bosch, 1778-1876. 1777-1796 as Oekonomische taks der Hollandsche Maatschappije. OTHER TITLE: Resolutionem genomen by de Allgemeen Vergaderingen des Oeconomischen taks van de Hollandsche Maatschappije der Wettenschappen te Haarlem. LOCATIONS: BUC, ULS SUBJECT: Economics - Societies - Haarlem.

715 Handelingen van het geneeskundig genootschapp onder de zinspreuk: servandis civibus. - v.1-16, 1776-1792. Amsterdam: Peter Condrad, 1776-1792. SOURCES: G39 LOCATIONS: BUC, CCF, CCN, ULS SUBJECT: Medicine - Societies - Amsterdam.

716 Handlingar / Göteborgska vetenskaps och vitterhets samhälle. - Ser.1-2, v.1-4, in both series, 1778-1797. Göteborg: 1778-1788. OTHER TITLE: Wetenskaps afdelningen, Göteborgska vetenskaps och vitterhets samhälle, 1778-1788; Witterhets afdelningen, Göteborgska vetenskaps och vitterhets samhälle, 1778-1797. CONTINUED BY: Nya handlingar, Göteborgska vetenskaps och vitterhets samhälle. SOURCES: G473, S604 LOCATIONS: BUC, ULS SUBJECT: Science - Societies - Göteborg.

717 Handlingar / Kungliga Svenska Vetenskapsakademien. - v.1-v.40, 1741-1779. Stockholm: Lorentz Ludvig Grefing, 1741-1779. TRANSLATED AS: Abhandlungen aus der Naturlehre, Haushaltungkunst und Mechanik. TRANSLATED AS: Analecta transalpina seu epitome commentariorum, Regia scientiarum academia suecica. TRANSLATED AS: Auserlesene Sammlung zum vortheil der Staats- wirthschaft, der Naturforschung, des Feldbaues. TRANSLATED AS: Medical, chirurgical and anatomical cases and experiences. TRANSLATED AS: Oeconomiske, physiske og mechaniske afhandlingar. TRANSLATED AS: Geneeskundige verhandelingen. TRANSLATED AS: Mémoires de l'Académie royale des sciences de Stockholm. TRANSLATED AS: Recueil des mémoires les plus interessans de chimie et d'histoire naturelle. Imprint varies: v.1-8, 1741-47, L.L. Grefing, v.9-36, 1748-1775, Lars Salvius, v.37-40, 1776-1779, J.G. Lange. French translation in Collection académique v.11. CONTINUED BY: Nya handlingar, Kungliga Svenska vetenskapsakademien. INDEXES: v.1-15 (1755), 16-30 (1770), 31-40(1789) also A.J.Stahl, Register öfer Kongl. Vetenskapsacademiens

handlinger ifrån deras borjam år 1739 til och med år 1825 Stockholm, 1825. **MICROFORMS:** RM(3). **SOURCES:** G477, OCLC, S693u **LOCATIONS:** BUC, CCF, ULS, ZDB **SUBJECT:** Science - Societies - Stockholm.

718 Handlingar / Physiographiska sällskapet, Lund. - v.1-3, 1776-1785, ns. v.1, 1786. Stockholm: 1776-1786. **SOURCES:** G486, S685 **LOCATIONS:** BUC, ULS, ZDB **SUBJECT:** Natural history - Societies - Lund. **SUBJECT:** Economics - Societies - Lund.

719 Handlingar / Svenska academie, Stockholm. - v.1-5 (1786-1795) 1801-1815, s.2, v.1-62 (1798-1885) 1801-1885, s.3, v.1- 1886- . Stockholm: C.D. Delën and J.G. Forsgren, 1801-1885. Some volumes were issued in new editions. **INDEXES:** There is a cumulative index for 1786-1895. **SOURCES:** G519, OCLC, S690 **LOCATIONS:** BUC, CCF, ULS, ZDB **SUBJECT:** General - Societies - Stockholm.

720 Handlingar / Svenska patriotiska sällskakpets. - v.1, 1770. Stockholm: Joh. Georg Lange, 1779. **LOCATIONS:** BUC **SUBJECT:** Economics - Societies - Stockholm.

721 Handlingar / Svenska vitterhetsacademien, Stockholm. - v.1-5, 1753-1788, ns. v.1-20, 1789-1852. Stockholm: Kongl. Tryckeriet, 1753-1952. **INDEXES:** Index 1753-1852 in ns. v.20. **SOURCES:** G520, OCLC **LOCATIONS:** BUC, CCF, ULS **SUBJECT:** Science - Societies - Stockholm.

722 Handlungsbibliothek. - v.1-3, 1785-1797. Hamburg: 1785-1797. **EDITORS:** Johann Georg Büsch and Christoph Daniel Eberling. **CONTINUED BY:** Neue Handlungsbibliothek. **SOURCES:** K2737 **LOCATIONS:** ZDB **SUBJECT:** Commerce - Periodicals - Germany.

723 Handlungszeitung, oder, wöchentliche Nachrichten von dem Handel, dem Manufakturwesen und der Ökonomie. - v.1-16, 1784-1799. Gotha: Ettinger, 1784-1799. Weekly. **EDITOR:** Johann Adolph Hildt. **CONTINUED BY:** Neue Handlungszeitung. **CONTINUED BY:** Neue Zeitung, oder, wöchentliche Nachrichten für Handel. **SOURCES:** K2738 **LOCATIONS:** CCF, ZDB **SUBJECT:** Commerce - Periodicals - Germany.

724 Hannoverische Beiträge zum Nutzen und Vergnügen. - v.1-4, 1759-1763. Hannover: E.C. Schlüter, 1759-1763. Weekly. **CONTINUATION OF:** Hannoverische gelehrte Anzeigen. **CONTINUATION OF:** Nützliche Sammlungen. **CONTINUED BY:** Hannoverisches Magazin. **CONTINUED BY:** Neues Hannoverisches Magazin. **SOURCES:** B4185b, G658, K219, OCLC, S2781 **LOCATIONS:** CCF, CCN, ULS, ZDB **SUBJECT:** General - Periodicals - Germany - Reviews.

725 Hannoverische gelehrte Anzeigen. - v.1-5, 1750-1754. Hannover: H.E.C. Schlüter, 1750-1754. **OTHER TITLE:** Sammlung kleiner Ausführungen aus verschiedenen Wissenschaften. **CONTINUED BY:** Nützliche Sammlungen. **CONTINUED BY:** Hannoverische Beiträge zum Nutzen und Vergnügen.

CONTINUED BY: Hannoverisches Magazin.   CONTINUED BY: Neues Hannoverisches Magazin.   SOURCES: B4185, K167   LOCATIONS: BUC, CCF, ZDB   SUBJECT: General - Periodicals - Germany - Reviews.

726   Hannoverisches Magazin, worin kleine Abhandlungen einzelne Gedanken, Nachrichten, Vorschläge und Erfahrungen, so die Verbesserung des Nahrungstandes aufbewahret sind. - v.1-28, 1764-1791.   Hannover: H.E.C. Schlüter, 1764-1791.   CONTINUATION OF: Hannoverische gelehrte Anzeigen.   CONTINUATION OF: Nützliche Sammlungen.   CONTINUATION OF: Hannoverische Beiträge zum Nutzen und Vergnügen.   EDITOR: Ludolf von Guckenberger.   CONTINUED BY: Neues Hannoverisches Magazin. INDEXES: Indexed in Allgemeines Sachregister (q.v.).   SOURCES: B4185c, G659, K243, OCLC, S2782   LOCATIONS: BUC, CCF, CCN, ULS, ZDB SUBJECT: General - Periodicals - Germany - Reviews.

727   Hanseatisches Magazin. - v.1-6, 1799-1804.   Bremen; Hamburg: Seyffert, 1799-1804. Imprint varies.   EDITOR: Johann Schmid.   SOURCES: B2037, G660, K1316, OCLC, S2430   LOCATIONS: BUC, CCF, ULS, ZDB SUBJECT: General - Periodicals - Germany.

728   Haushaltungszeitung, oder, Tagebuch vom Feldbau, von der Haushaltung und von einigen Hülfsmitteln für die Landleute in Ermangelung eines Arztes und Wundarztes. - St.1-12, 1781.   Heilbron: Eckebrecht, 1781.   SOURCES: K2932   SUBJECT: Home economics - Periodicals - Germany.   SUBJECT: Medicine - Periodicals - Germany - Popular.

729   Der Hausvater, eine ökonomische Schrift. - v.1-6, St.1, 1764-1773.   Hannover: Helwig, 1764-1773. Imprint varies. New edition in 1766.   EDITOR: Otto von Munchausen.   SOURCES: K2879   LOCATIONS: CCF, ULS, ZDB SUBJECT: Economics - Periodicals - Germany.

730   Heelkundige mengelschriften. - v.1-3, 1789. Haarlem: 1789.   Translated by Johann Daam.   TRANSLATION OF: Vermischte chirurgische Schriften. SOURCES: G434a, OCLC   LOCATIONS: CCN, DSG   SUBJECT: Surgery - Periodicals - Germany - Translations.

731   Helvetische Monatsschrift. - Heft 1-8, 1799-1802.   Winterthur: Steiner, 1799-1802.   CONTINUATION OF: Magazin für die Naturkunde Helvetiens. CONTINUATION OF: Allgemeines Helvetisches Magazin.   EDITOR: Johann Georg Albrecht Höpfner.   SOURCES: B2046, G536, K1321, S2187 LOCATIONS: CCF, ULS, ZDB   SUBJECT: Natural history - Periodicals - Switzerland.

732   Hermetischer Briefe wider Vorurtheile und Betrüggereyen. - v.1, 1788. Leipzig: Beer, 1788.   Johann Salomo Semler.   SOURCES: K3324   SUBJECT: Alchemy - Periodicals - Germany.

733 Hermetisches Museum, allen Liebhabern der wahren Weisheit gewidmet. - v.1-3, 1782-1785. Reval; Leipzig: Albrecht, 1782-1785. **SOURCES:** K3289, OCLC **LOCATIONS:** ZDB **SUBJECT:** Alchemy - Periodicals - Germany.

734 Hertha, et ugeblad af blandet inhold. - v.1, 1789. Copenhagen: 1789. **CONTINUATION OF:** Sundstidende. **CONTINUATION OF:** Nye sundstidende. **CONTINUATION OF:** Sundhed og underholdning. **CONTINUATION OF:** Hygaea og muserne. **CONTINUATION OF:** Museum for sunheds og kundskabs elskere. **EDITOR:** Johann Clemens Tode. **CONTINUED BY:** Sundhedsbog. **CONTINUED BY:** Medicinalblade. **CONTINUED BY:** Sundhedsjournal. **CONTINUED BY:** Sundhedsraad. **CONTINUED BY:** Sundhedsjournal et maanedskrift. **CONTINUED BY:** Nyeste sundhedsblade. **SOURCES:** G187 **SUBJECT:** Medicine - Periodicals - Denmark.

735 Hessische Beiträge zur Gelehrsamkeit und Kunst. - St.1-8, 1784-1788. Frankfurt: Varrentrapp and Wenner, 1784-1788. **EDITORS:** Georg Forster, Samuel Thom, S.T.Sommering, Dietrich Tiedemann and others. **SOURCES:** B2057, K374, OCLC, S2594 **LOCATIONS:** BUC, CCF, ULS, ZDB **SUBJECT:** General - Periodicals - Germany - Reviews.

736 Der Hessischer Arzt. - St.1-6, 1776. Leipzig: Kummer, 1776. **SOURCES:** K3575 **SUBJECT:** Medicine - Periodicals - Germany.

737 Hippocrates ridens, or, joco-serious reflections on the impudence and mischiefs of quacks and illiterate pretenders to physick. - No.1-4, 1686. London: Walter Davis, 1686. **SOURCES:** GI:8 **LOCATIONS:** BUC, ULS **SUBJECT:** Medicine - Periodicals - England - Popular.

738 Histoire de la Société royale de médecine avec les mémoires et de physique médicale. - T.1(1776) 1779, T.2 (1777-78) 1780, T.3 (1779) 1782, T.4 (1780-81) 1785, T.5 (1782-83) 1787, T.6 (1783) 1784, T.7(1784-85) 1788, T.8 (1786) 1790, T.9 (1787-88) 1790, T.10 (1789) 1798. Paris: Didot, 1779-1798. **TRANSLATED AS:** Sammlung der gemeinnützigsten practischen Aufsätze und Beobachtungen. **TRANSLATED AS:** Practische Abhandlungen aus der Schriften der Kóniglichen medicinischen Societát zu Paris. Publisher varies. **OTHER TITLE:** Histoire et mémoires de la Société royale de médecine. **SOURCES:** G41a, OCLC **LOCATIONS:** BUC, CCF, CCN, ULS, ZDB **SUBJECT:** Medicine - Societies - Paris.

739 Histoire de la Société royale des sciences établie à Montpellier, avec les mémoires de mathematiques et de physique tirées des registres de cette société. v.1(1706-1730)-v.2(1731-1745), 1764-1778. Lyons and Montpellier: Benoit Duplain, 1764-1778. **SOURCES:** G414, S1241b **LOCATIONS:** BUC, CCF, CCN, ULS **SUBJECT:** Science - Societies - Montpellier.

740 Histoire de l'Académie royale des sciences, avec les mémoires de mathematiques et de physiques pour le mêmes années, tirez des registres de cette Académie. - v.1-94, covering 92 years, 1699 to 1790, with two vols. each in 1718 and

1772. Paris: Imprimerie royale, 1702-1790. Annual, usually two to four years in arrears of proceedings for that year. CONTINUATION OF: Histoire et mémoires, Académie royale des science. CONTINUATION OF: Historia in quae praeter ipsius academiae originem. CONTINUATION OF: Mémoires de mathematiques et de physique tirez des registres de l'Académie royales des science. There were numerous editions both authorized and unauthorized as well as translations into English, German and other languages for particular years, or selections taken from the series. See title and name index. A 2nd ed. for the years 1699-1708, 1710 and 1718 are recorded, as well as a 3rd ed. for 1699 and 1732, including an edition in Amsterdam in 48 v.in 1706-1755 covering the years 1699-1746. An Italian translationin 1739 under the title: Istoria dell' Accademia realedelle scienze has been cited (Ref.60,p,548). Abstracts are included in the Recueil des mémoires, ou collection des pièces académique. INDEXES: Table alphabétique générale des matières contenues dans l'histoire et les mémoires, par Godin, Demours et Cotte. Paris, Imprimerie royale, 1719- 1809 in 10 v. (several editions of at least the first six volumes were published, the different editions of v.1 bearing the dates 1729, 1734 and 1778). IDC(8),RM(3). SOURCES: G17c, OCLC, S1276d LOCATIONS: BUC, CCF, CCN, ULS, ZDB SUBJECT: Science - Societies - Paris.

741   Histoire de l'Académie royale des sciences et des belles-lettres de Berlin, avec les mémoires pour le même année. - v.1-25, 1746 (1745)-1771 (1769). Berlin: Haude et Spener, 1745-1771. TRANSLATED AS: Anatomische, chymische und botanische Abhandlungen. CONTINUATION OF: Miscellanea Berolinensis. TRANSLATED AS: Physikalische und medicinische Abhandlungen. TRANSLATED AS: Physikalische, chemische, anatomische und medizinische Abhandlungen. Abstracted in Collection Académique, v.8-9,12. Publisher varies. CONTINUED BY: Nouveaux mémoires de l Académie royale des sciences et belles-lettres, Berlin. CONTINUED BY: Sammlung der deutschen Abhandlungen. CONTINUED BY: Abhandlungen der königlichen Preussischen Academie der Wissenschaften in Berlin. INDEXES: Kohnke, Otto, ed. Gesamtregister über die in den Schriften der Akademie von 1700-1899 erscheinen wissenschaftlichen Abhandlungen und Festreden. Berlin, 1900 (Ref.126). MICROFORMS: IDC(8), SMS(1), RM(3). SOURCES: G487b, K148, OCLC, S2262a LOCATIONS: BUC, CCF, CCN, ULS, ZDB SUBJECT: Science - Societies - Berlin.

742   Histoire de l'Académie royale des sciences, Paris. - v.1-11, (1666-1699) 1729-1734. Paris: Gabriel Martin; Jean Baptiste Coginar; Hippolyte Louis Guerin, 1729-1734. CONTINUATION OF: Historia in quae praeter ipsius academiae originem. A retrospective publication of the society's activities covering 1666 to 1699. v.1-2 as Histoire and v.3-10 as Mémoires, v.11 is an index. There was a reduced edition published in 6 vols. under the title: Mémoires, contenant les ouvrages adoptés par cette académie avant son renouvellement en 1699, at La Haye in 1731, and Amsterdam 1735-1736. OTHER TITLE: Mémoires de l'Académie royale des sciences, Paris. CONTINUED BY: Histoire de Í'Académie royale des sciences avec les mémoires. MICROFORMS: IDC(8), RM(3). SOURCES: G17b, OCLC,

S127f **LOCATIONS:** BUC, CCF, CCN, ULS, ZDB **SUBJECT:** Science - Societies - Paris.

743 Histoire des ouvrages des sçavans. - v.1-24, Sept.1687-June, 1709. Rotterdam: Reiner Leers, 1687-1709. Sept.1690-Aug.1694 was published at Amsterdam. There were at least 4 editions of some of the volumes (Ref.13). **OTHER TITLE:** Histoire des ouvrages des savans. **EDITOR:** Henri Basnage de Beauval. **MICROFORMS:** RP(1). **SOURCES:** B2060, OCLC, S900 **LOCATIONS:** BUC, CCF, CCN, ULS, ZDB **SUBJECT:** General - Periodicals - Netherlands - Reviews.

744 Histoire et mémoires de l'Académie royale des sciences, inscriptions et belles lettres de Toulouse. - Ser. 1, v.1-4, 1782-1790, ser. 2, v.1-6, 1807-1841. Toulouse: Desclassan, 1782-1841. Publisher varies. Ser. 2, v.1-6 also as v.5-10. **OTHER TITLE:** Mémoires de l'Académie des sciences, inscriptions et belles lettres de Toulouse. **INDEXES:** Table alphabétique des matières contenus dans les seize premieres tomes des mémoires (ser. 1-3, 1782-1850) des Mémoires. Toulouse, 1854. **MICROFORMS:** RM(3). **SOURCES:** G453, OCLC, S1679c **LOCATIONS:** BUC, CCF, CCN, ULS, ZDB **SUBJECT:** Science - Societies - Toulouse.

745 Histoire et mémoires / Société des sciences physiques de Lausanne. - v.1 (1782) 1784, 2 (1784-1786) 1789, 3 (1787-1788) 1790. Lausanne: 1784-1790. v.1 as Mémoires. **OTHER TITLE:** Mémoires, Société des sciences physiques de Lausanne. **SOURCES:** G515, S2171 **LOCATIONS:** BUC, CCF, CCN, ULS, ZDB **SUBJECT:** Science - Societies - Lausanne.

746 Histoire et mémoires / Société formée à Amsterdam en faveur des noyés. - v.1-3, 1768-1784. Amsterdam: 1768-1784. **TRANSLATED AS:** Kurzer Unterricht, ertunkene Menschen wieder lebendig zu machen. **CONTINUED BY:** Historie en gedanschriften van de Maatschappij tot redding van drenkelingen. **SOURCES:** G510 **LOCATIONS:** BUC, CCF **SUBJECT:** Resuscitation - Societies - Amsterdam.

747 Historia / Academia electoralis moguntinae quae Erfurti est. - v.1 (1778-1779), 1780. Erfurt: 1780. **SOURCES:** G439a **SUBJECT:** Science - Societies - Erfurt.

748 Historia / Accademia Pisana. - v.1-3, 1791-1795. Pisa: 1791-1795. **SOURCES:** S1999 **SUBJECT:** General - Societies - Pisa.

749 Historia et commentationes academiae electoralis scientiarum et elegantiarorum literarum Theodoro-Palatinae. - v.1-7, 1766-1794. Mannheim: Typis Academicus, 1766-1794. v.1-2, pars historica and pars physica in same volume, v.3ff in separate volumes, v.7 pars historica only (Ref.79,p.49-50). **OTHER TITLE:** Acta Academia Theodoro-Palatinae. **INDEXES:** Indexes to v.1-3 in 3 pt.1; 4-6 in 6 pt.1 and 2. **SOURCES:** K256, G440a, OCLC, S3067 **LOCATIONS:** BUC, CCF, CCN, ULS, ZDB **SUBJECT:** Science - Societies - Mannheim.

750   Historia in quae praeter ipsius Academiae originem et progressus, Académie royale des sciences, Paris. - 1 v.(1667-1696)1698. Paris: S. Michallet, 1698. CONTINUATION OF: Mémoires de mathématiques et de physique. A retropsective history written by Jean Baptiste Du Hamel. A 2nd ed. covering the academy's history to 1700 was issued in Paris: J.B. Delespine, 1701. CONTINUED BY: Histoire de l'Académie royale des sciences. CONTINUED BY: Histoire et mémoires de l'Academie royale des sciences. CONTINUED BY: Mémoires de mathématiques et de physique présentés à l'Académie royale des sciences. SOURCES: GII:17a   SUBJECT: Science - Societies - Paris.

751   Historia morborum / Academia Caesararea-Leopoldina Carolina naturae curiosorum. - Pt.1-3 (1699-1702) 1706-1710. Breslau: Christian Bauch, 1706-1710. MICROFORMS: RP(1).   SOURCES: GII:7   LOCATIONS: BL, DSG   SUBJECT: Medicine - Societies - Breslau.

752   Historie en gedanschriften van de Maatschappij tot redding van drenkelingen. - v.1-6, 1768-1822. Amsterdam: Pieter Meyer, 1774-1822. TRANSLATED AS: Kurzer Unterricht, ertunkene Menschen wieder lebendig zu machen. CONTINUATION OF: Histoire et mémoires, Société formée à Amsterdam en faveur des noyés. Suspended 1811-1821. SOURCES: G510a   LOCATIONS: CCF, DSG   SUBJECT: Resuscitation - Societies - Amsterdam.

753   Historisch-geographisches Journal. - St.1-6, 1789-1790. Halle; Leipizig; Jena: 1789-1790.   CONTINUATION   OF: Geographisches   Magazin. CONTINUATION OF: Neues geographisches Magazin.   EDITORS: Johann Ernst Fabri and Carl Hammerdörfer.   CONTINUED BY: Beiträge zur Geographie, Geschichte und Staatenkunde. SOURCES: B2061, K214, S2715 SUBJECT: Geography - Periodicals - Germany.

754   Historisch-ökonomische Abhandlungen von einer Gesellschaft Gelehrten. - St.1-15, 1790. Giessen: Krieger, 1790. SOURCES: K1221   LOCATIONS: BUC   SUBJECT: Economics - Periodicals - Germany.

755   Historische und geographische Monatsschrift. - Bd.1(St.1-12), 1788. Halle: 1788. Monthly.   CONTINUATION   OF: Geographisches   Magazin. CONTINUATION OF: Neues geographisches Magazin.   EDITOR: Johann Ernst Fabri and Carl Hammerdörfer.   CONTINUED BY: Beiträge zur Geographie, Geschichte und Staatenkunde. CONTINUED BY: Magazin für die Geographie, Staatenkunde und Geschichte.   SOURCES: K1205 LOCATIONS: CCF, ZDB   SUBJECT: Geography - Periodicals - Germany.

756   History of cradle-convulsions, vulgarly called black and white fits, or, monthly observations on the weekly bills of mortality. - No.1, 1701. London: 1701. SOURCES: (Ref.80,p.44).   LOCATIONS: BUC   SUBJECT: Medicine - Periodicals - England.

757   History of inventions and discoveries. - v.1-v.4, 1797- . London: J. Bell, 1797. There have been at least four numbered editions, the 4th of which appeared in 5 revisions by the same publisher, 1797-1892. EDITOR: Johann Beckmann.

TRANSLATION OF: Beiträge zur Geschichte der Erfindungen. **LOCATIONS:** BUC **SUBJECT:** Economics - Periodicals - Germany - Translations. **SUBJECT:** Manufactures - Periodicals - Germany - Translations.

758 History of the works of the learned, containing impartial accounts and accurate abstracts of the most valuable books published in Great Britain and foreign parts. - v.1-14, 1737-1743. London: Jacob Robinson, 1737-1743. Monthly. **CONTINUATION OF:** Present state of the republic of letters. **EDITOR:** Ephraim Chambers. **MICROFORMS:** UM(1). **LOCATIONS:** BUC, ULS **SUBJECT:** General - Periodicals - England - Reviews.

759 History of the works of the learned, or, an impartial account of books lately printed in all parts of Europe. - v.1-14, 1699-1712. London: H. Rhodes, 1699-1712. Monthly. Suspended Jan.-June, 1711. **EDITOR:** George Redpath. **MICROFORMS:** RP(1), UM(1). **SOURCES:** OCLC **LOCATIONS:** BUC, ULS, ZDB **SUBJECT:** General - Periodicals - England - Reviews.

760 Holländisches Museum, für Deutschlands Ärzte und Wundärzte. - v.1, 1794. Breslau: Korn, 1794. **EDITOR:** Christian Friedrich Niceus. **SOURCES:** G289, K3691 **LOCATIONS:** BUC, ULS **SUBJECT:** Medicine - Periodicals - Germany. **SUBJECT:** Surgery - Periodicals - Germany.

761 Hollands en stigt Utrechts collegie van de heeren dijkgraafen hooge heemraden zeeburg en diemer mits gaders muyder zeedijk. - Amsterdam: 1733-1738. **CONTINUED BY:** Hollands en Utrechts hoogheemraadschap van den zeeburg en diemerdijk. **LOCATIONS:** CCN **SUBJECT:** Engineering - Periodicals - Netherlands.

762 Hollands en Utrechts hoogheemraadschap van den zeeburg en diemerdijk. - Amsterdam: 1740-1809. **CONTINUATION OF:** Hollands en stigt Utrechts collegie van de heeren dijkgraafen hooge heemraden zeeburg en diemer mits gaders muyder zeedijk. **LOCATIONS:** CCN **SUBJECT:** Engineering - Periodicals - Netherlands.

763 Hollands magazijn voorzien van aardrijkskundige historische philosophische aanmerkcungen. - v.1-2, 1750-1761. Haarlem: 1750-1761. **LOCATIONS:** CCN,ULS **SUBJECT:** Geography - Periodicals - Netherlands.

764 Hushallnings-journal, utgifen genom Kongl, Patriotiska sälskaptes föranstaltande. - Stockholm: J.G. Lange, 1776-1789. **CONTINUED BY:** Ny journal uti hushallningen. **SOURCES:** (Ref 54:23). **SUBJECT:** Economics - Periodicals - Sweden.

765 Hygaea og muserne og kundskabs elskere. - Nos.1-24, 1788. Copenhagen: 1788. **CONTINUATION OF:** Sundstidende. **CONTINUATION OF:** Nye sundstidende. **CONTINUATION OF:** Sundhed og underholdning. **EDITOR:** Johann Clemens Tode. **CONTINUED BY:** Museum for sunheds og kundskabs elskere. **CONTINUED BY:** Hertha. **CONTINUED BY:** Sundhedsbog.

CONTINUED BY: Medicinalblade. **CONTINUED BY:** Sundhedsjournal. **CONTINUED BY:** Sundhedsraad. **CONTINUED BY:** Sundhedsjournal et maanedskrift. **CONTINUED BY:** Nyeste sundhedsblade. **SOURCES:** G185 **LOCATIONS:** ULS **SUBJECT:** Medicine - Periodicals - Denmark.

766   Hygea, eine heilkundige Zeitschrift, dem weiblichen Geschlecht vom Stande vorzüglich gewidmet. - Heft 1-4, 1793-1794. Eisenach: Wittekindt, 1793-1794. **EDITOR:** Wilhelm Julius Augustin Vogel. **SOURCES:** G192, K3683 **SUBJECT:** Medicine - Periodicals - Germany - Popular. **SUBJECT:** Women - Periodicals - Germany.

767   Ideenmagazin für Liebhaber von Gärten, Englischen Anlagen und für Besitzer von Landgärtern. - v.1-5 (no.1-60) 1796-1806. Leipzig: F.G. Baumgärtner, 1796-1806. There were three editions. **OTHER TITLE:** Recueil d'idées nouvelles pour la decoration des jardins. **EDITOR:** Johann Gottfried Grohmann. **CONTINUED BY:** Neues Ideen-Magazin für Liebhaber von Gärten. Supplemented by: Literarischer Anzeiger für das Ideenmagazin, 1800-1805. **SOURCES:** K4099, OCLC **LOCATIONS:** BUC, CCF, ULS **SUBJECT:** Gardening - Periodicals - Germany.

768   Die in der Medicin siegende Chymie. - St.1-8, 1743-1750. Erfurt: Jungnicol, 1743-1749. St.8 has title: Zugabe zu der in der Medicin noch immer und immer siegenden Chymie. This may be considered a book issued in parts. **EDITOR:** Hieronymus Ludolf. **SOURCES:** K3193, OCLC **LOCATIONS:** BN **SUBJECT:** Medicine - Periodicals - Germany. **SUBJECT:** Chemistry - Periodicals - Germany.

769   Institution of the Humane Society of the Commonwealth of Massachusetts. - Boston: 1786-1788. Title varies (Ref.81, p.10). **SOURCES:** OCLC **LOCATIONS:** DSG **SUBJECT:** Resuscitation - Societies - Boston.

770   Institution and proceedings of the Society for the encouragement of arts, manufacture and commerce of Barbados. - Barbados: John Anderson, 1781-1784. **LOCATIONS:** BUC **SUBJECT:** Economics - Societies - Barbados.

771   Instructions et observations sur les maladies des animaux domestique, avec les moyens de les guerir, de les preserver, de les conserver en santé, de les multiplier, de les élever avec avantage, et de n'etre point trompe dans leurs achat. - T.1-5, 1791-1795. Paris: 1791-1795. **CONTINUATION OF:** Almanach vétérinaire. **SOURCES:** G229 **LOCATIONS:** ZDB **SUBJECT:** Veterinary medicine - Periodicals - France.

772   Introduction aux observations sur la physique, sur l'histoire naturelle et sur les arts. - v.1-18, 1771-1772. Paris: Le Jay, 1771-1772. Monthly. **TRANSLATED AS:** Sammlung brauchbarer Abhandlungen aus Rozier's Beobachtungen über die Natur und Kunst. **CONTINUATION OF:** Observations périodique sur la physique, l'histoire naturelle. **TRANSLATED AS:** Osservazioni spettanti alla fisica, alla storia naturali ed alla arti. A 2nd ed. in

2 v. was published in 1777. Volume 2 also has the title Tableau du travail annuel de toutes les académies de l'Europe. **OTHER TITLE:** Observations sur la physique, sur l'histoire naturelle et les arts. **EDITOR:** Jean Baptiste François Rozier. **CONTINUED BY:** Observations et mémoires sur la physique. **CONTINUED BY:** Journal de physique, de chimie, d'histoire naturelle et des arts. **INDEXES:** There is an index in v.10 of Journal de physique, de chimie. **MICROFORMS:** RM(3). **SOURCES:** B2201, G661, OCLC, S1398 **LOCATIONS:** BUC, CCF, ULS, ZDB   **SUBJECT:** Science - Periodicals - France.

773   Ionian antiquities / Society of dilletanti, London. - v.1 (1769), 2(1797), 3(1840, 4(1881), 5(1915) (Ref.82). London: 1915. **OTHER TITLE:** Antiquities of Ionia. **SOURCES:** G516, OCLC   **LOCATIONS:** BUC, CCF, ULS **SUBJECT:** Archeology - Societies - London.

774   Italienische bibliothek, oder, Sammlung der merkwürdigsten kleinen Abhandlungen zur Naturgeschichte, Ökonomie und zum Fabrikwesen aus den neuesten italienischen Monatsschriften. - v.1-2, 1778-1779. Leipzig: Caspar Fritsch, 1778-1779. J.J. Volkmann.   **SOURCES:** B2234, K3264, S2938 **LOCATIONS:** BL, ZDB **SUBJECT:** Natural history - Periodicals - Germany - Reviews.   **SUBJECT:** Economics - Periodicals - Germany - Reviews.

775   Italienische medicinisch-chirurgische Bibliothek. - v.1-5, 1793-1799. Leipzig: J.G. Müller, 1783-1799. **EDITORS:** Carl Gottlob Kühn and Carl Weigel. **SOURCES:** G290 **LOCATIONS:** CCN, ULS, ZDB **SUBJECT:** Medicine - Periodicals - Germany - Reviews.   **SUBJECT:** Surgery - Periodicals - Germany - Reviews.

776   Jahrbücher der Berg-und Hüttenkunde. - v.1-5, 1797-1801. Salzburg: Mayer, 1797-1801. **EDITOR:** Carl Ehrenbert von Moll. **CONTINUED BY:** Annalen der Berg-und Hüttenkunde. **CONTINUED BY:** Neue Jahrbücher der Berg-und Hüttenkunde.   **SOURCES:** B2271, K3368, S3529   **LOCATIONS:** BUC, CCF, CCN, ULS, ZDB   **SUBJECT:** Mineral industries - Periodicals - Austria.

777   Le Jardinier portatif. - v.1-14, 1799-1812. Paris: 1799-1812. **CONTINUATION OF:** Bon Jardinier portatif. **CONTINUED BY:** Almanach du bon jardinier. **CONTINUED BY:** Bon jardinier. **SOURCES:** B1011a **LOCATIONS:** BUC **SUBJECT:** Gardening - Periodicals - France.

778   Jenaische gelehrte Zeitungen. - v.1-9. 1749-1757. Jena: Cuno: (Cröcker and Melchoir 1751ff), 1749-1757. **EDITORS:** Georg Erhard Hamburger and others. **CONTINUED BY:** Jenaische Zeitungen von gelehrten Sachen. **CONTINUED BY:** Jenaische gelehrte Anzeigen. **SOURCES:** B2312, K166, S2822 **LOCATIONS:** ULS **SUBJECT:** General - Periodicals - Germany - Reviews.

779   Jenaische Zeitungen von gelehrten Sachen. - v.1-20, 1765-1786. Jena: Cröcker, 1765-1786. **CONTINUED BY:** Jenaische gelehrte Zeitungen. **CONTINUED**

BY: Jenaische gelehrte Anzeigen.   SOURCES: B2314, K253, S2823
LOCATIONS: CCF, CCN, ULS, ZDB   SUBJECT: General - Periodicals -
Germany - Reviews.

780   Jenaische gelehrte Anzeigen. - St.1-102, 1787.   Jena: 1787.   Biweekly.
CONTINUATION OF: Jenaische Zeitungen von gelehrten Sachen.
CONTINUED BY: Jenaische gelehrte Zeitungen.   SOURCES: K407
SUBJECT: General - Periodicals - Germany - Reviews.

781   Journal aller Journale, oder Geist der vaterländischen Zeitschriften, nebst
Anzeigen aus den periodischen Schriften und besten Werken der Ausländer. -
Heft. 1-12, 1786-87, Heft. 13-18, 1788.   Hamburg: Chaidron, 1786-1788.
Imprint and subtitle vary.   EDITOR: Jonas Ludwig von Hess.   CONTINUED
BY: Neues Journal aller Journale.   SOURCES: G664, K4558   LOCATIONS:
ULS, ZDB   SUBJECT: General - Periodicals - Germany - Reviews.

782   Journal anglais contenant les déscouvertes faites dans les sciences, les arts libéraux
et mecanique, dans les trois royaumes. - v.1-7, 1775-1778.   Paris: Lacombe,
1775-1778.   Imprint and subtitle vary.   LOCATIONS: CCF, ULS, ZDB
SUBJECT: General - Periodicals - France - Reviews.

783   Journal Britannique. - T.1-24, 1750-1757.   La Haye: H. Scheurleer, 1750-1757.
CONTINUATION OF: Bibliothèque Britannique, ou Histoire des ouvrages des
savans de la Grand Bretagne.   EDITOR: Matthew Maty.   MICROFORMS:
IDC(8).   SOURCES: B93a, S828   LOCATIONS: BUC, CCF, CCN, ULS,
ZDB   SUBJECT: General - Periodicals - Netherlands.

784   Journal d'agriculture à l'usage des campagnes. - Paris: 1790.   OTHER TITLE:
Journal d'agriculture à l'usage des cultivateurs.   EDITOR: Louis Reynier.
SOURCES: B2342   LOCATIONS: CCF   SUBJECT: Agriculture -
Periodicals - France.

785   Journal d'agriculture à l'usage des habitants des campagnes. - No.1-24, Apr.1,
1791-Mar.15, 1792.   Paris: 1791-1792.   EDITOR: Henri Alexander Tessier.
SOURCES: B2343   LOCATIONS: CCF   SUBJECT: Agriculture -
Periodicals - France.

786   Journal d'agriculture, du commerce, des arts et des finances. - v.1-24 (tomes
1-72) 1778-1783.   Paris: 1778-1783.   CONTINUATION OF: Journal de
commerce et d'agriculture.   CONTINUATION OF: Journal d'agriculture, du
commerce et des finances. Reprinted in 1970 (Milan, Feltenelli).   EDITOR:
Hubert Pascal Ameilhon.   CONTINUED BY: Feuille du cultivateur.
SOURCES: B2463b, OCLC   LOCATIONS: CCF, ULS   SUBJECT:
Agriculture - Periodicals - France.   SUBJECT: Economics - Periodicals -
France.

787   Journal d'agriculture, du commerce et des finances. - v.1-48 (Tome 1-114)
1765-1774.   Paris: 1765-1774.   CONTINUATION OF: Journal de commerce
et d'agriculture. Reprinted: Milan, Feltrinelli. 1970.   EDITORS: Dupont de

Nemours, Nicolas Baudeau, P.J.A. Roubaud and others. **CONTINUED BY:** Journal d'agriculture, du commerce des arts et des finances. **CONTINUED BY:** Feuille du cultivateur.   Microfilm: Clearwater Publishing Co. **SOURCES:** B2463a, OCLC   **LOCATIONS:** CCF, ULS   **SUBJECT:** Agriculture - Periodicals - France.   **SUBJECT:** Economics - Periodicals - France.

788   Journal d'agriculture et de prosperité publique. - No.1-3, Apr.1, 1793-Mar.21, 1794. Paris: 1793-1794. **LOCATIONS:** CCF, ULS **SUBJECT:** Agriculture - Periodicals - France.

789   Journal d'agriculture et d'économie rurale, contenant les observations sur toutes les parties de l'agriculture. - No.1-7, 1794-1795. Paris: 1794-1795. **EDITOR:** Borelly. **SOURCES:** B2347 **LOCATIONS:** CCF **SUBJECT:** Agriculture - Periodicals - France.

790   Journal de chirurgie. - T.1-t.4, 1791-1792.   Paris: Desault, 1791-1792. **TRANSLATED AS:** Parisian chirurgical journal.   **TRANSLATED AS:** Auserlesene chirurgische Wahrnemungen. **EDITOR:** Pierre Joseph Desault. **SOURCES:** G97, OCLC   **LOCATIONS:** BUC, CCF, CCN, ULS, ZDB **SUBJECT:** Surgery - Periodicals - France.

791   Journal de commerce et d'agriculture. - v.1-24, 1759-1762. Brussels: J. van den Berhen, 1759-1762.   1759-1761 as Journal de commerce or Journal du commerce. **OTHER TITLE:** Journal de commerce; Journal du commerce. **EDITORS:** Louis Florent Le Camus and P.J.A. Roubard. **CONTINUED BY:** Journal d'agriculture, du commerce des arts et des finances. **CONTINUED BY:** Journal d'agriculture, du commerce et des finances. **CONTINUED BY:** Feuille d'agriculture et d'économie rurale.   **MICROFORMS:** RP(1) (Goldsmith-Kress Library). **SOURCES:** B2463 **LOCATIONS:** CCF, ULS, ZDB   **SUBJECT:** Economics - Periodicals - Belgium.   **SUBJECT:** Agriculture - Periodicals - Belgium.

792   Journal de la Société des pharmaciens de Paris, ou, recueil d'observations de chimie et de pharmacie. - v.1-3, 1797-1800. Paris: Bernard, 1799-1800. Monthly.   **OTHER TITLE:** Journal de pharmacie.   **EDITORS:** Antoine François de Fourcroy, Baptiste Nicolas Louis Vaquelin, Antoine Augustin Parmentier, Nicolas Déyeux and Edme Jean Baptiste Bouillon-Lagrange (Ref.83). **CONTINUED BY:** Annales de chimie. **SOURCES:** B6745, G220, OCLC **LOCATIONS:** BUC, CCF, ULS **SUBJECT:** Pharmacy - Societies - Paris.   **SUBJECT:** Chemistry - Societies - Paris.

793   Journal de l'art de conserver la santé et de prolonger la vie. - v.1, 1798. Angers; Paris: 1798. **EDITOR:** Joseph Charles Gilles de la Tourette.   **SOURCES:** G204 **SUBJECT:** Medicine - Periodicals - France.

794   Journal de l'école polytechnique, ou, bulletin du travail fait à cette école. - v.1-31 (No.1-64), 1795-1894.   Paris: École polytechnique, 1795-1815. **CONTINUATION OF:** Journal polytechnique. **EDITORS:** Joseph Louis de

Lagrange, Pierre Simon Laplace and others. **INDEXES:** Index to v.1-37 (1795-1858) in v.37. **MICROFORMS:** BHP(9). **SOURCES:** B2334a, G663, OCLC, S1379b **LOCATIONS:** BUC, CCF, CCN, ULS, ZDB **SUBJECT:** Science - Societies - Paris.

795   Journal de littérature, des sciences et des arts. - v.1-4, 1779-1783. Paris: 1779-1783. **CONTINUATION OF:** Mémoires pour servir à l'histoire des sciences et des beaux arts. **CONTINUATION OF:** Journal des sciences et des beaux-arts. Reprints issued in the 18th century. **EDITOR:** Jean Baptiste Grosier. **SOURCES:** B2924b, S1412 **LOCATIONS:** CCF, ULS **SUBJECT:** General - Periodicals - France - Reviews.

796   Journal de Lycée de Londres, ou, tableau de l'état present des sciences et des arts en Angleterre. - v.1-2, 1784. Paris: Peresse, 1784. **EDITOR:** J.B. Brisset de Warville. **LOCATIONS:** ULS **SUBJECT:** General - Periodicals - France.

797   Journal de marine, ou, bibliothèque raisonnée de la science du navigateur. - No.1-7, 1778-1780. Brest: 1778-1780. **EDITOR:** E. Blondeau (Ref.85). **SOURCES:** B2371, S1090 **LOCATIONS:** CCF **SUBJECT:** Naval art and science - Periodicals - France.

798   Journal de médecine. - Dijon: Edme Bidault, 1781-1787. Translated by G. Masuyer. **TRANSLATION OF:** London medical journal. **SOURCES:** G46a **LOCATIONS:** ULS **SUBJECT:** Medicine - Periodicals - England - Translations.

799   Journal de médecine, chirurgie, et de pharmacie. - v.1-30, 1754-1794. Paris: Vincent, 1754-1794. Monthly. **TRANSLATED AS:** Sammlung auserlesenen Wahrnehmungen aus der Arzneiwissenschaft, der Wundarznei-und der Apothekerkunst. **TRANSLATED AS:** Neue Sammlung auserlesenen aus allen Theilen der Arzneiwissenschaft. Published in both octavo and quarto editions. A 2nd ed. of v.1-58 was published in Paris, 1783-1786. v.1-7, 1754- 1757 under the title: Recueil périodique d'observations de médecine, de chirurgie et de pharmacie. **OTHER TITLE:** Recueil périodique d'observations de médecine, de chirurgie, et de pharmacie. **EDITORS:** Charles August Vandermonde (1754-June 1762), Augustin Roux (June, 1762-Sept.1776), Nicolas Louis de LaCaille (June 1776-Sept.1776), Demagen (Sept.1776-Dec. 1790), Bernard Bacher (Sept.1776-1794), Laurent Charles Pierre Leroux (Dec. 1790-1794) (Ref.14, p.66). **INDEXES:** Two index volumes were issued. **SOURCES:** G12, OCLC **LOCATIONS:** BUC, CCF, CCN, ULS, ZDB **SUBJECT:** Medicine - Periodicals - France. **SUBJECT:** Surgery - Periodicals - France. **SUBJECT:** Pharmacy - Periodicals - France.

800   Journal de médecine, chirurgie, pharmacie (Corvisart). - v.1-40, 1800-1817. Paris: 1800-1817. Sometimes identified with title of the same name edited by Vandermonde, Roux and others. Appeared in both duodecimo and octavo formats. **EDITORS:** Jean Nicolas Corvisart, Laurent Charles Pierre Leroux and Alexis Boyer. **CONTINUED BY:** Nouveau journal de médecine,

chirurgie, pharmacie. **SOURCES:** G174 **LOCATIONS:** BUC, CCF, CCN, ULS **SUBJECT:** Medicine - Periodicals - France.

801   Journal de médecine, d'education et de'économie. - v.1, 1799. Paris: 1799. **EDITOR:** J. Duverdier. **SOURCES:** G205 **SUBJECT:** Medicine - Periodicals - France - Popular. **SUBJECT:** Economics - Periodicals - France.

802   Journal de médecine, ou, observations des plus fameux médicins, chirurgiens et anatomistes de l'Europe tirées des journaux des pays étrangers et autres mémoires particuliers. - No.1 (Mar. 1681), 2(Jan-June 1683), 3(May 1685), 4(Apr. 1686). Paris: F. Lambert and J. Cusson, 1681-1686. Publisher varies. **EDITORS:** Jean Paul de la Roque and Claude Brunet. **SOURCES:** GI:6 **LOCATIONS:** ULS **SUBJECT:** Medicine - Periodicals - France.

803   Journal de médecine militaire. - Paris: Imprimerie Royale, 1782-1789. Vol. 1-7, 1782-1789. **TRANSLATED AS:** Journal of the practice of medicine, surgery and pharmacy.   Cover title: Journal de médicine militaire, chirurgie et pharmacie militaire.   **EDITOR:** Jacques de Horne. **SOURCES:** G50 **LOCATIONS:** BUC, CCF, ULC **SUBJECT:** Medicine, Military - Periodicals - France.

803a   Journal de médecine preservative.   No.1-2,Jan.-Feb.1794.   Paris: 1794. **EDITOR:** F. Pinglin (Ref.152,p.425). **LOCATIONS:** CCF **SUBJECT:** Medicine - Periodicals - France - Popular.

804   Journal de physique, de chimie, d'histoire naturelle et des arts. - v.1-53, 1794-1823. Paris: 1794-1823. **CONTINUATION OF:** Observations périodique sur la physique, l'histoire naturelle. **CONTINUATION OF:** Introduction aux observations sur la physique, sur l'histoire naturelle et sur les arts. **CONTINUATION OF:** Observations et mémoires sur la physique, sur l'histoire naturelle et sur les arts et métiers.   There is a gap of 4 years between v.2 and 3.   **EDITORS:** Henri Marie Ducrotay de Blainville and Louis Cotte. **INDEXES:** Table générale des articles contenus dans les vingt-six derniers volumes du Journal de Physique depuis 1787 jusqu'en 1802. Paris, n.d. **MICROFORMS:** RM(3).   **SOURCES:** B2201c, G670, OCLC, S1421 **LOCATIONS:** BUC, CCF, CCN, ULS, ZDB **SUBJECT:** Science - Periodicals - France.

805   Journal de santé et d'histoire naturelle / Société de santé et d'histoire naturelle. - v.1-3, 1796-1798. Bordeaux: 1796-1798. **EDITOR:** Jean Felix La Capelle. **SOURCES:** B2389, G201, S1067, 1073 **LOCATIONS:** CCF **SUBJECT:** Medicine - Societies - Bordeaux. **SUBJECT:** Natural history - Societies - Bordeaux.

806   Journal der Erfindungen, Theorien und Widersprüche in der Natur und Arzneiwissenschaft. - v.1-37, 1794-1804.   Gotha: Perthes, 1794-1804. **OTHER TITLE:** Journal der Erfindungen, Zweiful und Widersprüche in der gesammten Natur und Arzneiwissenschaft. **EDITOR:** August Friedrich Hecker. **CONTINUED BY:** Neues Journal der Erfindungen. **CONTINUED BY:**

Neuestes Journal der Erfindungen. Intelligenzblätter were issued. **INDEXES:** Index to the first six volumes in Das achzehnten Jahrhunderts Geschichte der Erfindungen. Gotha, Perthes, 1800. **SOURCES:** B2405, G110, K3675, OCLC, S2665 **LOCATIONS:** BUC, CCF, CCN, ULS, ZDB **SUBJECT:** Science - Periodicals - Germany. **SUBJECT:** Technology - Periodicals - Germany.

807   Journal der Pharmacie für Aerzte und Apotheker und Chemisten. - v.1-26, 1793-1817. Leipzig: Siegried Lebrecht Crusius, 1794-1817. **EDITOR:** Johann Bartholma Trommsdorf. **CONTINUED BY:** Neues Journal der Pharmacie für Aerzte, Apotheker und Chemiker. **INDEXES:** Index to v.1-14 (1794-1806) in v.15, v.16-25 in 26. **SOURCES:** B2409, G216, K3685, OCLC **LOCATIONS:** BUC, CCF, CCN, ULS, ZDB **SUBJECT:** Pharmacy - Periodicals - Germany.

808   Journal der Physik. - Bd.1-8 (each with 3 Hefte). Leipzig und Halle: Ambros. Barth, 1790-1794. **EDITOR:** Friedrich Albrecht Gren. **CONTINUED BY:** Neues Journal der Physik. **CONTINUED BY:** Annalen der Physik. **MICROFORMS:** RM(3). **SOURCES:** B2410, G669, K3337, OCLC **LOCATIONS:** BUC, CCF, CCN, ULS, ZDB **SUBJECT:** Science - Periodicals - Germany.

809   Journal der practischen Arzneykunde und Wundarzneykunst. - Bd 1-85, 1795-1837. Jena: Academische Buchhandlung, 1795-1837. Place and publisher vary. A 6 v.series of abstracts was published in 1808 under the title: Auszug aus C.W. Hufeland's Journal der practischen Arzneykunde und Wundarzneykunst. From v.8 also under the title: Neues Journal der practischen Arzneykunde und Wundarzneykunst. **EDITORS:** Christoph Wilhelm Hufeland (the entire series), v.28-39 with Carl Himly, v.40-47 with J.C.F. Harless, and 58-85 with E. Osann. Supplements were issued each year for the years 1822 and 1824-1830. **INDEXES:** There are indexes to v.1-20, 21-40 and 41-60. **SOURCES:** G129, K3696, OCLC **LOCATIONS:** BUC, CCF, CCN, ULS, ZDB **SUBJECT:** Medicine - Periodicals - Germany. **SUBJECT:** Surgery - Periodicals - Germany.

810   Journal der practischen Rossarznei- und Reitkunst. - Heft 1-2, 1800. Leipzig: Wienbrack, 1800. Imprint varies. **EDITOR:** Christian Ehrenfried Syfert von Tennecker.**INDEXES:** Indexed in Wimmel (Ref.7). **SOURCES:** K3738 **LOCATIONS:** CCN, ZDB **SUBJECT:** Veterinary medicine - Periodicals - Germany.

811   Journal der theoretischen und praktischen Ökonomie. - No.1-8, 1800. Leipzig: Leo, 1800. **EDITOR:** Friedrich Gottlieb Leonhardi. **SOURCES:** K3031 **SUBJECT:** Economics - Periodicals - Germany.

812   Journal des arts, des sciences et de la litterature. - No. 1-502, Jul. 23, 1799-1808, ser. 2, v.1-19 (no. 1-336), Apr. 15, 1810-1814. Paris: 1799-1814. Title varies. **CONTINUED BY:** Nain jaune, ou Journal des arts, de litterature et de

commerce. **SOURCES:** OCLC **LOCATIONS:** BUC, CCF, ULS, ZDB **SUBJECT:** General - Periodicals - France.

813 Journal des arts et manufactures, Publié sous la direction de la commission exécutive d'agriculture et des arts. - v.1-3, 1797-1797. Paris: 1795-1797. **MICROFORMS:** RP(1). (Goldsmiths-Kress Library of Economic literature). **SOURCES:** B2411 **LOCATIONS:** CCF, ULS **SUBJECT:** Economics - Periodicals - France.

814 Journal des inventions et découvertes. - No.1-7, 1793-1795. Paris: 1793-1795. Ser.1, no.1-14, 15 Apr-25 Sept., 1793. Ser.2, no.1-14, Jul., 1795-Nov. 1795 (Ref.87, p.17). **OTHER TITLE:** Journal du Lycée des arts, inventions et découvertes. **EDITOR:** Charles Gaullard Desaudray (Ref.88). **CONTINUED BY:** Annuaire du Lycée des arts. **CONTINUED BY:** Annuaire de l'Athenée des arts. **SOURCES:** B2431, G666 **LOCATIONS:** CCF **SUBJECT:** Technology - Societies - Paris.

815 Journal des mères de famille, ouvrage périodique entiérement consacré à celles que se destinent à élever leurs enfans dans l'orde de la nature. - Bordeaux: 1797-1798. **EDITOR:** I.M. Cailleau. **LOCATIONS:** CCF **SUBJECT:** Pediatrics - Periodicals - France.

816 Journal des mines, publié par l'agence (from v.4 le conseil) des mines de la république. - v.1-38 (nos.1-228), 1794-1815. Paris: Agence des mines de la république, 1794-1815. Subtitle and Publisher vary. Publication suspended Feb. 1799-Apr. 1801. **CONTINUED BY:** Annales des mines. **INDEXES:** Indexes to v.1-28, 1794-1809; and v.29-38, 1810-1815. **MICROFORMS:** RM(3). **SOURCES:** B2434, G667, OCLC, S1418 **LOCATIONS:** BUC, CCF, CCN, ULS, ZDB **SUBJECT:** Mineral industries - Periodicals - France.

817 Journal des observations minérologiques. - v.1, 1782. Nancy: H.Haenes, 1782. Prize essay for the Société royale des sciences, belles lettres et arts de Nancy by Esprit Pierre de Sivry. **SOURCES:** B2437 **LOCATIONS:** BL **SUBJECT:** Mineralogy - Periodicals - France.

818 Journal des observations physiques, mathématiques et botanique faites sur les cotes orientale de l'Amerique méridionale et dans le Indes occidentales. - Paris: 1707-1725. **SOURCES:** B2438 **LOCATIONS:** CCN **SUBJECT:** Science - Periodicals - France. **SUBJECT:** Mathematics - Periodicals - France.

819 Journal des officiers de santé de St. Dominique. - No.1, 1800. St. Dominque: 1800. Dates of publication vary, e.g. NUC:No.1, 1803. **EDITOR:** Victor Bally. **SOURCES:** G:p.343 **LOCATIONS:** CCF, NUC **SUBJECT:** Medicine - Periodicals - St. Dominique.

820 Journal des savans, augmenté de divers articles. - v.1-170 (1665-1753), v.1-79 (1754-1763), v.1-84 (1764- 1775), v.1-12 (1776-1782). Amsterdam: 1665-1782. **CONTINUATION OF:** Mémoires pour servir à l'histoire des sciences et des beaux arts. **CONTINUATION OF:** Journal des sçavans. Subtitle varies:

Combiné avec les Mémoires de Trévoux (1754- 1763), Avec des extraits des meilleurs journaux (1764-1775), Combiné avec les meilleurs journaux (1776-1782). Includes Journal des sçavans and Mémoires de Trévoux. **SOURCES:** S746 **LOCATIONS:** CCF, CCN, ZDb **SUBJECT:** General - Periodicals - Netherlands - Reviews.

821 Journal des savants. - v.1-. Paris: 1797- . **CONTINUATION OF:** Journal des sçavans. **SOURCES:** B2449a, S1425 **LOCATIONS:** BUC, CCF, ULS, ZDB **SUBJECT:** General - Periodicals - France.

822 Journal des savants d'Italie. - v.1-3, 1748-1749. Amsterdam: 1748-1749. **MICROFORMS:** ACR(1). **SOURCES:** S747 **LOCATIONS:** CCF, CCN, ZDB **SUBJECT:** General - Periodicals - Netherlands.

823 Journal des sçavans. - v.1-111, 1665-1792. Paris: Jean Cussan, 1665- . **TRANSLATED AS:** Journal des sçavans, hoc est ephemerides eruditorum. There was a counterfeit edition in Amsterdam 1665-1792. Reprinted in Paris by Editions Klincksieck. Frequency and imprint vary. **EDITORS:** Denis de Sallo, Jean Gallois, Jean Paul de la Roque and others (Ref.89). **CONTINUED BY:** Journal des savants. **INDEXES:** Cornelius à Beughem. La France Sçavant. Amsterdam, 1683; J.B. Robinet, Table générale alphabetique pour l'Editon de Holland. Amsterdam, Rey, 1765; Abbé Andre de Claustre. Table générale des matières contenus dans le Journal des Savans de l'edition de Paris (1665-1750). Paris, Briasson, 1753-1764. **MICROFORMS:** AACR(1), GO(9), RM(3). **SOURCES:** B2449, GII:3, OCLC, S1425 **LOCATIONS:** BUC, CCF, CCN, ULS, ZDB **SUBJECT:** General - Periodicals - France - Reviews.

824 Journal des sçavans, hoc est ephemrides eruditorum. - v.1-5, 1667-1671. Leipzig: Johann Eric Hohn, 1667-1671. Translated by Johann Friedrich Nietsche. **TRANSLATION OF:** Journal des sçavans. **SOURCES:** GII:3a, S2949 **LOCATIONS:** BUC, ULS **SUBJECT:** General - Periodicals - France - Translations.

825 Journal des sciences, arts et métiers, par une société de gens de lettres et d'artistes. - No.1-2, 1792. Paris: 1792. **EDITORS:** Jean Henri Hassenfratz and Alexander Brogniart. **SOURCES:** B2450, S1427 **LOCATIONS:** CCF **SUBJECT:** Science - Periodicals - France.

826 Journal des sciences et des beaux-arts. - v.1-50, 1768-1778. Paris; Trévoux: 1768-1778. **CONTINUATION OF:** Mémoires pour servir à l'histoire des sciences et des beaux arts. **OTHER TITLE:** Journal des beaux arts et des sciences. **EDITORS:** Jean Louis Castilhon, Jean Castilhon and Jean Louis Aubert (Ref.84). **CONTINUED BY:** Journal de littérature, des sciences et des arts. **INDEXES:** P.S. Sommervogel, Table methodique des Mémoires de Trevoux (1701-1775) Paris, 1864. **MICROFORMS:** ACR(1). **SOURCES:** B2924a, G662, S1428 **LOCATIONS:** BUC, CCF, ULS **SUBJECT:** General - Periodicals - France - Reviews.

827 Journal des sciences utiles, par une société de gens de lettres. - No.1-12, 1790, no.13-24, 1791. Paris: Perisse, 1790-1791. **EDITOR:** Pierre Bertholon. **LOCATIONS:** CCF **SUBJECT:** Science - Periodicals - France - Reviews.

828 Journal d'histoire naturelle (Berthollet). - v.1-3, 1787-1789. Paris, 1787-1789. **OTHER TITLE:** Nature considérée sur les animaux. **EDITORS:** Jean Berthollet and Alexis Boyer. **SOURCES:** B2358, G538, S1410 **LOCATIONS:** BUC, CCF **SUBJECT:** Natural history - Periodicals - France.

829 Journal d'histoire naturelle (Lamarck). - v.1, nos. 1-22, 1792. Paris: Directeurs de l'Imprimerie du Cercle social, 1792. Also issued in a quarto edition under the title: Choix de mémoires. Johnson reprint. **OTHER TITLE:** Choix de mémoires sur divers objets d'histoire naturelle. **EDITORS:** Jean Baptiste Lamarck, Jean Guillaume Brugnière, Guillaume Antoine Olivier, René Just Hauy, Bertrand Pelletier. **SOURCES:** B1289, G539, OCLC, S1410a **LOCATIONS:** BUC, CCF, ULS, ZDB **SUBJECT:** Natural history - Periodicals - France.

830 Journal d'instruction sur toutes les parties de l'art de guérir. - v.1, 1792. Montpellier: Jean Martel, 1792. **CONTINUATION OF:** Feuilles hebdomadaire sur la médecine, la chirurgie, la pharmacie et les science qui y ont rapport. **SOURCES:** G106 **LOCATIONS:** CCF **SUBJECT:** Medicine - Periodicals - France.

831 Journal d'observations de médecine, chirurgie et pharmacie. - v.1, 1763-1786. Calais: Moreaux, 1763-1786. **EDITOR:** Le Sieur Denis. **SOURCES:** GIII:17, OCLC **LOCATIONS:** ULS **SUBJECT:** Medicine - Periodicals - France. **SUBJECT:** Surgery - Periodicals - France. **SUBJECT:** Pharmacy - Periodicals - France.

832 Journal du point central des arts et métiers. - No.1-3, 1791. Paris: 1791. **SOURCES:** Ref.91,p.213. **LOCATIONS:** CCF **SUBJECT:** General - Periodicals - France.

833 Journal économique, ou, mémoires, notes et avis sur les arts, l'agriculture, la commerce, et tout ce qui peur avoir rapport à la santé, ansi qu'à la conservation et à l'augmentation des biens des familles. - Ser. 1, v.1-28 (4 v.a year), 1751-1757, Ser.2, v.1-15 (1 v.a year) 1758-1772. Paris: Antoine Bourdet, 1751-1772. **TRANSLATED AS:** Select essays on commerce, agriculture, mines, fisheries and other useful subjects. **OTHER TITLE:** Journal oeconomique. **EDITORS:** Badeau, Boudet, Jean Goulin and others (Ref.10, p.459). **MICROFORMS:** RP(1). (Goldsmiths' Kress Library). **SOURCES:** B2328 **LOCATIONS:** BUC, CCF, ULS **SUBJECT:** Science - Periodicals - France. **SUBJECT:** Technology - Periodicals - France. **SUBJECT:** Agriculture - Periodicals - France. **SUBJECT:** Medicine - Periodicals - France.

834 Journal encyclopédique ou universel. - v.1-304, 1756-1793. Liège; Bouillon; Paris: 1756-1793. **TRANSLATED AS:** Giornale enciclopedico di Liege.

849   Journal für Medicin, Chirurgie und Geburtshülfe, vorzügliche mit Rücksicht auf
      Aetiologie und Semiotik. - v.1-2, 1799-1803. Herborn; Hadamar: 1799-1803.
      Imprint varies. OTHER TITLE: Aetiologisches und semiotisches Journal.
      EDITOR: Johann Friedrich Sigmund Posewitz. SOURCES: G162, K3728
      LOCATIONS: BUC, CCF, ULS   SUBJECT: Medicine - Periodicals -
      Germany. SUBJECT: Surgery - Periodicals - Germany. SUBJECT:
      Obstetrics - Periodicals - Germany.

850   Journal of natural philosophy, chemistry and the arts. - v.1-5, 1797-1801.
      London: G.G. and J. Robinson, 1797-1801. Imprint varies. OTHER TITLE:
      Nicholson's Journal. EDITOR: William Nicholson. CONTINUED BY:
      Philosophical magazine. MICROFORMS: RM(3), RP(1). SOURCES:
      B2518, G668, OCLC, S324 LOCATIONS: BUC, CCF, CCN, ULS, ZDB
      SUBJECT: Science - Periodicals - England.

851   Journal of the practice of medicine, surgery and pharmacy in the military hospitals
      of France. - No.1, 1790. New York: J.M. McLean, 1790. Translated from the
      French by Joseph Browne. TRANSLATION OF: Journal de médecine
      militaire. SOURCES: G50a LOCATIONS: ULS SUBJECT: Medicine,
      Military - Periodicals - France - Translations.

852   Journal physico-médicale des eaux de plombières. - Nancy: H. Haener,
      1791-1799. OTHER TITLE: Journal, ou, recueil périodique d'observations sur
      les effets des eaux de plombières dans plusieurs malades. EDITOR: Martinet.
      LOCATIONS: CCF   SUBJECT: Medicine - Periodicals - France.
      SUBJECT: Plumbism - Periodicals - France.

853   Journal polytechnique, ou, bulletin du travail fait à cette école. - No.1, 1794.
      Paris: École centrale des travaux publics, 1794. EDITORS: Joseph Louis de
      Lagrange, Pierre Simon Laplace and others. CONTINUED BY: Journal de
      l'école polytechnique.   SOURCES: B2334, G663, OCLC, S1379a
      LOCATIONS: BUC, CCF, CCN, ULS   SUBJECT: Science - Societies -
      Paris.

854   Journal polytype des sciences et des arts. - No.1-152, Feb. 1786- Feb. 1787.
      Paris: Imprimerie polytype, 1786-1787. Originally issued in 3 parts: Partie des
      sciences, Partie des arts utile, and Partie des arts agreeable, each section once
      a week. Combined into one periodical with no. 58 (Ref.90). EDITORS:
      François Ignace Hoffmann and Bailly de Benfeld.   SOURCES: B2335
      LOCATIONS: BUC, CCF SUBJECT: Science - Periodicals - France.

855   Kabinet der natuurlyke historien, wetenschappen, konsten en handwerken. - v.1-9,
      1719-1727. Amsterdam: Hendrik Stuts, 1719-1727. 2nd ed. 1730, 3rd 1758.
      OTHER TITLE: Natuur-en konst-kabinet. EDITOR: Willem van Ranouw.
      INDEXES: Index published 1732.   SOURCES: B2533, G673, S748
      LOCATIONS: BUC, CCF, CCN, ULS, ZDB   SUBJECT: Natural history -
      Periodicals - Netherlands. SUBJECT: Technology - Periodicals - Netherlands.

856 Karlsruher meteorologische ephemeriden. - v.1, 1779. Mannheim: 1779. **EDITOR:** Johann Lorenz Böckmann. **SOURCES:** (Ref.71,p.2). **SUBJECT:** Meteorology - Periodicals - Germany.

857 Kleine Abhandlungen einiger Gelehrten in Schweden über verschiedene in die Physik, Chemie und Minerologie laufende Materien. - v.1-2, 1766-1768. Copenhagen; Leipzig: F.C. Mummen, 1766-1768. **SOURCES:** B2562, K3238, OCLC, S623 **SUBJECT:** Science - Periodicals - Sweden - Translations.

858 Kleine Beobachtungen für Taubstumme mit Anmerkungen von J.S. Beister und J.A.H. Reimarus. - No.1, 1799. Berlin: W.Viewig, 1799. **EDITORS:** Justus Arnemann, Jacob Albericht Reimarus and Johann Erich Biester. **SOURCES:** G367 **SUBJECT:** Deaf - Periodicals - Germany.

859 Kleine Gartenbibliothek. - v.1, 1790. Kiel: 1790. **CONTINUATION OF:** Taschenbuch für Gartenfreunde. **CONTINUATION OF:** Gartenkalender. **EDITOR:** Christian Cay Lorenz Hirschfeld. **SOURCES:** B4420, K2970, OCLC **LOCATIONS:** ZDB **SUBJECT:** Gardening - Periodicals - Germany.

860 Kleine medicinisch-chirurgische Abhandlungen, aus verschiedenen Sprache übersetzt. - v.1-2 (1779-1780) 1781. Leipzig: Weygand, 1781. **SOURCES:** G368, K3586 **SUBJECT:** Medicine - Periodicals - Germany - Reviews. **SUBJECT:** Surgery - Periodicals - Germany - Reviews.

861 Kleine physikalisch-chemische Abhandlungen. - v.1-8, 1785-1797. Leipzig: 1785-1797. **EDITOR:** Johann Friedrich Westrumb. **SOURCES:** B2566 **LOCATIONS:** BL **SUBJECT:** Science - Periodicals - Germany.

862 Kleine Schriften zur Land-und Stadt-wirthschaft von der ökonomischen Gesellschaft in Bern einzeln herausgegeben. - v.1-13 (1762-1789) 1791. Bern: Typographische Societät, 1791. A collection of separately printed monographs each with its own title page published between 1762 and 1789. Includes translations from other languages. **SOURCES:** K2982 **LOCATIONS:** BUC **SUBJECT:** Agriculture - Societies - Bern. **SUBJECT:** Economics - Societies - Bern.

863 Kongelige Danske Videnskabers selskab skrifter. - v.1-6 (1800-1809) 1801-1818. Copenhagen: Munksgard, 1801-1812. **CONTINUATION OF:** Skrifter, som udi det Kjöbenhavnske selskab af laerdoms og videnskabs elsster. **INDEXES:** Fortegnelse over de Konelige danske Videnskabers selskab. **SOURCES:** S615d **LOCATIONS:** BUC, CCF, CCN, ULS, ZDB **SUBJECT:** Science - Societies - Copenhagen.

864 De Koopman, of, bydragen ten opbouw van Neêderlands koophandel en zeevaard. - v.1-6, 1768-1770. Amsterdam: 1768-1770. **MICROFORMS:** RP(1) (Goldsmiths' Kress Library). **LOCATIONS:** BUC, CCN, ULS, ZDB **SUBJECT:** Economics - Periodicals - Netherlands.

865   Kopenhagener Magazin von öconomischen, cameral- polizei- handlungs-
        manufactur- mechanischen und Berwerksgesetzes, Schriften und kleinen
        Abhandlungen welche die königlichen Danischen Reiche und Lände betreffen.
        - v.1-3, 1757-1768. Copenhagen; Leipzig: Friedrich Christian Pelt, 1757-1768.
        Translated by Christian Gottlob Mengel. **OTHER TITLE:** Oeconomische
        Gedanken zu weiterem Nachdenken eröffnet. **TRANSLATION OF:** Danmark
        og Norges oeconomiske Magasin. **LOCATIONS:** BUC, CCF **SUBJECT:**
        Economics - Periodicals - Denmark - Translations.

866   Kosmographische Nachrichten und Sammlungen, von den Mitgliedern der
        kosmographischen Gesellschaft zusammengetragen. - v.1 (1748) 1750. Vienna:
        1750. **SOURCES:** B2574, S3596 **LOCATIONS:** CCF, ULS, ZDB
        SUBJECT: Geography - Societies - Vienna.

867   Der Kranke, eine Wochenschrift. - Danzig: 1764. Weekly. **EDITOR:** Gottfried
        Stolterforth. **SOURCES:** K3542 **SUBJECT:** Medicine - Periodicals -
        Germany - Popular.

868   Kratkoe opisaïe kommentariev / Akademïi nauk, St. Petersburg. - No.1, 1728. St.
        Petersburg: 1728. **CONTINUED BY:** Soderzhanïe uchenykh razsuzhdenïi,
        Academia scientiarum imperiales Petropolitana. **CONTINUED BY:**
        Ezhemiesiachnyia sochinenïia uchenykh dielaks, Academia scientiarum
        imperiales Petropolitana. **LOCATIONS:** ULS **SUBJECT:** General -
        Societies - St. Petersburg.

869   Kritik og antikritik, eller anmeldelser og bedommelser af de nysste indenlandske
        skrifter. - v.1-18, 1787-1796. Copenhagen: S. Poppe, 1787-1796. **EDITORS:**
        Johann Clemens Tode and others. **LOCATIONS:** ULS **SUBJECT:** General
        - Periodicals - Denmark - Reviews.

870   Kritische Beiträge zur neuesten Geschichte der Gelehrsamkeit. - v.1-10,
        1786-1791. Leipzig: Christian Gottlob Hertel, 1786-1791. **EDITOR:** Johann
        Peter Andreas Müller. **SOURCES:** B2580, K398, S2954 **LOCATIONS:**
        ULS, ZDB **SUBJECT:** General - Periodicals - Germany - Reviews.

871   Kritische Bemerkungen über Gegenstände aus dem Pflanzenreiche. - v.1, 1793.
        Mannheim: Schwan and Götz, 1793. **EDITOR:** Friedrich Casimir Medicus.
        SOURCES: B2581, G563, S3074 **SUBJECT:** Botany - Periodicals -
        Germany.

872   Kritische Nachrichten von kleinen medicinischen Schriften in- und ausländischer
        Akademien in Auszüge und kurzen Urtheilen dargelegt. - v.1 (1780) 1783, 2
        (1780-1781) 1784, 3 (1780-1783) 1788. Leipzig: Adam Friedrich Böhme; Jena:
        Christian Heinrich Cuno, 1783-1788. Volume 3 also issued as Neue kritische
        Nachrichten. **OTHER TITLE:** Neue kritische Nachrichten von kleinen
        medicinischen Schriften. **EDITOR:** Christian Gottfried Gruner. **SOURCES:**
        G291, K3603, OCLC **LOCATIONS:** BN **SUBJECT:** Medicine - Periodicals
        - Germany - Dissertations.

873   Kunst-oeffeningen over verscheide nuttige onderwerpten der wiskunde /
       Genootschap der mathematische wetenschappen, onder de spreuk: een
       onvermoeide arbeid komt alles te boven. - v.1, 1782, v.2, 1788. Amsterdam:
       F. Houttuyn, 1782-1788. **MICROFORMS:** RM(3). **SOURCES:** OCLC,
       S735 **LOCATIONS:** BUC, CCN   **SUBJECT:** Mathematics - Societies -
       Amsterdam.

874   Kurelle chemische Versuche und Erfahrungen. - St.1, 1756. Berlin: Haude undd
       Spener, 1756. **SOURCES:** K3225 **SUBJECT:** Chemistry - Periodicals -
       Germany.

875   Kurze Nachricht von der Entstehung und Einrichtung der Gesellschaft von
       Freunden der Entbindungswissenschaft. - No.1-2, 1796-1798. Göttingen:
       Johann Georg Rosenbusch, 1796-1798. No.2 as Zweite Nachricht. **OTHER
       TITLE:** Zweite Nachricht von der Verhandlungen der Gesellschaft von
       Freunden der Entbindungswissenschaft. **EDITOR:** Friedrich Benjamin
       Osiander. **LOCATIONS:** GB **SUBJECT:** Obstetrics - Societies - Göttingen.

876   Kurzer Unterricht, ertrunkene Menschen wieder lebendig zu machen. - v.1, 1775.
       Warsaw: 1775. **TRANSLATION OF:** Histoire et mémoires, Société formée
       à Amsterdam en faveur des noyés. **SOURCES:** G510b **SUBJECT:**
       Resuscitation - Societies - Amsterdam - Translations.

877   Kweekschool der genees- heel- ontleed- natuur-ziekte- schei- en vroekunde. - 10
       v., 1759-1773. Amsterdam: David Weige, 1759-1771. Issued in 7 series: (1)
       Kweekschool der artzenykunde 1v.1773; (2) Kweekschool der heelkunde,
       3v.1771; (3) Kweekschool der natuurkunde, 1v.1773; (4) Kweekschool der
       ontleedkunde 2v.1772; (5) Kweekschool der Scheikunde, 1v.1773; (6)
       Kweekschool der vroedkunde, 1v.1773; (7) Kweekschool der ziektekunde,
       1v.1772. **SOURCES:** G292 **LOCATIONS:** CCN, DSG **SUBJECT:**
       Science - Periodicals - Netherlands.

878   The ladies' diary, or, the Woman's almanack. - nos.1-180, 1704-1876. London:
       1704-1876. Imprint varies. **OTHER TITLE:** Palladium, or, appendix to the
       Ladies' diary; Diarian miscellany; Companion to the Ladies' diary; Ladies'
       diary, supplement; Diarian repository; Mathematical questions proposed in the
       Ladies' diary. **EDITORS:** John Tipper, Henry Beighton, Robert Heath,
       Thomas Simpson, Edward Rollinson, Charles Hutton, and Olinthus Gilbert
       Gregory. **CONTINUED BY:** Ladies' and gentleman's diary. Supplements
       issued as: The Palladium, or appendix to the Ladies' diary, 1748-1779; Diarian
       miscellany, 1705-1773 (1775); Ladies' diary, supplement, 1788-1789;
       Companion to the Ladies' diary, 1777-1782; Diarian repository or mathematical
       register, 1704-1760 (1771-1774). Mathematical questions proposed in the
       Ladies' diary, 1704-1816 (1817). **SOURCES:** B2601, B8577, OCLC, S332
       LOCATIONS: BUC, ULS **SUBJECT:** Mathematics - Periodicals - England
       - Popular.

879   The Lady's and gentleman's diary, or, Royal almanac. - No.1-4, 1776-1779.
       London: Thomas Carnon, 1776-1779. At least three supplements were

published 1779-1781. (Ref.12,no.12). **OTHER TITLE:** Carnon's diary. **EDITOR:** Reuben Burrow. **CONTINUED BY:** Ladies' diary. **SOURCES:** B8541 **LOCATIONS:** BUC **SUBJECT:** Mathematics - Periodicals - England - Popular.

880   Lady's and gentleman's scientific repository, containing enigmas, rebuses, paradoxes, philosophical and other useful queries; arithmetical and mathematical questions and problems with their respective solutions. - No.1-10, 1783-1784. Newark, England: J. Tomlinson, 1783-1784. **EDITORS:** William Spalton and Joseph Gibbs. **SOURCES:** OCLC **LOCATIONS:** BUC, ULS **SUBJECT:** Mathematics - Periodicals - England - Popular.

881   Laeger et medicinst ugeskrift. Copenhagen: Johann Christian Berling and Georg Christopher Berling, 1766. A translation of Unzer's Der Arzt, eine medicinische Wochenschrift by Niels Friberg (Ref.42). **TRANSLATION OF:** Arzt, eine medicinische Wochenschrift. **SOURCES:** (Ref.42) **SUBJECT:** Medicine - Periodicals - Germany - Popular - Translations.

882   Läkare och naturforskaren. - v.8-15, 1787-1807. Stockholm: Strangnäs, 1787-1807. **CONTINUATION OF:** Veckoskrift för läkare och naturforskare. **EDITORS:** Andreas Johan Hagström, Johan Kraak, Johann Lorens Odhelius and Gabriel Lund. **SOURCES:** B4657, G47, S701 **LOCATIONS:** CCF, ULS **SUBJECT:** Medicine - Periodicals - Sweden. **SUBJECT:** Natural history - Periodicals - Sweden.

883   Der Landarzt, eine medicinische Wochenschrift. - St.1-52, 1765-1766. Mitau: Christian Liedtke, 1765-1766. Reprinted Frankfurt and Leipzig, 1769. **EDITOR:** Peter Ernst Wilde. **CONTINUED BY:** Liefländische Abhandlungen von der Arzneywissenschaft. **SOURCES:** GIII:18, K3543 **LOCATIONS:** CCF, ULS, ZDB **SUBJECT:** Medicine - Periodicals - Russia - Popular.

884   Der Landarzt, oder, Archiv für das Landvolk. - v.1, 1794. Augusburg: Matthäus Rieger, 1794. **EDITOR:** Johann Gottfried Essich. **SOURCES:** G194 **LOCATIONS:** DSG **SUBJECT:** Medicine - Periodicals - Germany - Popular.

885   Der Landwirth. - v.1-2, 1779-1782. Warsaw; Dresden: Michael Gröll, 1779-1782. **SOURCES:** K2925 **LOCATIONS:** BL **SUBJECT:** Agriculture - Periodicals - Poland.

886   Landwirthschafliche Erfahrungen zum Besten des Landsmänner, eine Wochenschrift. - Quartal 1-3, 1768-1769. Altona; Lübeck: David Iverson, 1768-1769. Weekly. **EDITOR:** Johann Heinrich Pratje. **SOURCES:** B2627, K2892 **SUBJECT:** Agriculture - Periodicals - Germany.

887   Landwirthschaftliche Nachrichten für Bündten. - Genf: 1799. **SOURCES:** (Ref.95,p.8). **SUBJECT:** Agriculture - Periodicals - Switzerland.

888   Landwirthschaftliches Magazin. - v.1-2, 1788-1791. Leipzig: Siegfried Crusius, 1788-1791. **EDITOR:** Sebastian Georg Friedrich Mund. **SOURCES:** K2960

LOCATIONS: BUC, CCF, ULS   SUBJECT: Agriculture - Periodicals - Germany.

889   Landwirthschaftliches Wochenblatt. - Chur: 1775. Weekly. SOURCES: (Ref.95,p.8). SUBJECT: Agriculture - Periodicals - Switzerland.

890   Landwirtschaftliche Monatsschrift. - St.1-2, 1799-1800. Leipzig: Rein, 1799-1800. Monthly. EDITOR: Lüders Hermann Hans Engel. SOURCES: K3026 SUBJECT: Agriculture - Periodicals - Germany.

891   Lausitzisches Wochenblatt zu Ausbreitung nützlicher Kenntnisse aus der Natur-, Haushaltungs-, Staats- und Völker- Kunde der Ober- und Niederlausiz. - Zittau: Schöps, 1790-1791. Began as a weekly but became a monthly in the second quarter. CONTINUATION OF: Provinzialblätter oder Sammlungen zur Geschichte, Naturkunde, Moral und anderen Wissenschaften. OTHER TITLE: Beiträge zur natürlichen, ökonomischen und politischen Geschichte der Ober-und Niederlausitz. EDITOR: Christian August Pescheck. CONTINUED BY: Lausitzische Monatsschrift, Oberlausitzische Gesellschaft der Wissenschaften, Görlitz. SOURCES: K1248 SUBJECT: General - Periodicals - Germany.

892   Lausitzische Monatsschrift. - v.1-32, 1793-1808. Görlitz: Johann Rudolf Unger, 1793-1808. Monthly. CONTINUATION OF: Provinzialblätter oder Sammlungen zur Geschichte, Naturkunde, Moral und anderen Wissenschaften. CONTINUATION OF: Lausitzisches Wochenblatt zu Ausbreitung nützlicher Kenntnisse aus der Natur-, Haushaltungs-, Staats- und Völker- Kunde der Ober- und Niederlausiz. Publisher varies. After 1799 as Neues Lausitzische Monatsschrift. OTHER TITLE: Neues Lausitzische Monatsschrift. SOURCES: G485, K1280   LOCATIONS: BUC, CCP, ULS, ZDB SUBJECT: General - Periodicals - Germany.

893   Lausitzisches Magazin, oder, Sammlung verschiedener Abhandlungen und Nachrichten zum Behuf der Natur-, Kunst-, Welt-, und Vaterlands-Geschichte. - v.1-25 (24 issues each) 1768-1792. Görlitz: Johann Friedrich Fickelscherer, 1768-1792. EDITOR: Carl Gottlob Dietmann. CONTINUED BY: Neues Lausitzische Magazin. SOURCES: B2678, K268, S2650 LOCATIONS: BUC, CCF, ZDB SUBJECT: General - Periodicals - Germany.

894   Leipziger gelehrte Zeitungen. - v.1-112, 1715-1797. Leipzig: Breitkopf, 1715-1797. Subtitles and imprints vary. OTHER TITLE: Neue Zeitungen von gelehrten Sachen; Neue Leipziger Zeitung von gelehrten Sachen; Leipziger gelehrte Anzeigen; Neue Leipziger gelehrte Anzeigen; Neue Leipziger gelehrte Zeitungen. EDITORS: Johann Christoph Adelung, Johann Gottlieb Krause, Johann Burchard Mencke and others. INDEXES: There are indexes to 1715-1784 and 1785-1797. SOURCES: B4925, G703, K41, 396, 425, S2958, 2981 LOCATIONS: ULS, ZDB SUBJECT: General - Periodicals - Germany - Reviews.

895   Leipziger gelehrtes Tagebuch. - v.1-26, 1780-1807. Leipzig: Johann Gottlob
       Immanuel Breitkopf, 1780-1807. Imprint varies. EDITOR: Johann Georg
       Eck. INDEXES: Index to 1780-1802 published 1804. SOURCES: B2684,
       K347, OCLC, S2957 LOCATIONS: ULS, CCN, ZDB SUBJECT: General
       - Periodicals - Germany.

896   Leipziger Magazin für reine und angewandte Mathematik. - v.1-3 (4 issues each),
       4 (St.1-2) 1786-1788. Leipzig: J.G. Müller, 1786-1788. OTHER TITLE:
       Leipziger Magazin für die Mathematik. EDITORS: Carl Friedrich Hindenburg
       and Johann Bernouilli. CONTINUED BY: Archiv der reinen und angewandten
       Mathematik. SOURCES: B2686, K3311, OCLC, S2961 LOCATIONS:
       CCF, CCN, ULS, ZDB SUBJECT: Mathematics - Periodicals - Germany.

897   Leipziger Magazin zur Naturgeschichte und Ökonomie. - v.1-2 (St.1-4 each), 3
       (St.1-2) 1786-1788. Leipzig: Johann Gottfried Müller, 1786-1788.
       CONTINUATION OF: Leipziger Magazin zur Naturkunde, Mathematik und
       Ökonomie. EDITOR: Nathaniel Gottfried Leske. INDEXES: Index 1-4 in 4
       of Leipziger Magazin zur Naturkunde. SOURCES: G674, K3312
       LOCATIONS: BUC, CCF, CCN, ULS, ZDB SUBJECT: Natural history -
       Periodicals - Germany. SUBJECT: Economics - Periodicals - Germany.

898   Leipziger Magazin zur Naturkunde, Mathematik und Ökonomie. - v.1-5,
       1781-1784. Leipzig; Dessau: Buchhandlung der Gelehrten, 1781-1784.
       EDITORS: Carl Friedrich Hindenburg, Christian Benedict Fund and Nathaniel
       Gottfried Leske. CONTINUED BY: Leipziger Magazin zur Naturgeschichte
       und Ökonomie. INDEXES: Index v.1-4 in 4. SOURCES: B2687, G674,
       K3280, S2960 LOCATIONS: BUC, CCF, CCN, ULS, ZDB SUBJECT:
       Natural history - Periodicals - Germany. SUBJECT: Economics - Periodicals
       - Germany. SUBJECT: Mathematics - Periodicals - Germany.

899   Leipziger Sammlungen von Wirthschaftlichen-, Polizei-, Cameral- und
       Finanzsachen. - v.1-16, 1742-1767; v.13. Leipzig: Carl Ludwig Jacobi,
       1742-1767. OTHER TITLE: Sammlungen von allerhand zum Land- und
       Stadtwirthschaftlichen, Polizei- Finanz- und Cameral-Wesen dienlilchen
       Nachrichten. EDITOR: Georg Heinrich Zincke. INDEXES: Index v.1-12
       (1742-1757) in 12, v.13-16 (1758-1767) in 18. LOCATIONS: NUC, ZDB
       SUBJECT: Economics - Periodicals - Germany.

900   Lesebibliothek für die Liebhaber der Apotheker- und Wundarzneykunst. - v.1-2,
       1788-1789. Regensburg: 1788-1789. EDITOR: Johann Jacob Kohlhaas.
       SOURCES: K3646   SUBJECT: Pharmacy - Periodicals - Germany.
       SUBJECT: Surgery - Periodicals - Germany.

901   Letters and essays on the small-pox and innoculation, the measles, the dry
       belly-ache, the yellow and remitting, and intermitting fevers of the West Indies.
       - v.1, 1778. London; Edinburgh: J.Murray, 1778. EDITOR: Donald Monro.
       SOURCES: OCLC   SUBJECT: Medicine - Periodicals - England.
       SUBJECT: Smallpox - Periodicals - England.

902   Letters and papers on agriculture, planting &c., selected from the correspondence
       book of the Society instituted at Bath for the encouragement of agriculture, arts,
       manufactures, and commerce. - Volume 1-14, 1779-1816. Bath: R. Crutwell,
       1779-1816. Imprint varies. There were several editions of the early volumes.
       OTHER TITLE: Letters to the Bath and West of England Agricultural Society.
       CONTINUED BY: Journal, Bath and West and Southern counties society.
       MICROFORMS: RM(3), MIM(1), RP(1). (Goldsmiths'-Kress Library).
       SOURCES: OCLC, S34a   LOCATIONS: BUC, CCN, ULS   SUBJECT:
       Agriculture - Societies - Bath.

903   Lettres hebdomadaires sur l'utilité minéraux dans la société civil. - v.1-2, 1770.
       Paris: Durand, 1770. EDITOR: Pierre Joseph Buc'hoz. CONTINUED BY:
       Nature considérée sous ses différent aspects.   SOURCES: B2691c, OCLC,
       S1433   LOCATIONS: CCF   SUBJECT: Mineralogy - Periodicals - France.

904   Lettres périodique sur la méthode de s'enrichir promptement et conserver sa santé
       par la culture des végetaux. - v.1-2, 1768-1770. Paris: Durand, 1768-1770.
       EDITOR: Pierre Joseph Buc'hoz. CONTINUED BY: Nature considérée sous
       ses différent aspects.   SOURCES: B2691a, S1435   LOCATIONS: CCF
       SUBJECT: Botany - Periodicals - France.   SUBJECT: Medicine - Periodicals
       - France.

905   Lettres périodiques, curieuses, utiles et intéresantes sur les avantages que la
       société peut retirer de la connaissance des animaux. - v.1-4, 1769-1770. Paris:
       Durand, 1769-1770. EDITOR: Pierre Joseph Buc'hoz. CONTINUED BY:
       Nature considérée sous ses différent aspects.   SOURCES: B2691b, S1434
       LOCATIONS: CCF   SUBJECT: Zoology - Periodicals - France.

906   Liefländische Abhandlungen von der Arzneywissenschaft. - St.1-13, 1766-1767.
       Schloss-Oberpahlen: 1766-1767.   CONTINUATION OF: Landarzt, eine
       medicinische Wochenschrift.   Reprinted along with Landarzt, St.1-52 with
       index, Schloss- Oberphlen in 1782 (Ref.94). EDITOR: Peter Ernst Wilde.
       SOURCES: K3545   SUBJECT: Medicine - Periodicals - Russia - Popular.

907   Lineae medicae singulos per menses quotidie ductae continentes, observationes,
       historias, experimenta. - v.1-6, Jan. 1695-Dec. 1700. Augsburg: 1695-1700.
       EDITOR: Viti Riedlini. MICROFORMS: RP(1).   LOCATIONS: BUC,
       CCF   SUBJECT: Medicine - Periodicals - Germany.

908   Literatur der Ökonomie, Technologie, Polizei und Cameralwissenschaften. -
       St.1-6, 1790-1791. Leipzig: Johann Gottlob Hamann, 1790-1791. SOURCES:
       K2979   LOCATIONS: BUC, ULS   SUBJECT: Economics - Periodicals -
       Germany - Reviews.   SUBJECT: Technology - Periodicals - Germany -
       Reviews.

909   The London medical journal, by a Society of physicians. - v.1-11, 1781-1790.
       London: J. Johnson; Edinburgh: Elliot; Dublin: Byrne, 1781-1790.
       TRANSLATED AS: Journal de médecine. TRANSLATED AS: Sammlung der
       neuesten Beobachtungen englischer Aerzte und Wundärzte. TRANSLATED

AS: Geneeskundig journal van London. EDITOR: Samuel Foart Simmons. CONTINUED BY: Medical facts and observations. SOURCES: G46 LOCATIONS: BUC, CCF, ULS, ZDB SUBJECT: Medicine - Periodicals - England.

910 London medical review and magazine, by a society of physicians and surgeons. - v.1-6 (in 2 series), 1799-1802. London: Cadell and Davies, 1799-1802. Monthly. OTHER TITLE: Medical review. EDITOR: William Blair. Absorbed by: Medical and physical journal. SOURCES: G166 LOCATIONS: BUC, CCF, CCN, ULS, ZDB SUBJECT: Medicine - Periodicals - England.

911 Lucina, oder, Magazin für Geburtshelfer. - St.1, 1787. Marburg: Neue akademische Buchhandlung, 1787. EDITORS: Johann David Busch and Heinrich Daum. SOURCES: G69, K3640 LOCATIONS: CCF SUBJECT: Obstetrics - Periodicals - Germany.

912 Maandelijke uitreksels of Boekzaal der geleerde wereld. - No.1-194, 1715-1811. Amsterdam: 1715-1811. CONTINUATION OF: Boekzaal van Europe. CONTINUATION OF: Twee-maandellijke uitreksel. CONTINUATION OF: Boekzaal der geleerde wereld of tijdschrift voor leteerkundigen. INDEXES: Index, 1715-June 1730; July. 1730-June. 1745; July. 1745-Dec. 1759. LOCATIONS: CCF, ULS SUBJECT: General - Periodicals - Netherlands - Reviews.

913 Machines et inventions aprouvées par Académie des sciences, depuis son établissement jusqu'à présent, avec leur description dessinée & publiées du consentement de l'Académie, par M. Gallon. - v.1(1666-1701), v.2(1702-1712), v.3(1713-1719, v.4(1720-1721), v.5(1727-1731), v.6(1732-1734), v.7(1735-1754), 1735-1777 (Ref.15, p.462). Paris: Gabriel Martin, 1735-1777. Printer varies. V.7, which was published after Gallon's death (1775), has an amended title page which reads "depuis 1734 jusqu'à 1754." INDEXES: Table alphabétique at end of v.6. MICROFORMS:RP(1). SOURCES: B1726, G17h, OCLC LOCATIONS: BUC, CCN, ULS, ZDB SUBJECT: Manufactures - Societies - Paris. SUBJECT: Technology - Societies - Paris.

914 Magasin encyclopédique, ou, journal des sciences, des lettres et des arts. - v.1-122, 1795-1816. Paris: 1795-1816. EDITORS: Aubin Louis Millin, François Noel and Israel Warrens. CONTINUED BY: Annales encyclopdique. CONTINUED BY: Revue encyclopédique. INDEXES: Table générale des matières des 122 volumes. Paris, J.B. Sajou, 1819. MICROFORMS: RM(3). SOURCES: B2728, G675, OCLC, S1437 LOCATIONS: BUC SUBJECT: General - Periodicals - France.

915 Magazin aller neuen Erfindungen, Entdeckungen und Verbesserungen für Fabrikanten, Künstler, Handwerker. - v.1-8, 1797-1808. Leipzig: Baumgärtner, 1797-1808. EDITORS: Sigismund Friedrich Hermbstädt, Christian Ludwig Sebass, and Friedrich Gotthelf Baumgärtner. CONTINUED BY: Magazin der neuesten Erfindungen, Entdeckungen. CONTINUED BY: Neuer Magazin aller

neuen Erfindungen. **SOURCES:** B2742, K3917 **LOCATIONS:** BUC, CCF, CCN, ULS, ZDB **SUBJECT:** Manufactures - Periodicals - Germany. **SUBJECT:** Technology - Periodicals - Germany.

916   Magazin auserlesener medicinischer Abhandlungen von berühmten französischen Aerzten. - v.1, 1797. Berlin: Gottlieb August Lange, 1797. **EDITOR:** Hermann Wilhelm Lindemann. **SOURCES:** G369 **LOCATIONS:** NUC **SUBJECT:** Medicine - Periodicals - France - Translations.

917   Magazin der Bergbaukunde. - v.1-13, 1785-1799. Dresden: Walther, 1785-1799. **OTHER TITLE:** Magazin für die Bergbaukunde. **EDITOR:** Johann Friedrich Lempe. **MICROFORMS:** RM(3). **SOURCES:** B2746, K3307, S2509 **LOCATIONS:** BUC, CCF, CCN, ULS, ZDB **SUBJECT:** Mineral industries - Periodicals - Germany.

918   Magazin der italienischen Litteratur und Künste. - v.1-8, 1780-1785. Weimar: Carl Ludolf Hoffmann, 1780-1785. Imprint varies. **EDITOR:** Christian Joseph Jagemann. **SOURCES:** G678, K4513, OCLC **LOCATIONS:** BUC, CCF, ULS, ZDB **SUBJECT:** General - Periodicals - Germany - Reviews.

919   Magazin der neuesten ausländischen Insekten. - v.1, 1794. Erlangen: Heyder, 1794. **EDITOR:** Eugen Johann Christoph Esper. **SOURCES:** B2750, G583, K3356, S2569 **LOCATIONS:** ZDB **SUBJECT:** Entomology - Periodicals - Germany.

920   Magazin der patriotisch-ökonomischen Gesellschaft in Franken. - v.1-2, 1771. Nürnberg: 1771. **CONTINUATION OF:** Gesammelte Nachrichten der oekonomischen Gesellschaaft in Francken. **SOURCES:** (Ref.6,p.289). **SUBJECT:** Economics - Societies - Anspach.

921   Magazin der Regierungskunst, der Staats-und Landwirthschaft. - St.1-3, 1775-1779. Leipzig: 1775-1779. **EDITOR:** Carl Friedrich Dacheröden. **SOURCES:** K2015 **LOCATIONS:** ULS, ZDB **SUBJECT:** Agriculture - Periodicals - Germany.

922   Magazin der theoretischen und praktischen Arzneikunst, für Freunde und Feinde der neuen Lehre. - v.1, 1796. Heilbronn: J.D. Closs, 1796. **EDITOR:** Melchoir Adam Weickard. **SOURCES:** G253, K3703, OCLC **LOCATIONS:** ZDB **SUBJECT:** Brunonianism - Periodicals - Germany. **SUBJECT:** Medicine - Periodicals - Germany.

923   Magazin der Viehartzneikunst. - v.1, 1784. Vienna: Friedrich August Hartmann; Leipzig: Böhme, 1784. **INDEXES:** Indexed in Wimmel (Ref.7). **SOURCES:** G225, K3609 **LOCATIONS:** ZDB **SUBJECT:** Veterinary medicine - Periodicals - Austria.

924   Magazin des Pflanzenreichs. - Volume 1, 1793-1796. Erlangen: 1793-1796. **EDITOR:** Gottfried Christian Reich. **SOURCES:** B2756, G565, S2570 **LOCATIONS:** CCF, ULS, ZDB **SUBJECT:** Botany - Periodicals - Germany.

925   Magazin des Thierreichs. - v.1 (no.1-3) 1793-1795. Erlangen: Wolfgang Walther, 1793-1795. **EDITOR:** Gottfried Christian Reich. **SOURCES:** G572, K3352, OCLC, S2571 **LOCATIONS:** CCF, ULS, ZDB **SUBJECT:** Zoology - Periodicals - Germany.

926   Magazin für die Thiergeschichte, Thieranatomie und Thierarzneikunde. - v.1 (St.1-2) 1790-1794. Göttingen: Johann Christian Dieterich, 1790-1794. **EDITOR:** Friedrich Albrecht Anton Meyer. **CONTINUED BY:** Zoologische Annalen. **CONTINUED BY:** Zoologisches Archiv. **INDEXES:** Indexed in Wimmel (Ref.7). **SOURCES:** B2791, G571, K3339, S2689 **LOCATIONS:** BUC, ULS, ZDB **SUBJECT:** Veterinary medicine - Periodicals - Germany.

927   Magazin für allgemeine Natur und Thiergeschichte. - v.1-6, 1788-1796. Göttingen und Leipzig: Brose, 1788-1796. **EDITOR:** Carl Friedrich August Müller. **SOURCES:** B2762, G541, K3327, S2688 **LOCATIONS:** BUC, CCF, ZDB **SUBJECT:** Natural history - Periodicals - Germany. **SUBJECT:** Zoology - Periodicals - Germany.

928   Magazin für Apotheker, Materialisten und Chemisten. - Heft. 1-3, 1785-1787. Nürnberg; Schmidner, 1785-1787. There was a second edition in 1796. **OTHER TITLE:** Magazin für Apotheker, Chemisten und Materialisten. **EDITOR:** Johann Caspar Philipp Elwert. **CONTINUED BY:** Repertorium für Chemie, Pharmacie und Arzneimittelkunde. **SOURCES:** B2763, G212, K3619, S3163 **SUBJECT:** Pharmacy - Periodicals - Germany. **SUBJECT:** Chemistry - Periodicals - Germany.

929   Magazin für Arzneikunde. - Heft 1, 1785. Erfurt: 1785. **EDITOR:** Johann Christoph Fahner. **SOURCES:** K3720 **LOCATIONS:** ZDB **SUBJECT:** Medicine - Periodicals - Germany.

930   Magazin für das neueste aus der Physik und Naturgeschichte. - v.1-12, 1781-1799. Gotha: Carl Wilhelm Ettinger, 1781-1799. There was a 2nd ed. of v.1-2, 1785-1787. **EDITORS:** Ludwig Christian Lichtenberg (v.1-3) and Johann Heinrich Voigt (v.4-12). **CONTINUED BY:** Magazin für den neuesten Zustand der Naturkunde. **INDEXES:** v.12 is an index to the first 11 v. **MICROFORMS:** RM(3). **SOURCES:** B2775, G679, K3281, OCLC, S2666 **LOCATIONS:** BUC, CCF, CCN, ULS, ZDB **SUBJECT:** Science - Periodicals - Germany.

931   Duplicates no. 930.

932   Magazin für den neuesten Zustand der Naturkunde in Rücksicht auf die dazu gehörigen Hülfswissenschaften. - Volumes 1-12, 1797-1806. Jena; Weimar: Industrie Comptoir, 1797-1806. **CONTINUATION OF:** Magazin für das Neueste aus der Physik. **EDITOR:** Johann Heinrich Voigt. **SOURCES:** B2776, G680, K3370, S2827 **LOCATIONS:** CCF, CCN, ULS, ZDB **SUBJECT:** Science - Periodicals - Germany.

933   Magazin für die Arzneimittellehre, aus verschiedenen Sprachen übersetzt. - St.1,
      1794. Chemnitz: Mauke, 1794. **EDITOR:** Carl Gottlob Kühn. **SOURCES:**
      G217, K3689   **LOCATIONS:** ZDB   **SUBJECT:** Pharmacy - Periodicals -
      Germany.

934   Magazin für die Botanik. - St. 1-12, 1787 in 3v. each with 4 issues. **OTHER
      TITLE:** Botanisches Magazin. Zürich: Johann Caspar Fuessli, 1787-1790.
      EDITORS: Johann Jacob Römer and Paul Usteri. **CONTINUED BY:** Annalen
      der Botanik. **CONTINUED BY:** Neue Annalen der Botanik. **CONTINUED
      BY:** Neues Magazin für die Botanik in ihrem ganzen Umfange. **INDEXES:**
      Index included in St. 12. **MICROFORMS:** RM(3), IDC(8). **SOURCES:**
      B2778, G564, K3318, OCLC, S2200 **LOCATIONS:** BUC, CCF, CCN, ULS,
      ZDB   **SUBJECT:** Botany - Periodicals - Germany.

935   Magazin für die Geographie, Staatenkunde und Geschichte. - Bd.1-Bd.3, 1797.
      Nürnberg: Bauer, 1797. **CONTINUATION OF:** Geographisches Magazin.
      CONTINUATION OF: Neues geographisches Magazin. **CONTINUATION
      OF:** Historische und geographische Monatsschrift. **CONTINUATION OF:**
      Beiträge zur Geographie, Geschichte und Staatenkunde. **EDITOR:** Johann
      Ernst Fabri and Carl Hammerdörfer. **SOURCES:** B2779, K1303
      LOCATIONS: BUC, CCF, ZDB   **SUBJECT:** Geography - Periodicals -
      Germany.

936   Magazin für die gerichtliche Arzneikunde und medicinische Polizei. - v.1-2 (4
      issues each) 1782-1784. Stendal: Christian Franzen und Johann Christian
      Grosse, 1781-1784. **EDITORS:** Johann Theodor Pyl and Conrad Friedrich
      Uden. **CONTINUED BY:** Neues Magazin für die gerichtliche Arzneikunde.
      CONTINUED BY: Repertorium für die öffentliche und gerichtliche
      Arzneiwissenschaft. **SOURCES:** G49, K3591 **LOCATIONS:** BUC, CCF,
      CCN, ULS, ZDB **SUBJECT:** Medical jurisprudence - Periodicals - Germany.

937   Magazin für die gesammte populäre Arzneikunde, besonders für die sogennante
      Hausmittel. - v.1-2 (St.1-12) 1785-1786. Frankenhausen: Cöler, 1785-1786.
      Monthly. Imprint varies. **EDITOR:** Johann Christoph Fahner. **SOURCES:**
      G181, K3621   **LOCATIONS:** ZDB   **SUBJECT:** Medicine - Periodicals -
      Germany - Popular.

938   Magazin für die höhere Naturwissenschaft und Chemie. - v.1-2, 1784-1787.
      Tübingen: Jacob Friedrich Heerbrandt, 1784-1787. **SOURCES:** B2782, G677,
      K3300, OCLC, S3279 **LOCATIONS:** ULS, ZDB   **SUBJECT:** Alchemy -
      Periodicals - Germany. **SUBJECT:** Chemistry - Periodicals - Germany.

939   Magazin für die Liebhaber der Entomologie. - Bd.1-2, each with 2 issues,
      1778-1779. Zürich; Winterthur: Heinrich Steiner, 1778-1779. Publisher varies.
      OTHER TITLE: Magazin für Liebhaber der Entomologie. **EDITOR:** Johann
      Caspar Fuessli. **CONTINUED BY:** Neues Magazin für die Liebhaber der
      Entomologie. **CONTINUED BY:** Neuestes Magazin für die Liebhaber der
      Entomologie. **MICROFORMS:** RM(3). **SOURCES:** B2784, G582, K3266,

OCLC, S2201   **LOCATIONS:** BUC, CCF, ULS, ZDB   **SUBJECT:** Entomology - Periodicals - Switzerland.

940   Magazin für die Mineralogie und mineralogische Technologie. - v.1-2, 1789-1790. Halle: Johann Jacob Gebauer, 1789-1790. **EDITOR:** Johann Hermann Pfingston. **SOURCES:** B2785, K3331, S2721 **LOCATIONS:** CCF, ULS, ZDB **SUBJECT:** Mineral industries - Periodicals - Germany. **SUBJECT:** Mineralogy - Periodicals - Germany.

941   Magazin für die Naturgeschichte des Menschen. - Bd.1-3 each with two issues, 1788-1791. Zittau; Leipzig: Johann David Schöps, 1788-1791. **MICROFORMS:** RM(3). **SOURCES:** B2786, G243, K3328, OCLC, S3339 **LOCATIONS:** BUC, CCF. ULS, ZDB **SUBJECT:** Anthropology - Periodicals - Germany.

942   Magazin für die Naturkunde Helvetiens. - Bd.1-4, 1787-1789. Zürich: Otell, Gessner, Füssli, 1787-1789. **EDITOR:** Johann Georg Albrecht Höpfner. **CONTINUED BY:** Allgemeines Helvetisches Magazin. **CONTINUED BY:** Helvetische Monatschrift. **MICROFORMS:** RM(3). **SOURCES:** B2787, G542, K3319, OCLC, S2202 **LOCATIONS:** BUC, CCF, CCN, ULS, ZDB **SUBJECT:** Natural history - Periodicals - Switzerland.

943   Magazin für die Naturkunde und Ökonomie Mecklenburgs. - v.1-2, 1791-1792. Schwerin; Leipzig: William Bättensprung, 1791-1792. **EDITOR:** Adolf Christian Siemssen. **SOURCES:** B2788, G543, K2980, S3216 **LOCATIONS:** BUC, ULS   **SUBJECT:** Natural history - Periodicals - Germany. **SUBJECT:** Economics - Periodicals - Germany.

944   Magazin für die neue Historie und Geographie. - v.1-23, 1767-1793. Hamburg: F.C. Ritter, 1767-1793. Imprint varies. **EDITOR:** Anton Friedrich Büsching. **CONTINUED BY:** Neues Magazin für die neuere Geschichte, Erd- und Völkerkunde. **INDEXES:** Register über sammtliche Theile in v.23. **SOURCES:** B6919, K1044, OCLC **LOCATIONS:** BUC, CCF, CCN, ULS, ZDB **SUBJECT:** Geography - Periodicals - Germany.

945   Magazin für die pathologische Anatomie und Physiologie. - Heft. 1, 1796. Altona: Verlagsgesellschaft, 1796. **EDITOR:** August Friedrich Hecker. **SOURCES:** B2790, G131, K3702 **LOCATIONS:** BUC, ULS, ZDB **SUBJECT:** Pathology - Periodicals - Germany. **SUBJECT:** Medicine - Periodicals - Germany.

946   Magazin für die Thierarzneikunde. - v.1-4, 1799-1802. Berlin: Friedrich Maurer, 1799-1802. Quarterly. **EDITOR:** Johann Nicolaus Rohlwes. **INDEXES:** Indexed in Wimmel (Ref.7). **SOURCES:** G233, K3730 **SUBJECT:** Veterinary medicine - Periodicals - Germany.

Magazin für die Thiergeschichte, see no. 926.

947 Magazin für die Wundarzneiwissenschaft. - v.1-2 (St.1-4), 3(St.1-3), 1797-1803. Göttingen: Vandenhoek und Ruprecht, 1797-1803. Volume 3 issued as Allgemeines Magazin für die Wundarzneiwissenschaft. **OTHER TITLE:** Allgemeines Magazin für die Wundarzneiwissenschaft. **EDITOR:** Justus Arnemann. **SOURCES:** G142, K3710 **LOCATIONS:** BUC, CCF, CCN, ULS, ZDB **SUBJECT:** Surgery - Periodicals - Germany.

948 Magazin für Forst und Jagdwesen. - Heft. 1-14, 1797-1805. Leipzig: Magazin für Industrie, 1797-1805. **EDITOR:** Friedrich Gottlieb Leonhardi. **SOURCES:** K3014 **LOCATIONS:** ZDB **SUBJECT:** Forestry management - Periodicals - Germany. **SUBJECT:** Hunting - Periodicals - Germany.

949 Magazin für Freunde der Naturlehre und Naturgeschichte, Scheidekunst, Land-, und Stadtswirthschaft, Volks- und Staatsarznei. - v.1-8, 1794-1797. Griefswald: Lenge, 1794-1797. Imprint varies. **EDITOR:** Christian Ehrenfried Weigel. **SOURCES:** B2766, G676, K3355, S2345 **LOCATIONS:** BUC, CCN, ULS, ZDB **SUBJECT:** Natural history - Periodicals - Germany. **SUBJECT:** Medicine - Periodicals - Germany. **SUBJECT:** Medical jurisprudence - Periodicals - Germany.

950 Magazin für Geburtshelfer. - v.1 (St.1-2) 1794. Frankfurt; Leipzig: Gebhard, 1794. **EDITOR:** Christian Ludwig Schweickhard. **SOURCES:** G119, K3690 **LOCATIONS:** ULS, ZDB **SUBJECT:** Obstetrics - Periodicals - Germany.

951 Magazin für gemeinnützige Arzneikunde und medicinische Polizey. Heft 1-Heft 2, 1799-1801. Zürich: Orell, 1799-1801. **CONTINUATION OF:** Gazette de Santé, oder, gemeinnütziges medizinisches Magazin für Leser aus allen Ständen. **CONTINUATION OF:** Archiv gemeinnützger physischer und medicinischer Kenntnisse. **CONTINUATION OF:** Gemeinnütziges Wochenblatt physischen und medizinischen Kenntnisse. **OTHER TITLE:** Magazin für medizinische Polizey und gemeinnützige Arzneikunde. **EDITOR:** Johann Heinrich Rahn. **SOURCES:** G161, K3729 **LOCATIONS:** ULS **SUBJECT:** Medicine - Periodicals - Switzerland.

952 Magazin für Ingenieure und Artilleristen. - v.1-12, 1777-1795. Giessen: 1777-1795. **EDITOR:** Andreas Böhm. **SOURCES:** K3961 **LOCATIONS:** CCN, ZDB **SUBJECT:** Engineering - Periodicals - Germany. **SUBJECT:** Military art and science - Periodicals - Germany.

953 Magazin für Pharmazie, Botanik und Materia Medica. - v.1-2, 1782-1783. Halle: J.C. Hendel, 1782-1783. **EDITOR:** Johann Hermann Pfingsten. **CONTINUED BY:** Repertorium für Chemie (G214). **SOURCES:** B2771, G211, K3598, S2722 **LOCATIONS:** ZDB **SUBJECT:** Pharmacy - Periodicals - Germany.

954 Magazin fürs volk, medicinischen, ökonomischen und historischen Inhalts. - v.1-2, 1789. Weimar: Hoffmann, 1789. Includes medical articles primarily from Unzer's Der Arzt. **SOURCES:** G293, K5816 **LOCATIONS:** ZDB

SUBJECT: Medicine - Periodicals - Germany - Popular.   **SUBJECT:**
Economics - Periodicals - Germany.

955   Magazin kleiner gemeinnütziger und unterhaltender Reisebeschreibungen. - v.1-2,
1794-1795. Görlitz: 1794-1795.   **SOURCES:** B2744, S2651   **SUBJECT:**
Travel - Periodicals - Germany.

956   Magazin, Physiographiska sällskpaket / Lund. - Lund: 1781.   **SOURCES:** S674c
LOCATIONS: ULS   **SUBJECT:** Science - Societies - Lund.

957   Magazin von merkwürdigen neuen Reisebeschreibungen, aus fremden Sprachen
übersetzt und mit erläuternden Anmerkungen gegleitet. - v.1-38, 1790-1839.
Berlin: Boss, 1790-1839. v.23-39 also as Neues Magazin etc. **OTHER TITLE:**
Neues Magazin von merkwürdigen neuen Reisebeschreibungen.   **EDITOR:**
Johann Reinhold Forster and others.   **SOURCES:** B2792, S2346
LOCATIONS: BUC, CCN, ULS, ZDB   **SUBJECT:** Travels - Periodicals -
Germany - Translations.

958   Magazin vor Aerzte. - Bd.1-4 (St.1-12) 1776-1778.   Cleve: J.C. Bärstecher;
Leipzig: F.G. Jacobäer, 1775-1778. Imprint varies. **OTHER TITLE:** Magazin
für Aerzte.   **EDITOR:** Ernst Gottfried Baldinger.   **CONTINUED BY:** Neues
Magazin für Aerzte.   **SOURCES:** G36, K3574   **LOCATIONS:** BUC, CCF,
ULS, ZDB   **SUBJECT:** Medicine - Periodicals - Germany.

959   Magazin zur nähere Kenntniss des physischen und politischen Zustandes von
Europa und dessen merkwürdigen Kolonien. - v.1-3, 1792-1794.   Berlin:
Akademmische Buchhandlung, 1792-1793.   **EDITOR:** Friedrich Leopold
Brunn.   **SOURCES:** K1722   **LOCATIONS:** BUC   **SUBJECT:** Science -
Periodicals - Germany.   **SUBJECT:** Geography - Periodicals - Germany.

960   Magazin zur Vervollkommung der theoretischen und practischen Heilkunde. -
v.1-10 (3 nos. each) 11(no.1) 1799-1809.   Frankfurt: Andreae, 1799-1809.
v.8-10 also as Magazin für Physiologie und Medizin.   **OTHER TITLE:**
Magazin für Physiologie und Medizin.   **EDITOR:** Andreas Röschlaub.
SOURCES: G167, K3731   **LOCATIONS:** BUC, CCF, ULS, ZDB
SUBJECT: Medicine - Periodicals - Germany.   **SUBJECT:** Physiology -
Periodicals - Germany.

961   Magnetisches Magazin für Niederdeutschland. - v.1-8, 1787-1900.   Bremen:
Johann Heinrich Cramer, 1787-1790. Imprint varies. **MICROFORMS:** RM(3).
SOURCES: B2811, G238, K3641, OCLC, S2431   **LOCATIONS:** BUC, ULS
SUBJECT: Animal magnetism - Periodicals - Germany.

962   Man, a paper for the ennobling of the species. - No.1-53, 1 Jan.-31 Dec. 1755.
London: J. Hberkorn, 1755.   Weekly.   **EDITOR:** Peter Shaw.
MICROFORMS: RP(1).   **SOURCES:** OCLC, (Ref.17no.7)   **LOCATIONS:**
BUC, ULS   **SUBJECT:** Anthropology - Periodicals - England.

963 Mannigfaltigkeiten aus der Natur und dem Menschenleben. - No.1, 1792. Altona: 1792 **SOURCES:** B2830, S2238 **SUBJECT:** Natural history - Periodicals - Germany.

964 Mannigfaltigkeiten, eine gemeinnützige Wochenschrift. - v.1-4, 1770-1773. Berlin: 1770-1773. Weekly. **EDITOR:** Friedrich Heinrich Wilhelm Martini. **CONTINUED BY:** Neue Mannigfaltigkeiten. **CONTINUED BY:** Neueste Mannigfaltigkeiten. **CONTINUED BY:** Allerneueste Mannifgaltigkeiten. **SOURCES:** B2831a, K5425, S2347 **LOCATIONS:** BUC, CCF, ULS, ZDB **SUBJECT:** General - Periodicals - Germany.

965 Manufacturer, or, the British trade truly stated. - No.1-31, Oct.1719-17 Feb.1720. London: 1719-1720. **EDITOR:** Daniel Defoe. **MICROFORMS:** RP(1). (Goldsmiths' Kress Library). **LOCATIONS:** BUC, ZDB **SUBJECT:** Commerce - Periodicals - England.

966 The Massachusetts magazine, or, monthly museum of knowledge and rational entertainment. - v.1-8, 1789-1796. Boston: Isaiah Tomas, 1789-1796. Monthly. Publication suspended Jan.-Mar. 1796. **EDITORS:** Isaiah Thomas, Thaddeus Mason Harris and William Biglow. **MICROFORMS:** UM(1). **SOURCES:** B2848, OCLC, S3998 **LOCATIONS:** ULS, ZDB **SUBJECT:** General - Periodicals - United States.

967 Materialen für die Anthropologie. - St.1-2, 1791-1793. Tübingen: Heilbronn und Rothenburg, 1791-1793. **CONTINUATION OF:** Untersuchungen den thierischen Magnetismus. **CONTINUATION OF:** Neue Untersuchungen über den thierischen Magnetismus. **EDITOR:** Eberhard Gelin. **SOURCES:** G246 **SUBJECT:** Anthropology - Periodicals - Germany.

968 Materialen für die Staatsarzneikunde und Jurisprudenz. - St.1, 1792; St.2, 1795. Königsberg: 1792-1795. **CONTINUATION OF:** Medicinisch-gerichtliche Bibliothek. **CONTINUATION OF:** Bibliothek für Physiker. **EDITOR:** Johann Daniel Metzger. **CONTINUED BY:** Annalen der Staats-arzneikunde. **SOURCES:** G108, OCLC **LOCATIONS:** ZDB **SUBJECT:** Medical jurisprudence - Periodicals - Germany.

969 Materialen für die Staatsarzneiwissenschaft und practische Heilkunde. - v.1-11, 1800-1824. Jena; Meiningen: Johann Christian Gottfried Göpferdt, 1800-1824. Imprint varies. 1819-1824 as Neue Materialen. **OTHER TITLE:** Neue Materialen für die Staatsarzneiwissenschaft und practische Heilkunde. **EDITOR:** Julius Heinrich Gottlob Schlegel. **SOURCES:** G176, K3739 **LOCATIONS:** CCF, CCN, ULS, ZDB **SUBJECT:** Medical jurisprudence - Periodicals - Germany. **SUBJECT:** Medicine - Periodicals - Germany.

970 Materialen für Electiker. - v.1, 1788. Halle: Hemmerde, 1788. **SOURCES:** (Ref.3,no.254). **SUBJECT:** Electricity - Periodicals - Germany. **SUBJECT:** Physics - Periodicals - Germany.

971   Materialen zur theoretischen und practischen Heilkunde. - Pt.1-2, 1799-1800. Breslau: Hirschberg und Lissa, 1799-1800. **EDITOR:** J.G. Knebel. **SOURCES:** G158   **SUBJECT:** Medicine - Periodicals - Germany.

972   Mathematicae commentationes / Academia electoralis moguntinae scientiarum utilium quae Erfurti est. - Erfurt: Keyser, 1778-1780. **LOCATIONS:** ZDB **SUBJECT:** Mathematics - Societies - Erfurt.

973   Mathematical exercises. - Nos. 1-6, 1750-1753. London; Wrexham: James Morgan and R. Marsh, 1750-1753. Reprinted in one vol. (London, J.F.C. Rivington (Ref.16,v.50:267). **EDITOR:** John Turner. **SOURCES:** B8562, OCLC   **LOCATIONS:** ULS   **SUBJECT:** Mathematics - Periodicals - England.

974   Mathematical, geometrical and philosophical delights, containing essays, problems, solutions, theorems &c. - No.1-11, 1792-1798. London: T.N. Longman, 1792-1798. **EDITOR:** Thomas Whiting. **LOCATIONS:** ULS, ZDB **SUBJECT:** Mathematics - Periodicals - England - Popular.

975   Mathematical magazine and philosophical repository, containing a variety of original pieces in all parts of mathematical science. - v.1, Apr.-Aug., 1761. London: J. Walker, 1761. Monthly. **EDITORS:** George Witchell and Thomas Moss.   **SOURCES:** B8563   **LOCATIONS:** BUC, ULS   **SUBJECT:** Mathematics - Periodicals - England - Popular.

976   Mathematical post. - London: Myles Davies, 1703.   **SOURCES:** Offered by Kraus in 1974.   **SUBJECT:** Mathematics - Periodicals - England.

977   Mathematical repository (Dobson). - v.1-3, 1748-1753. London: 1748-1753. Sometimes cited as ser. 1 of Mathematical repository (Leybourn) which was published 1795ff. **EDITOR:** James Dobson. **CONTINUED BY:** Mathematical repository (Leybourn).   **SOURCES:** B2864, S357   **LOCATIONS:** ULS **SUBJECT:** Mathematics - Periodicals - England.

978   Mathematical repository (Leybourn). - Ser.1, v.1-3 (nos.1-14), 1795-1804; ser. 2, v.1-6 (nos 1-24), 1806-1835. London: Glendenning, 1795-1835. Ser. 1, no.2-8, 1795-1799 as Mathematical and philosophical repository, and no. 9-14, 1800-1804 as Mathematical and philosophical repository and review. The three parts had separate title pages: as Mathematical repository, 3 v.1798-1804, Philosophical repository, 2 v.1801-1810, and Review of mathematical and philosophical books, 1 v.1804. There was a second edition of v.1 in 1799. Issued irregularly.   **OTHER TITLE:** Mathematical and philosophical repository. **EDITOR:** Thomas Leybourn.   **SOURCES:** B2864, OCLC **LOCATIONS:** BUC, ULS **SUBJECT:** Mathematics - Periodicals - England.

979   Mathematical transactions and collections, with a methodical history of mathematics. - No.1, 1762. London: 1762. **EDITOR:** Samuel Clark. **LOCATIONS:** ZDB **SUBJECT:** Mathematics - Periodicals - England.

980   The Mathematician, containing many curious dissertations on the rise, progress
       and improvement of geometry. - No.1-6, 1745-1750. London: John Wilcox,
       1745-1750. (Ref.12no.6). Paged consecutively. **EDITORS:** Edward Rollinson,
       Thomas Simpson and John Turner.   **SOURCES:** B2866, OCLC, S358
       LOCATIONS: BUC, ULS, ZDB   **SUBJECT:** Mathematics - Periodicals -
       England.

981   Mathematische Abhandlungen der Akademie nützlicher Wissenschaften zu Erfurt.
       - v.1, 1799. Erfurt: Otto, 1799. **OTHER TITLE:** Commentationes
       mathematicae, Academiae scientiarum uatilium quae Erfurti est; Mathematicae
       commentationes, Academiae scientiarum uatilium quae Erfurti est.
       SOURCES: K3375   **LOCATIONS:** ZDB   **SUBJECT:** Mathematics -
       Societies - Erfurt.

982   Mathematische Liefhebberye met het nieuws der fransche en duytsche schoolen in
       Nederland. - v.1-18, 1754-1769. Purmerende: P. Jordann, 1754-1769.
       OTHER TITLE: Maandelijkse mathematische Liefhebberye. **SOURCES:**
       B2870, S754   **LOCATIONS:** CCN, ULS   **SUBJECT:** Mathematics -
       Periodicals - Netherlands.

983   La Médecine éclairée par les sciences physiques, ou, Journal des découvertes
       relatives aux différentes parties de l'art de guérir. T.1-t.4, 1791-1782. Paris:
       Buisson, 1791-1792. Issued twice a month. **TRANSLATED AS:** Aufklärung
       der Arzneiwissenschaft. **OTHER TITLE:** Journal des découvertes relatives aux
       différentes parties de l'art de guérir; Bibliographie physique et médicinale.
       EDITOR: Antoine Franĉois de Fourcroy.   **SOURCES:** G98, OCLC
       LOCATIONS: BUC,CCF,ULS   **SUBJECT:** Medicine - Periodicals - France.

984   The Medical and chirurgical review, or, compendium of medical literature,
       foreign and domestic v.1-15,16 no.1,(nos.1-94), May 1794-Dec.1807, v.16,
       no.1, 1808. London: T.Boosey, 1794-1808. Subtitle and imprint vary.
       EDITOR: Henry Clutterbuck. **INDEXES:** Index v.1-7 in v.8. **SOURCES:**
       G125, OCLC   **LOCATIONS:** BUC, CCF, ULS   **SUBJECT:** Medicine -
       Periodicals - England - Reviews.

985   Medical and philosophical commentaries by a Society in Edinburgh. - Ser. 1,
       v.1-10, 1773-1785; ser. 2, v.1-10, 1786-1795. Edinburgh: Kincaid; London:
       J. Murray; Dublin: T.Ewing, 1773-1797. Quarterly. **TRANSLATED AS:**
       Medicinische Commentarien von einer Gesellschaft der Aerzte in Edinburgh.
       CONTINUATION OF: Medical essays and observations. **CONTINUATION
       OF:** Essays and observations, physical and literary. From v.7 (1780) appeared
       as Medical commentaries. Several editions appeared of some of the volumes.
       An edition was printed in Philadelphia by Dobson from 1793-1797. **OTHER
       TITLE:** Medical commentaries. **EDITORS:** Andrew Duncan, Sr. and Andrew
       Duncan, Jr. **CONTINUED BY:** Annals of medicine. **CONTINUED BY:**
       Edinburgh medical and surgical journal. **INDEXES:** Index to ser. 1, v.6-10
       (1778-1785) in v.10, and ser. 2, v.1-10 in v.10.   **SOURCES:** G31, OCLC,
       S361   **LOCATIONS:** BUC, CCF, CCN, ULS, ZDB   **SUBJECT:** Medicine
       - Periodicals - Scotland.

986   The Medical and physical journal, containing the earliest information on subjects of medicine, surgery, pharmacy, chemistry and natural history. - v.1-32, 1799-1814. London: R. Phillips, 1799-1814. **TRANSLATED AS:** Physisch-medicinisches Journal. **CONTINUATION OF:** Medical and physical journal. **OTHER TITLE:** London physical and medical review. **EDITORS:** Thomas Bradley and A.F.M. Willich. **CONTINUED BY:** Medical quarterly review. **CONTINUED BY:** British and foreign medical review. **INDEXES:** Index to v.1-40 (1799-1813) published London, 1820. **SOURCES:** B2903, G168, OCLC, S362   **LOCATIONS:** BUC, CCF, CCN, ULS, ZDB **SUBJECT:** Medicine - Periodicals - England.   **SUBJECT:** Science - Periodicals - England.

987   Medical, chirurgical and anatomical cases and experiences, communicated by Dr. Haller and other eminent physicians to the Royal Academy of Sciences at Stockholm. London: A. Linde, 1758. **TRANSLATION OF:** Handlingar, Kungliga Svenska Vetenskapsakademien. **SOURCES:** OCLC **LOCATIONS:** NUC **SUBJECT:** Science - Societies - Stockholm - Translations. **SUBJECT:** Medicine - Societies - Stockholm - Translations.

988   Medical communications of the Society for promoting medical knowledge. - v.1-2, 1784-1790. London: Joseph Johnson, 1784-1790. **TRANSLATED AS:** Medicinische Beiträge, (Society for promoting medical knowledge, London). **SOURCES:** G61, OCLC **LOCATIONS:** BUC, CCF, ULS **SUBJECT:** Medicine - Societies - London.

989   Medical essays and observations, abridged from the Memoirs of the Royal Academy of Sciences of Paris. - v.1-4, 1764. London: J. Knox, 1764. An abridgment of the medical papers in the Histoire de l'Académie des Sciences, Paris, "from their establishment in 1699 to the year 1750 inclusive." **EDITOR:** Thomas Southwell. **TRANSLATION OF:** Histoire de l'Académie royale des sciences, Paris. **INDEXES:** Index to v.1-2, in v.2.   **SOURCES:** OCLC **LOCATIONS:** BUC **SUBJECT:** Medicine - Societies - France - Translations.

990   Medical essays and observations relating to the practice of physic and surgery, abridg'd from the Philosophical transactions from their first publication down to the present time. - v.1-2, 1745. London: 1745. There was also an edition in Lübeck in 1774-1778 (GII:4f). **OTHER TITLE:** Philosophical transactions. **EDITOR:** Samuel Mihles. **SOURCES:** GII:4f, OCLC **LOCATIONS:** BUC, NUC **SUBJECT:** Medicine - Societies - London - Abridgments.

991   Medical essays and observations / Society in Edinburgh. - v.1-6, 1731-1746. Edinburgh: William Monro, 1731-1746. **TRANSLATED AS:** Essais et observations. **TRANSLATED AS:** Medicinische Versuche und Bemerkungen. **TRANSLATED AS:** Neue Versuche und Bemerkungen. **TRANSLATED AS:** Geneeskundige proeven en aanmerkingen. **TRANSLATED AS:** Saggi ed osservazioni di medicina. It appeared in at least 5 editions and was translated into French, Dutch, Italian and German. Published in an abridged edition by W. Lewis (London: 1746). Imprint varies. **EDITOR:** Alexander Monro, primus. **CONTINUED BY:** Essays and observations, physical and literary.

CONTINUED BY: Medical and philosophical commentaries. **CONTINUED BY:** Annals of medicine. **CONTINUED BY:** Edinburgh medical and surgical journal. **SOURCES:** GIII7 **LOCATIONS:** BUC, CCF, ULS, ZDB **SUBJECT:** Medicine - Societies - Edinburgh.

992 Medical extracts on the natural laws of health and the laws of the nervous and fibrous systems by a friend to improvements. - v.1-4, 1794-1797. London: Johnson, 1794-1797. Imprint and subtitle vary. **EDITOR:** Robert John Thornton. **SOURCES:** G294, OCLC **LOCATIONS:** BUC **SUBJECT:** Medicine - Periodicals - England - Popular.

993 Medical facts and observations. - v.1-8, 1791-1800. London: J. Johnson, 1791-1800. **TRANSLATED AS:** Erläuterung der medicinischen und chirurgischen Praxis. **CONTINUATION OF:** London medical journal. **EDITOR:** Samuel Foart Simmons. **SOURCES:** G102 **LOCATIONS:** BUC, CCF, ULS **SUBJECT:** Medicine - Periodicals - England.

994 Medical magazine, or, general repository of practical physic and surgery by a society of gentlemen. - No.1, Jan. 1774. London: 1774. **LOCATIONS:** BUC, ULS **SUBJECT:** Medicine - Periodicals - England.

995 The Medical miscellany, or, a collection of cases, tracts and commentaries exhibiting a view of the present state of medical and chirurgical practice and literature in England. - v.1, 1768. London: W. Nicoll, 1768. **SOURCES:** G370, OCLC, (Ref.7,no.12). **LOCATIONS:** DSG **SUBJECT:** Medicine - Periodicals - England.

996 The Medical museum, or, a repository of cases, experiments, researches, and discoveries collected at home and abroad by gentlemen of the faculty. - v.1-3, 1763-1764. London: W. Bristow, 1763-1764. Monthly. A 2nd edition in 3v. published London, John Churchill, 1781. **SOURCES:** GIII:16, OCLC, (Ref.17,no.9). **LOCATIONS:** BUC, CCN, ULS **SUBJECT:** Medicine - Periodicals - England.

997 Medical observtions and inquiries by a Society of Physicians in London. - v.1-6, 1757-1784. London: William Johnson, 1757-1784. **TRANSLATED AS:** Medicinische Bemerkungen und Untersuchungen einer Gesellschaft von Aerzten in London .There were several editions of some of the volumes. Publisher varies. **SOURCES:** G11, OCLC **LOCATIONS:** BUC, CCF, CCN, ULS, ZDB **SUBJECT:** Medicine - Periodicals - England.

998 Medical papers communicated to the Massachusetts Medical Society. - v.1-24 (5-19 also as s2 v.1-15) 1790-1913. Boston: Thomas and Andrewes, 1790-1913. Imprint varies. **OTHER TITLE:** Medical Communications, Massachusetts Medical Society. **INDEXES:** Index to v.1-18, 1790-1901. **MICROFORMS:** UM(1). **SOURCES:** G95, OCLC **LOCATIONS:** BUC, ULS **SUBJECT:** Medicine - Societies - Boston.

999   Medical records and researches, selected from the papers of a private medical association. - Pt.1-2, 1798. London: T.Cox and G.G. and J. Robinson, 1798. A reprint was issued in 1813. **SOURCES:** G147, OCLC **LOCATIONS:** BUC, ULS, ZDB **SUBJECT:** Medicine - Societies - London.

1000   The Medical register. - London: J. Rose, 1779-1783. Publisher varies. Published in 3 editions. **EDITOR:** Samuel Foart Simmons. **SOURCES:** G:XIX **LOCATIONS:** BUC, CCF, DSG **SUBJECT:** Medicine - Directories - England.

1001   The Medical repository. - v.1-23, 1797-1824. New York: T.and J. Swords, 1797-1824. Quarterly. Subtitle varies. Three editions of the early volumes were published. **OTHER TITLE:** New York medical reporter. **EDITORS:** Samuel Latham Mitchill, Edward Miller and Elihu H. Smith. **SOURCES:** B2905, G144, OCLC **LOCATIONS:** BUC, CCF, ULS, ZDB **SUBJECT:** Medicine - Periodicals - United States.

1002   The Medical spectator. - v.1-3 (nos.1-48) 1791-1796. London: J. Nichols, 1791-1796. Weekly. Two "extraordinary" nos. were published. **SOURCES:** G101, OCLC **LOCATIONS:** BUC, ULS **SUBJECT:** Medicine - Periodicals - England.

1003   Medical transactions published by the Royal College of Physicians of London. - v.1(1768), 2(1777), 3(1785), 4(1813), 5(1815), 6(1820) (Ref.17:13). London: S. Baker and J. Dodsley, 1768-1820. **TRANSLATED AS:** Arzneykundige Abhandlungen herausgegeben von Collegio des Aerzte in London. Publisher varies. There were several editions of some of the volumes. **SOURCES:** G26, OCLC **LOCATIONS:** BUC, CCF, CCN, ULS, ZDB **SUBJECT:** Medicine - Societies - London.

1004   Medicina curiosa, or, a variety of new communications in physick, chirurgery and anatomy. - Nos.1-2, 1684, 1 no. 1695 (Ref.17, no.1). London: Thomas Basset, 1684-1695. **SOURCES:** GI:7, OCLC **LOCATIONS:** BUC **SUBJECT:** Medicine - Periodicals - England.

1005   Medicinalblade, en blandet ugeblad. - Nos. 1-23, 1791-1793. Copenhagen: 1791-1793. **CONTINUATION OF:** Sundstidende. **CONTINUATION OF:** Nye sundstidende. **CONTINUATION OF:** Sundhed og underholdning. **CONTINUATION OF:** Hygaea og muserne. **CONTINUATION OF:** Museum for sunheds og kundskabs elskere. **CONTINUATION OF:** Hertha. **CONTINUATION OF:** Sundhedsbog. **EDITOR:** Johann Clemens Tode. **CONTINUED BY:** Sundhedsjournal. **CONTINUED BY:** Sundhedsraad. **CONTINUED BY:** Sundhedsjournal et maanedskrift. **CONTINUED BY:** Nyeste sundhedsblade. **SOURCES:** G191 **SUBJECT:** Medicine - Periodicals - Denmark.

1006   Medicinisch-chirurgische Aufsätze, Krankengeschichten und Nachrichten. - v.1-3, 1791-1795. Altenburg: Richter, 1791-1795. **CONTINUATION OF:** Taschenbuch für die deutschen Wundärzte. **EDITOR:** Friedrich August Waitz

(Weiz)   SOURCES: G100   LOCATIONS: ULS   SUBJECT: Surgery - Periodicals - Germany - Reviews.   SUBJECT: Medicine - Periodicals - Germany - Reviews.

1007   Medicinisch-chirurgische Bibliothek. Bd.1-Bd.10, 1775-1787.   Copenhagen: Johann Gottlob Rothe, 1775-1787.   EDITOR: Johann Clemens Tode. CONTINUED BY: Arzneikundige Annalen.   CONTINUED BY: Medicinisch-chirurgisches Journal.   SOURCES: G34, K3573 LOCATIONS: BUC, CCF, ULS, ZDB   SUBJECT: Surgery - Periodicals - Denmark. SUBJECT: Medicine - Periodicals - Denmark.

1008   Medicinisch-chirurgische Zeitung. - In 53 Jahrgange each with 4 vol., 1790-1742. Salzburg: Mayr; Leipzig: Köhler, 1790-1842. Issued twice a week. Imprint varies. OTHER TITLE: Salzburgsche medicinische Zeitung. EDITORS: Johann Jakob Hartenkeil, Franz Xavier Mezler and J.N. Erhardt von Ehrenstein. CONTINUED BY: Neue medicinisch-chirurgische Zeitung. CONTINUED BY: Medicinisch-chirurgische Monatshefte.   Includes 39 supplementary volumes (Erganzungsband) issued irregularly. INDEXES: Three Universalrepertorium were issued in 1795(1790- 1794), 1801(1795-1800) and 1823(1801-1820).   SOURCES: G94, K3669, OCLC   LOCATIONS: BUC, CCF, CCN, ULS, ZDB   SUBJECT: Medicine - Periodicals - Germany - Reviews.   SUBJECT: Surgery - Periodicals - Germany - Reviews.

1009   Medicinisch-chirurgisches Journal. Bd.1-Bd.5, 1793-1804.   Copenhagen und Leipzig: Schubothe, 1793-1804.   CONTINUATION OF: Medicinisch-chirurgische Bibliothek. CONTINUATION OF: Arzneikundige Annalen.   Bd.1-Bd.3 was under the title: Medicinisches Journal.   OTHER TITLE: Medicinisches Journal. EDITOR: Johann Clemens Tode. SOURCES: G116, K3684 LOCATIONS: ULS, ZDB SUBJECT: Medicine - Periodicals - Denmark.   SUBJECT: Surgery - Periodicals - Denmark.

1010   Medicinisch-gerichtliche Bibliothek. - Bd.1-5 (each with 4 issues) 1784-1787. Königsberg: G.L. Hartung, 1784-1787.   EDITORS: Christoph Friedrich Elsner and Johann Daniel Metzger.   CONTINUED BY: Bibliothek für Physiker. CONTINUED BY: Annalen der Staats-arzneikunde.   CONTINUED BY: Materialen für die Staatsarzneikunde und Jurisprudenz.   SOURCES: G60, K3607   LOCATIONS: BUC, CCF, ULS, ZDB   SUBJECT: Medical jurisprudence - Periodicals - Germany.

1011   Medicinisch-physische Sammlungen. - v.1, 1782.   Vienna: Gerold, 1782. EDITOR: Friedrich August Wasserberg.   SOURCES: B2908, G374, K3290, S2908   SUBJECT: Medicine - Periodicals - Germany.   SUBJECT: Science - Periodicals - Germany.

1012   Medicinisch-praktische Bibliothek. - v.1 (St.1-3) 1785-1786.   Göttingen: Johann Christian Dieterich, 1785-1786.   EDITOR: Christian Friedrich Michaelis. SOURCES: G296, K3617, OCLC   LOCATIONS: ULS   SUBJECT: Medicine - Periodicals - Germany.

1013 Medicinisch-praktische Bibliothek. - v.1-3 (4 St. each) 1774-1780. Göttingen: Johann Christian Dieterich, 1774-1780. **CONTINUATION OF:** Medicinische Bibliothek (Vogel). **CONTINUATION OF:** Neue medicinische Bibliothek. **EDITOR:** Johann Andreas Murray. **CONTINUED BY:** Medicinische Bibliothek (Blumenbach). **SOURCES:** G295, K3565 **LOCATIONS:** BUC, CCF, ULS **SUBJECT:** Medicine - Periodicals - Germany - Reviews.

1014 Medicinisch-praktische Bibliothek für Ärzte und Wundärzte. - v.1-3 (3 St. each) 1789-1791. Münster; Hamm: Philipp Heinrich Perrenon, 1789- 1791. **EDITORS:** Carl Georg Theodor Kortun and Johann Christian Schäffer. **SOURCES:** G297, K3653 **LOCATIONS:** CCF, ZDB **SUBJECT:** Medicine - Periodicals - Germany - Reviews. **SUBJECT:** Surgery - Periodicals - Germany - Reviews.

1015 Medicinisch-praktische Bibliothek in Verbindung mit mehreren Mitgliedern der Helvetischen Gesellschaft correspondierender Ärzte und Wundärzte. - v.1 (St.1-2) 1795-1796. Zürich: Orell, 1795-1796. **EDITOR:** Johann Heinrich Rahn. **SOURCES:** G127, K3694 **LOCATIONS:** CCF, ULS **SUBJECT:** Medicine - Societies - Zürich.

1016 Medicinisch-praktisches Archiv.- v.1-2, 1799-1801/09. Göttingen: 1799-1809. **LOCATIONS:** ZDB **SUBJECT:** Medicine - Periodicals - Germany.

1017 Medicinische Annalen für Aerzte und Gesundheitsliebende. - v.1 (1779-1780) 1781. Leipzig: Weygand, 1781. Monthly. **EDITOR:** Johann Gottlieb Fritze. **SOURCES:** G43, K3588, OCLC **LOCATIONS:** BUC, CCF, ULS **SUBJECT:** Medicine - Periodicals - Germany.

1018 Medicinische Aufsätze für Aerzte. - Sammlung 1-2, 1791-1793. Wittenberg; Zerbst: 1791-1793. **EDITOR:** Johann Andreas Garn. **SOURCES:** G371 **SUBJECT:** Medicine - Periodicals - Germany.

1019 Medicinische Beiträge / Society for promoting medical knowledge, London. - v.1, 1785. Göttingen: Dieterich, 1785. Translated by Christian Friedrich Michaelis. **TRANSLATION OF:** Medical communications of the Society for promoting medical knowledge. **SOURCES:** G61a, K3616 **LOCATIONS:** BUC **SUBJECT:** Medicine - Societies - London - Translations.

1020 Medicinische Bemerkungen und Untersuchungen einer Gesellschaft von Aerzten in London. - v.1-7, 1759-1787. Altenburg: Richter, 1759-1787. Translated by Samuel Gottlieb Silchmüller and Johann Christian Friedrich Scherf. **TRANSLATION OF:** Medical observations and inquiries. **SOURCES:** GIII:11a, K3541, OCLC **LOCATIONS:** CCF, ULS, ZDB **SUBJECT:** Medicine - Societies - London - Translations.

1021 Medicinische Beobachtungen. - v.1-3 (Heft 1-7) 1784-1787. Quedlinburg; Blankenburg: Christoph August Reussner, 1784-1787. Imprint varies. **EDITOR:** Friedrich Christian Carl Krebs. **SOURCES:** G372, K3606 **SUBJECT:** Medicine - Periodicals - Germany.

1022 Medicinische Beobachtungen, eine Auswahl. - v.1, 1799. Göttingen: Johann Christian Dieterich, 1799. **EDITOR:** L.C.W. Cappel. **TRANSLATION OF:** Nova acta physico-medica academiae Caesareae Leopoldinae- Carolinae naturae curiosorum. **SOURCES:** GII:7g **SUBJECT:** Medicine - Societies - Halle - Translations.

1023 Medicinische Bibliothek (Blumenbach). - Bd.1-3 (each with 4 issues). Göttingen: Johann Christian Dieterich, 1783-1795. **CONTINUATION OF:** Medicinisch-practische Bibliothek (Murray). **CONTINUATION OF:** Medicinische Bibliothek (Vogel). **CONTINUATION OF:** Neue medicinische Bibliothek. **EDITOR:** Johann Friedrich Blumenbach. **SOURCES:** G229, K3602, OCLC **LOCATIONS:** BUC, CCF, CCN, ULS, ZDB **SUBJECT:** Medicine - Periodicals - Germany - Reviews.

1024 Medicinische bibliothek (Vogel). - v.1-2 (St.1-19) 1751-1753. Erfurt; Leipzig: Schröder, 1751-1753. **EDITOR:** Rudolf Austin Vogel. **CONTINUED BY:** Neue medicinische Bibliothek. **CONTINUED BY:** Mediccinisch-praktische Bibliothek (Murray). **CONTINUED BY:** Medicinische Bibliothek (Blumenbach). **SOURCES:** G298, K3533, OCLC **LOCATIONS:** BUC, CCF **SUBJECT:** Medicine - Periodicals - Germany - Reviews.

1025 Medicinische Chronik. - v.1-3 (3 Heft each) 1792-1795. Vienna: Meyer und Patzowsky, 1793-1795. Monthly. **EDITOR:** Joseph Eyerel. **SOURCES:** G113, K3682 **LOCATIONS:** BUC, ULS, ZDB **SUBJECT:** Medicine - Periodicals - Germany.

1026 Medicinische Commentarien von einer Gesellschaft der Aerzte in Edinburgh. - Decade 1-3 (in 21 v.) 1774-1799. Altenburg: Richter, 1774-1799. Translated by Georg Heinrich Königsdörfer and from v.8 by Friedrich Diel. **OTHER TITLE:** Medicinische Annalen Englischen Aerzte. **TRANSLATION OF:** Medical and philosophical commentaries. **SOURCES:** G31a, K3566 **LOCATIONS:** ULS, ZDB **SUBJECT:** Medicine - Societies - Edinburgh - Translations.

1027 Medicinische Ephemeriden. - v.1, 1794. Vienna: Alb. Ant.Patzowsky, 1794. A year by year account of the prevailing weather and accompanying morbidity published 1794. Not a periodical. Translated and augmented by Joseph Eyerel. **EDITOR:** Samuel Benkö. **TRANSLATION OF:** Ephemerides meteorologico-medicae annorum 1780-93. **SOURCES:** B853, G280a, S3572 **SUBJECT:** Meteorology - Periodicals - Germany. **SUBJECT:** Medicine - Periodicals - Germany.

1028 Medicinische Ephemeriden von Berlin. - v.1 (St.1-4) 1799-1800. Berlin: Nauck, 1799-1800. **EDITOR:** Johann Ludwig Formey. **SOURCES:** G152, K3727, OCLC **LOCATIONS:** BUC, ULS **SUBJECT:** Medicine - Periodicals - Germany. **SUBJECT:** Meteorology - Periodicals - Germany.

1029 Medicinische Litteratur des Jahres. - v.1-2, 1796-1797. Leipzig: Peter Philipp Wolff, 1796-1797. **CONTINUATION OF:** Repertorium der medicinischen

Litteratur. **EDITOR:** Paul Usteri. **SOURCES:** G300 **LOCATIONS:** CCF, DSG **SUBJECT:** Medicine - Periodicals - Germany - Reviews.

1030 Medicinische Litteratur für practische Aerzte. - v.1-12, 1781-1787. Leipzig: Carl Friedrich Schneider, 1681-1787. **EDITOR:** Johann Christian Traugott Schlegel. **CONTINUED BY:** Neue medicinische Litteratur. **CONTINUED BY:** Übersicht der neuesten medicinschen Litteratur. **SOURCES:** G301, K3590 **LOCATIONS:** BUC, CCF, ULS, ZDB **SUBJECT:** Medicine - Periodicals - Germany - Reviews.

1031 Medicinische nationalzeitung für Deutschland, und die mit selbigem zunächst verbundenen Staaten. - v.1-2, 1798-1799. Altenburg: 1798-1799. Weekly. **CONTINUED BY:** Allgemeine medizinischen Annalen. Supplement published under the title: Intelligenzblatt. **SOURCES:** GIII:150 **LOCATIONS:** BUC, CCF, ULS, ZDB **SUBJECT:** Medicine - Periodicals - Germany.

1032 Der Medicinische Richter, oder, Acta physico-medico- forensia. - v.1-4, 1755-1759. Ansbach: 1755-1759. **EDITOR:** Johann Georg Hasenest. **SOURCES:** (Ref.96,no.182) **SUBJECT:** Medical jurisprudence - Periodicals - Germany.

1033 Medicinische und chirurgische Berlinische wöchtentliche Nachrichten. - v.1-6, 1738-1748. Berlin: Rüdiger, 1739-1748. **OTHER TITLE:** Medicinische und chirurgische Nachrichten. **EDITOR:** Samuel Schaarschmidt. **SOURCES:** K3525 **LOCATIONS:** ULS **SUBJECT:** Medicine - Periodicals - Germany. **SUBJECT:** Surgery - Periodicals - Germany.

1034 Medicinische und physische Nachrichten. - v.1, 1762. Dresden: 1762. **SOURCES:** B2902, S2511 **LOCATIONS:** ULS **SUBJECT:** Medicine - Periodicals - Germany. **SUBJECT:** Science - Periodicals - Germany.

Medicinische Unterhaltungen, see no. 1049.

1035 Medicinische Unterweisungen, eine periodische Schrift. - Dessau: 1772. **EDITOR:** Friedrich Samuel Kretzchmar. **SOURCES:** K3561 **SUBJECT:** Medicine - Periodicals - Germany.

1036 Medicinische Verhandlungen des Kollegiums der Aerzte zu Philadelphia, aus dem Englischen übersetzt von Christian Friedrich Michaelis. - v.1, pt.1, 1795. Leignitz; Lleipzig: David Siegert, 1795. **TRANSLATION OF:** Transactions of the College of Physicians of Philadelphia. **SOURCES:** G111a, OCLC **SUBJECT:** Medicine - Societies - Philadelphia - Translations.

1037 Die Medicinischen Versuche und Bemerckungen / Gesellschaft in Edinburg. - v.1-7, 1749-1752. Altenburg: Richter, 1749-1752. Vol.6 and 7 also under the title: Zusätze zu den medicinischen Versuche. There was a second edition of v.1 (Altenburg: Richter, 1766). **TRANSLATION OF:** Medical essays and observations, Medical Society in Edinburgh. **CONTINUED BY:** Neue Versuche und Bemerkungen aus der Arztneykunst. **SOURCES:** GIII7b, K3531

LOCATIONS: ULS, ZDB SUBJECT: Medicine - Societies - Edinburgh - Translations.

1038 Medicinischer Almanach für das Landvolk. - v.1, 1782. Nurnberg: Stiebner, 1782. SOURCES: (Ref 21,v.22:113). SUBJECT: Medicine - Periodicals - Germany - Popular.

1039 Medicinischer Briefwechsel von einer Gesellschaft Aerzte. - St.1-2, 1785-1786. Halle: Gebauer, 1785-1786. August Gottlob Weber. SOURCES: G64, K3618 LOCATIONS: CCF, ULS SUBJECT: Medicine - Periodicals - Germany.

1040 Medicinischer Rathgeber für Aerzte und Wundärzte. - v.1-4, 1794-1797. Frankfurt: Jäger, 1794-1797. CONTINUATION OF: Medicinisches Wochenblatt für Aerzte, Wundärzte und Apotheker. CONTINUATION OF: Neues medicinisches Wochenblatt für Aerzte, Wundärzte und Apotheker. CONTINUATION OF: Frankfurter medicinische Annalen für Aerzte, Wundärzte, Apotheker und denkende Leser aus allen Ständen. EDITORS: Johann Valentin Müller and Georg Friedrich Hoffmann. CONTINUED BY: Medicinisches Repertorium für Gegenstände aus allen Fächern der Arzneiwissenschaft. SOURCES: G124, K3692 LOCATIONS: ULS SUBJECT: Medicine - Periodicals - Germany. SUBJECT: Surgery - Periodicals - Germany. SUBJECT: Pharmacy - Periodicals - Germany.

1041 Medicinisches Archiv von Wien und Oesterreich under der Enns. - v.1-5 (1798-1802) 1799-1803. Vienna: Schaumburg, 1799-1803. Subtitle and publisher vary. EDITOR: Paschal Joseph Ferro. SOURCES: G163, K3715 LOCATIONS: CCF, ULS, ZDB SUBJECT: Medicine - Periodicals - Austria.

1042 Medicinisches Journal. - v.1-9 (St.1-36) 1784-1796. Göttingen: J.C. Dieterich, 1784-1796. OTHER TITLE: Medicinisches und physisches Journal (v.6-9); Physisch-medicinisches Journal. EDITOR: Ernst Gottfried Baldinger. CONTINUED BY: Neues medicinisches und physisches Journal. SOURCES: B2907, G62, K3608, S3082 LOCATIONS: CCF, ULS,ZDB SUBJECT: Medicine - Periodicals - Germany.

1043 Medicinisches Journal über allerhand in die Arzneiwissenschaft und deren Ausübung einschlagende Materien. - v.1-5, 1767-1770. Chemnitz: Stössel, 1767-1770. EDITOR: Gottwald Schuster. CONTINUED BY: Vermischte Schriften, als eine Fortsetzung des medicinischen Journals. SOURCES: GIII:25, K3548 LOCATIONS: ULS SUBJECT: Medicine - Periodicals - Germany.

1044 Medicinisches Magazin der Holländischen Litteratur. - St.1, 1790. Leiden: Hornkoop; Marburg: Akademische Buchhandlung, 1790. OTHER TITLE: Magazin der medicinischen Litteratur. EDITORS: Franz Xaver Jansen and Johann Christian Jonas. SOURCES: G302, K3665 SUBJECT: Medicine - Periodicals - Netherlands - Translations.

1045 Medicinisches Repertorium für Gegenstände aus allen Fächern der Arzneiwissenschaft. - v.1, 1798-1799. Frankfurt: 1798-1799. **CONTINUATION OF:** Medicinisches Wochenblatt für Aerzte, Wundärzte und Apotheker. **CONTINUATION OF:** Neues medicinisches Wochenblatt für Aerzte, Wundärzte und Apotheker. **CONTINUATION OF:** Frankfurter medicinische Annalen für Aerzte, Wundärzte, Apotheker und denkende Leser aus allen Ständen. **CONTINUATION OF:** Medicinischer Rathgeber für Aerzte und Wundärzte. **EDITORS:** Johann Valentin Müller and Georg Friedrich Hoffmann. **SOURCES:** G151 **SUBJECT:** Medicine - Periodicals - Germany. **SUBJECT:** Surgery - Periodicals - Germany. **SUBJECT:** Pharmacy - Periodicals - Germany.

1046 Medicinisches Wochenblatt. - Vienna: 1781. Weekly. **SOURCES:** K3594 **SUBJECT:** Medicine - Periodicals - Austria.

1047 Medicinisches Wochenblatt für Aerzte, Wundärzte und Apotheker. - v.1-9, 1780-1788. Frankfurt: Jäger, 1780-1788. Weekly. **EDITORS:** Joseph Jacob Reichard (v.1-3) and Johann Valentin Müller (v.4-8). **CONTINUED BY:** Neues medicinisches Wochenblatt für Aerzte, Wundärzte und Apotheker. **CONTINUED BY:** Frankfurter medicinische Annalen für Aerzte, Wundärzte, Apotheker. **CONTINUED BY:** Medicinisches Wochenblatt oder forgesetzte medicinische Annalen. **CONTINUED BY:** Medicinischer Rathgeber für Aerzte und Wundärzte. **CONTINUED BY:** Medicinisches Repertorium für Gegenstände aus allen Fächern der Arzneiwissenschaft. **SOURCES:** G42, K3585 **LOCATIONS:** BUC, ULS, ZDB **SUBJECT:** Medicine - Periodicals - Germany. **SUBJECT:** Surgery - Periodicals - Germany. **SUBJECT:** Pharmacy - Periodicals - Germany.

1048 Medicinisches Wochenblatt, oder, forgesetzte medicinische Annalen. - v.1-3, 1790-1793. Frankfurt: Jäger, 1790-1793. Weekly. **CONTINUATION OF:** Medicinisches Wochenblatt für Aerzte, Wundärzte und Apotheker. **CONTINUATION OF:** Neues medicinisches Wochenblatt für Aerzte, Wundärzte und Apotheker. **CONTINUATION OF:** Frankfurter medicinische Annalen für Aerzte, Wundärzte, Apotheker und denkende Leser aus allen Ständen. **EDITORS:** Johann Valentin Müller and Georg Friedrich Hoffmann. **CONTINUED BY:** Medicinischer Rathgeber für Aerzte und Wundärzte. **CONTINUED BY:** Medicinisches Repertorium für Gegenstände aus allen Fächern der Arzneiwissenschaft. **SOURCES:** G93, K3668 **LOCATIONS:** ULS **SUBJECT:** Medicine - Periodicals - Germany. **SUBJECT:** Surgery - Periodicals - Germany. **SUBJECT:** Pharmacy - Periodicals - Germany.

1049 Medicinische Unterhaltungen, eine Wochenschrift für Gesunde und Kranke. - St.1-16, 1781-1781. Berlin: Johann Friedrich Bergmann, 1781-1781. Weekly. Reprinted in Dessau in 1783 as: Gemeinnützige Aufsäte etc. **OTHER TITLE:** Gemeinnützige Aufsätze für Gesunde und Kranke. **EDITOR:** Conrad Friedrich Uden. **SOURCES:** G54, K3592 **SUBJECT:** Medicine - Periodicals - Germany - Popular.

1050   Medicorum Silesicorum satyrae, quae varias observationes, casus, experimenta, tenamina ex omni medicinae ambitu petita exhibent. - v.1-8, 1736-1742. Breslau: Korn, 1736-1742.   **EDITOR:** Gottfried Heinrich Burghart. **SOURCES:** G503, K3523, OCLC   **LOCATIONS:** DSG, ZDB   **SUBJECT:** Medicine - Periodicals - Germany.

1051   Os Medicos perieitos, ou, novo methodo de curar todas as efermidades. - Lisbon: 1759-1760.   **OTHER TITLE:** Semana proveitosas ao vivente racional. **SOURCES:** (Ref.34,no.15).   **SUBJECT:** Medicine - Periodicals - Portugal.

1052   Meine Bemerkungen über den Entwurf zur Patriotischen Gesellschaft für Schlesien, ein Wochenblatt. - v.1-7, 1771-1779.   Breslau: Korn, 1771-1779. Weekly.   **EDITOR:** Carl Gottfried Tschirner.   **SOURCES:** K1056 **SUBJECT:** Economics - Societies - Breslau.

1053   Mélanges de philosophie et de mathématiques de la Société de Turin. - v 1-5 (1759-1773)   1759-1776.   Turin: Imprimerie   royale,   1759-1776. **TRANSLATED AS:** Mémoires concernant l'histoire naturelle, la physique, la chymie, la botanique.   Abstracted in Collection académique, v.13.   **OTHER TITLE:** Miscellanea taurinensia; Miscellanea philosophico-mathematica. **CONTINUED BY:** Mémoires de l'Académie des sciences, littérature et beaux-arts, Turin.   **INDEXES:** Table général des publications de la Société, 1759-1813/14, in v.13 of its Mémoires.   **SOURCES:** G497a, OCLC, S2075 **LOCATIONS:** BUC, CCF, CCN, ULS, ZDB   **SUBJECT:** Science - Societies - Turin.   **SUBJECT:** Mathematics - Societies - Turin.

1054   Mélanges physico-mathématiques, ou, Recueil de mémoires, contenant la description de plusiers machines et instrumens nouveaux, de physique, d'économie domestique etc. - v.1, 1800? Paris: A. Le Clerc, 1800.   **EDITOR:** Joseph Balthasar Bérard.   **SOURCES:** OCLC, S1442   **LOCATIONS:** ULS **SUBJECT:** Science - Periodicals - France.

1055   Mémoires / Académie Delphinale. - v.1-3, 1787-1789. Grenoble: J. Allier, 1787-1789.   **SOURCES:** S1152   **LOCATIONS:** BUC, CCF, ULS **SUBJECT:** General - Societies - Grenoble.

1056   Mémoires / Académie royale de marine. - v.1, 1773. Brest: R. Malassin, 1773. **SOURCES:** S1089   **LOCATIONS:** CCF, CCN   **SUBJECT:** Naval art and science - Societies - Brest.

1057   Mémoires analytique des travaux de l'Académie royale des sciences, arts et belles lettres de Caen. - T.1-5, 1754-1760. Caen: A. Hardel, 1754-1760.   Publisher varies.   **OTHER TITLE:** Mémoires de l'Académie de belles-lettres de Caen. **INDEXES:** Gaste, Armand. Table chronologique, methodique et alphabetique des travaux inserèes dans les Mémoires depuis 1754 jusqu'en 1883.   Caen, 1884.   **SOURCES:** G445, OCLC, S1095c   **LOCATIONS:** BUC, CCF, ULS **SUBJECT:** General - Societies - Caen.

1058 Mémoires / Cercle de Philadelphes. - v.1, 1788. Port au Prince: Mozard, 1788.
      SOURCES: S3920 **LOCATIONS:** CCF, ULS **SUBJECT:** Natural history
      - Societies - Port au Prince.

1059 Mémoires choises parmi ceux lus ou adressés à la Société libre d'agriculture, arts
      et commerce du départment des Ardennes. - v.1-3, 1799-1803. Mezieres:
      1799-1803. **LOCATIONS:** BUC, CCF, CCN, ULS **SUBJECT:** Agriculture
      - Societies - Ardennes.

1060 Mémoires concernant l'histoire naturelle, la physique, la chmie, la botanique etc.
      - v.1, 1779. Paris: 1779. Translated by Paul and others. **TRANSLATION
      OF:** Miscellanea philosophico mathematica. **SOURCES:** S1601 **SUBJECT:**
      Science - Societies - Turin - Translations.

1061 Mémoires concernat l'histoire, les science, les arts, les moeurs des chinois, par
      les missionaires de Pekin. - v.1-17, 1776-1815. Paris: Nyon, 1776-1815.
      OTHER TITLE: Mémoires concernant les chinois; Mémoires sur les chinois.
      EDITORS: Joseph Marie Amiot and others. **INDEXES:** Table génerale des
      matières. (v.1-10) in 10. **MICROFORMS:** IDC(8). **SOURCES:** B2919,
      G685, OCLC **LOCATIONS:** BUC, CCF, ULS, ZDB **SUBJECT:** China -
      Periodicals - France.

1062 Mémoires couronnés sur l'utilité des lichens dans la médecine et dans les arts /
      Académie des sciences, belles lettres et arts de Lyon. - Part 1-3, 1787. Lyon:
      Piestre et Delambellières, 1787. Contains prize essays of Georg Franz
      Hoffmann, Pierre Joseph Amoreux and Remi Willemet. An edition in
      Strassburg by König in 1788 is cited (Ref 21, 25:154). **SOURCES:** G451,
      OCLC, S1176 **LOCATIONS:** BUC **SUBJECT:** Botany - Societies - Lyons.

1063 Mémoires d'agriculture, d'économie rurale et domestique / Société royale
      d'agriculture de la géneralité Paris. - v.1-114, 1761-1916. Paris: Chucet,
      1761-1916. Imprint varies. **SOURCES:** OCLC **LOCATIONS:** BUC, CCF,
      ULS **SUBJECT:** Agriculture - Societies - Paris.

1064 Mémoires de la Société d'agriculture du département des Deux Sèvres. - v.1-3,
      1800-1807. Niort: P. Plisson, 1800-1807. **SOURCES:** OCLC **SUBJECT:**
      Agriculture - Societies - Deux Sèvres.

1065 Mémoires de la Société d'agriculture du département du Seine-et-Oise. - v.1-64,
      1797-1864. Versailles: 1797-1864. **SOURCES:** S1702 **LOCATIONS:**
      BUC, CCF, ULS **SUBJECT:** Agriculture - Societies - Seine-et-Oise.

1066 Mémoires de la Société d'histoire naturelle de Paris. - T.1, 1799. Paris: 1799.
      CONTINUATION OF: Actes de la société d'histoire naturelle de Paris.
      SOURCES: S1583b **LOCATIONS:** BUC, CCF, ULS, ZDB **SUBJECT:**
      Natural history - Societies - Paris.

1067 Mémoires de la Société du commerce et de l'industrie de la Seine-Infèriere. -
      v.1-3, 1798-1800. Rouen: 1798-1800. Continued to 1876 under various titles.

CONTINUED BY: Opuscules et rapports de la Société libre d'émulation de la Seine-infèrieure. **SOURCES:** S1646 **LOCATIONS:** CCF **SUBJECT:** Economics - Societies - Seine-infèriere.

1068 Mémoires de la Société établi à Geneve pour l'encouragement des arts et de l'agriculture. - v.1-2, 1778-1780. Geneva: Bonnant, 1778-1780. **SOURCES:** G509 **LOCATIONS:** BUC, CCF, ULS **SUBJECT:** Agriculture - Societies - Geneva. **SUBJECT:** Economics - Societies - Geneva.

1069 Mémoires de la société médical d'émulation séante à l'École de médecine de Paris. - T.1-t.9. 1797-1826. Paris: Maredan et Croulebois, 1797-1826. **TRANSLATED AS:** Auserlesene Beobachtungen der wetteifernden medicinischen Gesellschaft in Paris. T.9 also as Nouveaux mémoires. Publisher varies. **OTHER TITLE:** Nouveau mémoires de la société médicale d'émulation. **SOURCES:** G145 **LOCATIONS:** BUC, CCF, CCN, ULS, ZDB **SUBJECT:** Medicine - Societies - Paris.

1070 Mémoires de la Société royale d'émulation d'Abbeville. - Rouen: 1797-1810. Title and imprint vary. **SOURCES:** G513, S1002 **LOCATIONS:** BUC, CCF, ULS **SUBJECT:** Agriculture - Societies - Abbeville. **SUBJECT:** Economics - Societies - Abbeville.

1071 Mémoires de la Société royale des sciences et belles lettres de Nancy. - v.1-4, 1754-1759. Nancy: Pierre Antoine, 1754-1759. **INDEXES:** Table générale des publications de la Société, 1750-1900. **SOURCES:** G455, S1248a **LOCATIONS:** BUC, CCF, ULS **SUBJECT:** General - Societies - Nancy.

1072 Mémoires de l'Académie de Dijon. - v.1 (1768) 1769-v.2 (1772) 1774. Dijon: Causse, 1769-1774. Publisher varies. **CONTINUED BY:** Nouveaux mémoires de l'Académie de Dijon. **INDEXES:** Table générale et particuliere des travaux contenus dans les mémoires (1769-1913). Dijon, 1915. **MICROFORMS:** RM(3). **SOURCES:** G448, OCLC, S1137b **LOCATIONS:** BUC, CCF, CCN, ULS, ZDB **SUBJECT:** General - Societies - Dijon.

1073 Mémoires de l'Académie des sciences, inscriptions, belles lettres, beaux arts, etc. / Troyes. - v.1, 1756. Troyes (Paris): Libraire de l'Académie, 1756. A satire of a fictitious academy by Pierre Jean Grosley and others. **SOURCES:** GXX, OCLC **LOCATIONS:** BUC, ULS, ZDB **SUBJECT:** General - Societies - Troyes en Champagne.

1074 Mémoires de l'Académie des sciences, littérature et beaux-arts de Turin. - v.6-22 (1784-1814) 1786-1816. Turin: 1786-1816. **CONTINUATION OF:** Mélanges de philosophie et de mathématiques de la Société de Turin. **CONTINUED BY:** Memorie della Reale Accademia dell scienze, Turin. **INDEXES:** Table général des publications de la Société, 1759-1813/14, in v.13 of its Mémoires. **SOURCES:** G497c, S2051b **LOCATIONS:** BUC, CCF, CCN, ULS, ZDB **SUBJECT:** Science - Societies - Turin.

1075 Mémoires de l'Académie royale de chirurgie. - v.1-5, 1743-1774. Paris: Osmont, 1743-1774. **TRANSLATED AS:** Abhandlungen der königlichen Parisischen Academie der Chirurgie. **TRANSLATED AS:** Sammlung der Preisschriften, Académie royale de chirurgie, Paris. **TRANSLATED AS:** Memoirs of the Royal Academy of Surgery at Paris. Publisher varies. Also published in a duodecimo edition in Paris in 15 v, 1743-1774. Several other editions also appeared. **SOURCES:** G9, OCLC **LOCATIONS:** BUC, CCF, CCN, ULS, ZDB **SUBJECT:** Surgery - Societies - Paris.

1076 Mémoires de l'Académie royale de Prusse concernant l'anatomie, la physiologie, la physique, l'histoire naturelle, la botanique, la minerologie. - 9 v.in 10, (1745-1760) 1760-1769. Avignon; Paris: 1760-1769. First published in quarto and octavo, then in another edition in 2 vol. in quarto, and in another in 7 v. in octavo, published by Paul. An edition in 7 vol. in duodecimo covering the years 1760 to 1769 was published in 1770. A 3 vol. edition was published in 1774 with the title: Resumé retropective des Histoire de l'Académie royale et belles-lettres de Berlin. **INDEXES:**(Ref.126).   **SOURCES:** S1275a **LOCATIONS:** BUC, CCF, ULS, ZDB   **SUBJECT:** Science - Societies - Berlin.

1077 Mémoires de l'Académie royale des sciences de Stockholm. v.1 (1739-1767) 1772. Paris: 1772. **TRANSLATION OF:** Handlingar, Kungliga Svenska Vetenskapsakademien. **SOURCES:** G477d **LOCATIONS:** BUC, CCF, NUC **SUBJECT:** Science - Societies - Stockholm - Translations.

1078 Mémoires de l'Académie royale des sciences, des lettres et des beaux arts de Bruxelles. - T.1-5 (1772-1788) 1777-1788.   Brussels: 1777-1788. **TRANSLATED AS:** Beiträge zur burgerlichen Geschichte, zur Geschichte der Cultur, zur Naturgeschichte, Naturlehre und des Feldbau. A second ed. of v.1-4 was published in Brussels in 1780-1783. **CONTINUED BY:** Nouveaux mémoires de l'Académie royale des sciences et belles lettres de Bruxelles. **INDEXES:** Tables générale des Mémoires de l'Académie royale des sciences, lettres et des beaux arts de Belgique (1772-1897). Bruxelles, Hayez, 1898. **SOURCES:** G443, OCLC, S943e **LOCATIONS:** BUC, CCF, CCN, ULS, ZDB **SUBJECT:** General - Societies - Brussels.

1079 Mémoires de l'Académie royale des sciences et belles lettres depuis l'avénement de Fréderic Guillaume II au throne, avec l'histoire pour le même temps. - Ser. 1, t. 1-7 (1786-1797) 1792-1800; ser. 2, t. 1-6 (1798-1804) 1801-1807. Berlin: Georg Jacques Becker, 1792-1807. **TRANSLATED AS:** Abhandlungen der Königlich-Preussischen Akademie der Wissenschaften in Berlin. **CONTINUATION OF:** Histoire avec les mémoires Preussische-Königliche Akademie der Wissenschaften. **CONTINUATION OF:** Nouveaux mémoires de l'Académie royale des sciences et belle-lettres, Berlin. **CONTINUATION OF:** Sammlung der deutschen Abhandlungen. Publisher varies. **INDEXES:** Kohnke, Otto, ed. Gesamtregister über die in den Schriften der Akademie von 1700-1899 erscheinen wissenschaftlichen Abhandlungen und Festreden. Berlin, 1900. **MICROFORMS:** RM(3), IDC(8), RP(1). **SOURCES:** G487d, K443,

OCLC, S2262 **LOCATIONS: BUC, CCF, CCN, ULS, ZDB   SUBJECT:** General - Societies - Berlin.

1080 Mémoires de l'Institut des science, lettres et arts, sciences, mathematiques et physique, Paris. - v.1-14 (1796-1815) 1798-1818. Paris: Badouin, 1796-1818. **CONTINUATION OF:** Histoire de l'Académie royale des sciences, Paris. v.7-14 under the title: Mémoires de la classe des sciences, mathematiques et physique. **MICROFORMS:** IDC(8).   **SOURCES:** G17n, OCLC, S1396d **LOCATIONS:** BUC, CCF, CCN, ULS, ZDB **SUBJECT:** Science - Societies - Paris.

1081 Mémoires de mathématiques et de physique / Académie royale des science, Paris. - v.1-2, 11 issues in 1692 and 10 in 1693. Paris: Imprimerie Royale, 1692-1693. Monthly. **TRANSLATED AS:** Physical and mathematical memoirs. A new edition was issued in 1 vol. in Amsterdam in 1723. Issued monthly except for July, Sept.and Oct., with 2 issues in June and 2 in Dec. in 1692 and 2 issues in Dec. in 1693.   **SOURCES:** OCLC, S719c **LOCATIONS:** BUC, CCF, CCN **SUBJECT:** Science - Societies - Paris.

1082 Mémoires de mathématiques et de physique, présentés à l'Académie Royale des Sciences, par divers scavans (étrangers) et lus dans ses assemblées. - T.1 (1746) 1750;t.2, 1755; t.3, 1760; t.4, 1763; t.5, 1768; t.6, 1774; t.7, (1773), 1776; t.8, 1780, t.9, 1780; t.10, 1785; t.11, 1786. Paris: Imprimerie royale, 1746-1786. Publisher varies. **OTHER TITLE:** Savants étrangers, Académie royales des sciences, Paris. **INDEXES:** Table générale des travaux contenus dans les mémoires de l'Académie. Paris, Gauthier-Villars, 1787. **MICROFORMS:** RM(3), RP(1).   **SOURCES:** G17i, OCLC, S1276g **LOCATIONS:** BUC, CCF, CCN, ULS, ZDB **SUBJECT:** Science - Societies - Paris.

1083 Mémoires de mathematiques et de physique, redigées à l'Observatoire de Marseilles. - Avignon: Geraid, 1755-1756.   **LOCATIONS:** CCF, ZDB **SUBJECT:** Astronomy - Societies - Marseilles.   **SUBJECT:** Science - Societies - Marseilles.

1084 Mémoires de physique pure, sans mathématiques, de toutes les académies des sciences. - v.1 (1665-1667) 1754. Lausanne: Antoine Chapius, 1754. Translations of Philosophical transactions and Saggi de naturali esperienze, Accademia del cimento, Florence etc. Only 1 v.seems to have been published. **TRANSLATION OF:** Philosophical transactions. **TRANSLATION OF:** Saggi de naturali esperienze. **LOCATIONS:** CCF **SUBJECT:** Science - Societies - London - Translations.   **SUBJECT:** Science - Societies - Florence - Translations.

1085 Mémoires du nord pour le progrès de l'histoire naturelle,des sciences et des arts libéraux et mécaniques. - v.1-2, 1756. Altona, 1756. **SOURCES:** B2922, G686, S2239 **SUBJECT:** Science - Periodicals - Germany.

1086 Mémoires et observations recueillies par la Société oeconomique de Bern. -
      Jahrg.1-12 (in 34 issues). Bern: Société économique à Bern, 1762-1773.
      CONTINUATION OF: Recueil de mémoires concernant l'économie rurale.
      Published simultaneously in French and German.   OTHER TITLE:
      Abhandlungen und Beobachtungen durch die ökonomische Gesellschaft zu Bern
      gesammelt. EDITOR: Elie Bertrand (Ref.29,p.152). MICROFORMS: RP(1)
      (Goldsmiths' Kress Library).   SOURCES: S2141   LOCATIONS: BUC,
      CCF, CCN, ULS, ZDB   SUBJECT: Agriculture - Societies - Bern.
      SUBJECT: Economics - Societies - Bern.

1087 Mémoires et prix / Société de santé, Paris. - v.1 (1796-1817) 1817. Paris: 1817.
      CONTINUATION OF: Histoire de la société royale de médecine.
      CONTINUED BY: Journal générale de médecine, de chiururgie et de
      pharmacie.   SOURCES: G135   LOCATIONS: BUC, DSG   SUBJECT:
      Medicine - Societies - Paris.

1088 Mémoires / Institut d'Égypt.- v.1-4, 1800-1804. Paris: 1800-1804. SOURCES:
      S1393a   LOCATIONS: CCF   SUBJECT: Science - Societies - Egypt.

1089 Mémoires littéraires, critiques, philosophiques, biographiques et bibliographiques
      pour servir à l'histoire ancienne et moderne de la médecine. - v.1 (no.1-52)-v.2
      (no.1) 1775-1776.   Paris: Pyre & Bastien, 1775-1776.   Imprint varies.
      EDITOR: Jean Goulin.   SOURCES: OCLC   SUBJECT: Medicine -
      Periodicals - France - Reviews.

1090 Mémoires pour l'histoire générale du diocese de Béziers / Académie des sciences
      et belles lettres de Béziers. - Béziers: Estienne Barbut, 1728.   SOURCES:
      OCLC   LOCATIONS: BL   SUBJECT: General - Societies - Béziers.

1091 Mémoires pour servir à l'histoire des sciences et des beaux arts. - v.1-265,
      1701-1767. Trévoux; Paris, 1701-1767.   Bimonthly.   TRANSLATED AS:
      Memorie per la storia delle scienze e buone arti.   A counterfeit edition was
      issued in Amsterdam 1665-1767 under the title: Journal des savans augmenté de
      divers articles.   Translated into Italian at Pesaro in 1742 and Venice in 1748.
      Reprinted by Slatkine, Geneva, 1968.   Reprinted in Trévoux 1731-1733 and
      Lyons 1734-1767.   Imprint varies.   Suspended May-Dec. 1720.   OTHER
      TITLE: Mémoires de Trévoux; Journal de Trévoux.   EDITORS: François
      Catrou, René Joseph de Tournemine and others. (Ref.38,p.219-240 and Ref.84).
      CONTINUED BY: Journal des sciences et des beaux-arts.   CONTINUED BY:
      Journal de littérature, des sciences et des arts.   INDEXES: L'Esprit des
      journalistes de Trévoux (1701-1762) 1771. P.S. Sommervogel, Table
      methodique des Mémoires de Trevoux (1701- 1775). Paris, 1864-1865.
      MICROFORMS: DA(1), ACR(1), RP(1).   SOURCES: B2924, G684, OCLC,
      S1690   LOCATIONS: BUC, CCF, CCN, ULS, ZDB   SUBJECT: General
      - Periodicals - France - Reviews.

1092 Mémoires pour servir à l'histoire naturelle des animaux et des plantes / Académie
      royale des sciences, Paris. - v.1-2, 1671.   Paris: Imprimerie royale, 1671.
      TRANSLATED   AS: Memoirs   for   a   natural   history   of   animals.

TRANSLATED AS: Abhandlungen zur Naturgeschichte der Thiere und Pflanzen. Numerous other editions and translations were published in Paris, Amsterdam, La Haye, Leipzig and London. Edited by Claude Perrault. **SOURCES:** B2925, S830   **SUBJECT:** Natural history - Societies - Paris.

1093   Mémoires pour servir à l'histoire physique et naturelle de la Suisse. - v.1, 1788. Lausanne: J. Mourer; Paris: G. de Burelainé, 1788. **EDITORS:** Louis Reynier and Henri Struve. **TRANSLATION OF:** Magazin für die Naturkunde Helvetiens (in part). **SOURCES:** B2926, G687, OCLC, S2169 **LOCATIONS:** BL   **SUBJECT:** Science - Periodicals - Switzerland.

1094   Mémoires / Société d'emulation patriotique de Neuchâtel. - v.1, 1793-1795. **LOCATIONS:** ULS   **SUBJECT:** Agriculture - Societies - Neuchâtel. **SUBJECT:** Economics - Societies - Neuchâtel.

1095   Mémoires sur différentes parties des sciences et des arts. - v.1-5, 1768-1783. Paris: Laurent Prault, 1768-1783. Imprint varies. **OTHER TITLE:** Mémoires sur différentes parties de l'histoire naturelle, des sciences et arts. **EDITOR:** Jean Etienne Guettard. **CONTINUED BY:** Nouvelle collection des mémoires sur différents parties interessantes des sciences. **SOURCES:** OCLC **LOCATIONS:** BUC, ULS   **SUBJECT:** Science - Periodicals - France.

1096   Mémoires sur les question proposées par l'Académie imperiale et royale des sciences et belles-lettres, Bruxelles, qui ont remportés les prix. - Ser. 1, v.1-4, 1769-1787; new ser. 1, v.1-5, 1793-1825. Brussels: 1769-1825. Title varies. **OTHER TITLE:** Mémoires en response à la question proposée par l'Académie. Text in French Latin or Dutch. **INDEXES:** Tables générale des Mémoires de l'Académie Royale des sciences, lettres et des beaux arts de Belgique (1772-1897). Bruxelles, Hayez, 1898. **SOURCES:** G443, OCLC, S943e **LOCATIONS:** BUC, CCF, CCN, ULS, ZDB   **SUBJECT:** General Societies - Brussels - Prize essays.

1097   Mémoires sur les sujets proposées pour les prix, Académie royale de chirurgie, Paris. - T.1-5, 1753-1783.   Paris: Michel Lambert, 1753-1783. **TRANSLATED AS:** Sammlung der Preisschriften der königlichen Parischen Academie der Chirurgie. Publisher varies. Published in both a quarto edition in 12 v. and a duodecimo edition in 29 v. There was an edition in 13 v., 1757-1778, in 5 v.in 1770-1798, and in 5 v. again in 1819 and 1836. T.1-3(1732-1758) under the title Recueil des pièces qui ont concourru pour le prix. **OTHER TITLE:** Recueil des pièces qui ont concouru pour le prix, de l'Académie royale de chiururgie, Paris. **INDEXES:** The 1836 edition (Paris, Michel Fossone) is accompanied by "une table alphabétique des matières." **SOURCES:** G9b, OCLC   **LOCATIONS:** BUC, CCF, CCN, ULS **SUBJECT:** Surgery - Societies - Paris - Prize essays.

1098   Memoirs for a natural history of animals, being the anatomical descriptions of several animals dissected by the Royal Acadamy at Paris. To which is added an account of the measure of the earth published by the same academy. - London: J. Streater, 1688. A translation by Alexander Pitfield of Perrault's

Mémoires pour servir à l'historie naturelle des animaux et des plantes. Another edition was published in London in 1702, translated by Richard Waller under the title The Natural history of animals. **OTHER TITLE:** Natural history of animals. **TRANSLATION OF:** Mémoires pour servir à l'historie naturelle des animaux et des plantes. **LOCATIONS:** NUC **SUBJECT:** Natural history - Societies - Paris - Translations.

1099   Memoirs for the curious. - v.1, Jan. 1707-Dec. 1708. London: A. Baldwin, 1707-1708. 2nd ed. was published as A complete volume of the memoirs for the curious, London, J. Morphew, 1710. **OTHER TITLE:** Monthly miscellany or memoirs for the curious. **EDITOR:** James Petiver. **MICROFORMS:** UM(1). **SOURCES:** B1334, S255 **LOCATIONS:** BUC **SUBJECT:** General - Periodicals - England.

1100   Memoirs for the ingenious, containing curious observations in philosophy, mathematicks, physick, philology and other useful arts and sciences. - v.1 (no.1-12) 1693. London: D. Rhodes and J. Harris, 1693. Monthly. **EDITOR:** Jean Cornard de la Crose. **MICROFORMS:** RP(1), UM(1). **SOURCES:** OCLC **LOCATIONS:** BUC, ULS, ZDB **SUBJECT:** General - Periodicals - England.

1101   Memoirs for the ingenious, or, the Universal mercury. - v.1, no. 1, Jan. 1694. London: Randal Taylor, 1694. **OTHER TITLE:** Universal mercury. **LOCATIONS:** BUC, ULS, ZDB **SUBJECT:** General - Periodicals - England.

1102   Memoirs of agriculture and other economical arts / Society instituted at London for the encouragment of the arts, manufactures and commerce.. - v.1-v.3, 1768-1782. London: J. Nourse, 1768-1782. **CONTINUATION OF:** Museum rusticum et commerciale. **EDITOR:** Robert Dossie. **CONTINUED BY:** Transactions of the Society instituted at London for the encouragment of the arts, manufactures and commerce. **SOURCES:** OCLC **LOCATIONS:** BUC **SUBJECT:** Agriculture - Societies - London. **SUBJECT:** Economics - Societies - London.

1103   Memoirs of science and the arts, or, an abridgement of the transactions published by the principal learned and oeconomical societies established in Europe, Asia and America. - v.1-2, pt.1, 1793-1794; 2nd ed. pt.1-3, 1798. London: Robert Faulder, C. Dilly and J. Egerton, 1793-1798. **LOCATIONS:** BUC, ULS **SUBJECT:** Science - Periodicals - England - Reviews.

1104   Memoirs of the American academy of arts and sciences / Boston. - Pt.1 (1783) 1785. Boston: Adams and Nourse, 1785. Continued in 1804 with part 2. **SOURCES:** G456, S3974 **LOCATIONS:** BUC, CCF, CCN, ULS, ZDB **SUBJECT:** Science - Societies - Boston.

1105   Memoirs of the literary and philosophical society of Manchester. v.1-v.5, 1785-1802. Warrington and Manchester: 1785-1802. **TRANSLATED AS:** Physikalische und philosophische abhandlungen der Gesellschaft der Wissenschaften zu Manchester. **TRANSLATED AS:** Auserlesene

Abhandlungen für Aerzte Naturforscher und Psychologen. Imprint varies. **INDEXES:** An index to the first 17 vols. can be found in ser.2, v.12:285-385 of the memoirs. **MICROFORMS:** RM(3), BHP, GO(9). **SOURCES:** G480, OCLC, S494 **LOCATIONS:** BUC, CCF, CCN, ULS, ZDB **SUBJECT:** Science - Societies - Manchester.

1106 Memoirs of the medical society of London. - v.1-2, 1787-1789. London: C. Dilley, 1787-1789. **TRANSLATED AS:** Merkwürdige Abhandlungen der zu London 1773 errichteten medicinischesn Gesellschaft. A 2nd edition in 6 v.appeared 1792-1799. **SOURCES:** G78, OCLC **LOCATIONS:** BUC, CCF, CCN, ULS, ZDB **SUBJECT:** Medicine - Societies - London.

1107 Memoirs of the Royal academy of surgery at Paris. - v.1-3, 1750-1759. London: J. Rivington and J. Fletcher, 1750-1759. Translated by George Neale. Publisher varies. **TRANSLATION OF:** Mémoires de l'Académie royale de chirurgie. **LOCATIONS:** BUC, CCF, DSG **SUBJECT:** Surgery - Societies - Paris - Translations.

1108 Memoirs of the Royal Society, being a new abridgement of the Philosophical transactions. - v.1-10 (1665-1735) 1738-1741. London: G. Smith and T. Cooper, 1738-1741. **CONTINUATION OF:** Philosophical transactions. There was a 2nd ed. in 10v. covering 1665-1740, London, J. Nourse, 1745. **EDITOR:** Benjamin Baddam. **SOURCES:** GII:4e, OCLC, S434d **LOCATIONS:** BUC, CCF, CCN, NUC, ZDB **SUBJECT:** Science - Societies - London - Abridgments.

1109 Memorial literario, instructivo y curioso de la corte de Madrid. - v.1-21 (no.1-125) 1784-1790. Madrid: 1784-1790. Suspended Feb. 1791-Jul. 1793. **CONTINUED BY:** Continuacion del memorial literario. **SOURCES:** S1750a **LOCATIONS:** BUC, CCF, ULS **SUBJECT:** General - Periodicals - Spain.

1110 Memorias / Academia de medicina de Madrid. - v.1-10, 1797-1889. Madrid: Imprenta Real, 1797-1889. **SOURCES:** G146 **LOCATIONS:** BUC, CCF, ULS **SUBJECT:** Medicine - Societies - Madrid.

1111 Memorias / Academia medico-practical de la ciudad de Barcelona. - v.1, 1798. Barcelona: 1798. **LOCATIONS:** ULS **SUBJECT:** Medicine - Societies - Barcelona.

1112 Memorias academicas de la real sociedad de medicina y demas ciencias de Seville. - v.1-10, 1766-1819. Seville: 1766-1819 (Ref no.97). **SOURCES:** GIII:22 **LOCATIONS:** BUC, CCF **SUBJECT:** Medicine - Societies - Seville. **SUBJECT:** Science - Societies - Seville.

1113 Memorias de Acadamia real das sciencias de Lisboa. - v.1(1780-1788)1797-v.12, 1839. Lisbon: Typografia de Academia, 1797-1839. v.2 also under the title: Memorias de mathematica e phisica. v.4-12 also under the title: Historia e memorias. **OTHER TITLE:** Memorias de mathematica e phisica da Acadamia real das sciencias de Lisboa; Historia e memorias da Acadamia real das

154   SCIENTIFIC AND TECHNICAL PERIODICALS

sciencias de Lisboa. **INDEXES:** Index to the society's scientific publications from 1779-1924 in its: Memorias, Classe de Ciencias, 1936, 1:11-61. **MICROFORMS:** RM(3). **SUBJECT:** Science - Societies - Lisbon.

1114 Memorias de agricultura premidas pela Real Academia das sciencias de Lisboa. - v.1-2, 1788-1791. Lisbon: 1788-1791. **INDEXES:** Index to the society's scientific publications from 1779-1924 in its: Memorias, Classe de Ciencias, 1936, 1:11-61. **MICROFORMS:** RP(1)(Goldsmith' Kress Library). **LOCATIONS:** ULS   **SUBJECT:** Agriculture - Societies - Lisbon.

1115 Memorias de la junta general / Sociedad económica de Guatemala. - v.1-5, 1797-1798. Neuva Guatemala: Sebastien de Aravalo, 1797-1798. **OTHER TITLE:** Junta pública, Sociedad económica de Guatemala.   **LOCATIONS:** BUC, ULS   **SUBJECT:** Economics - Societies - Guatemala.

1116 Memorias economicas da Academia real das sciencias de Lisboa paro o adiantamento da agricultura, das artes e da industriae em Portugal e suas conquistas. - v.1-5, 1789-1815. Lisbon: 1789-1815. **INDEXES:** Index to the society's scientific publications from 1779-1924 in its: Memorias, Classe de Ciencias, 1936, 1:11-61.   **LOCATIONS:** BUC, CCF, CCN, ULS, ZDB **SUBJECT:** Economics - Societies - Lisbon.

1117 Memorias instructivas y curiosas sobre agricultura, comercio, industria, economia, chymica, botanica, historia natural etc. - v.1-12, 1778-1791. Madrid: P. Marin, 1778-1791.   Title varies slightly.   **EDITOR:** Miquel Gerónimo Suárez y Nuálunez.   **SOURCES:** B2934, OCLC, S1751   **LOCATIONS:** ULS **SUBJECT:** Economics - Periodicals - Madrid.   **SUBJECT:** Science - Periodicals - Madrid.

1118 Memorias / Real sociedad patriotica, Sevilla. - v.1-2, 1779. Seville: Varquez, 1779.   **SOURCES:** OCLC, S1776 **LOCATIONS:** BUC, ULS   **SUBJECT:** Economics - Societies - Seville.

1119 Memorias / Sociedad económica, Madrid. - v.1-5, 1780-1795. Madrid: A. de Sancha, 1780-1795. **MICROFORMS:** RP(1). (Goldsmiths' Kress Library). **SOURCES:** S1763   **LOCATIONS:** BUC, CCF, CCN, ULS, ZDB **SUBJECT:** Economics - Societies - Madrid.

1120 Memorias / Sociedad económica Mallorquina de Amigos del pais. - v.1, 1784. Palma de Mallorca: 1784.   **SOURCES:** OCLC, S1771   **SUBJECT:** Economics - Societies - Palma de Mallorca.

1121 Memorie di matematica e di fisica della Società italiana delle scienze / Verona. - v.1-25, 1778-1855. Verona: Dionigi Ramanzini, 1782-1855.   Imprint and society location vary i.e. Modena and Rome.   **SOURCES:** G504, OCLC, S2108 **LOCATIONS:** CCF, ULS, ZDB   **SUBJECT:** Science - Societies - Verona.

1122 Memorie della Reale Accademia di scienze, belle lettere ed arti / Mantua. - v.1, 1795. Mantua: Alberto Pazzoni, 1795. **SOURCES:** G496, S1871 **LOCATIONS:** BUC, CCF, ULS, ZDB **SUBJECT:** General - Societies - Mantua.

1123 Memorie delle reale società agraria / Turin. - v.1-11, 1788-1838. Turin: Giannuchele Biolo, 1788-1838. **LOCATIONS:** BUC, CCF, ULS **SUBJECT:** Agriculture - Societies - Turin.

1124 Memorie di varia erudizione della società columbaria fiorentina. - v.1-2, 1747-1752. Florence: L.S. Olschki, 1747-1752. **SOURCES:** OCLC, S1848 **LOCATIONS:** BUC, CCF, ULS, ZDB **SUBJECT:** General - Societies - Florence.

1125 Memorie ed osservazioni / Societá d'agricoltura pratica, Udine. - v.1, 1772. Udine: 1772. **LOCATIONS:** BUC **SUBJECT:** Agriculture - Societies - Udine. \

1126 Memorie per i curioisi di agricoltura e di economia rurale. - v.1-9, 1800-1802. Naples: 1800-1802. **LOCATIONS:** BUC, ULS **SUBJECT:** Agriculture - Periodicals - Italy.

1127 Memorie per la storia delle scienze e buone arti, comunciate ad imprimersi l'anno 1701, à Trevoux e l'anno 1756. - v.1-36, 1742-1753. Peraso: N. Gavelli, 1742-1753. **TRANSLATION OF:** Mémoires pour servir à l'histoire des sciences et des beaux arts. **LOCATIONS:** ULS **SUBJECT:** General - Periodicals - Paris - Translations.

1128 Memorie sopra la fisica e la storia naturale di diversi valent'uonimi. - v.1-4, 1743-1757. Lucca: 1743-1757. **SOURCES:** B2938, S1869 **LOCATIONS:** CCF, ULS **SUBJECT:** Science - Periodicals - Italy.

1129 The Merchant's and manufacturer's magazine of trade and commerce. - v.1-2, 1785-1786. London: 1785-1786. **MICROFORMS:** RP(1) (Goldsmiths' Kress Library). **LOCATIONS:** BUC, ULS **SUBJECT:** Economics - Periodicals - England.

1130 Mercure sçavant. - No.1-2, 1684. Amsterdam: Henry Desbordes, 1684. **EDITOR:** Nicolas de Blegny. **LOCATIONS:** BUC, CCF **SUBJECT:** General - Periodicals - Netherlands.

1131 Mercurio peruano de historia, literatura y noticias públicas / Sociedad académica de amantes de Lima. - v.1-12, 1791-1795. Lima: Jacinto Calero y Moiya, 1791-1795. Published in facsimile by Biblioteca Nacional de Peru. Lima, 1946. **SOURCES:** S3905 **LOCATIONS:** BUC, CCF, CCN, ULS **SUBJECT:** General - Periodicals - Guatemala.

1132 Mercurio volante, con noticias importantes i curiosas sobre various asuntos de fisica i medicina. - No. 1-16, 1772-1773. Mexico: Felipe de Zúñiga,

1772-1773. Reprinted by Biblioteca, Universidad de Mexico, 1979. **EDITOR:** Josef Ignacio Bartolache. **SOURCES:** G29, OCLC   **LOCATIONS:** ULS **SUBJECT:** General - Periodicals - Mexico.   **SUBJECT:** Medicine - Periodicals - Mexico.

1133   Merkwürdige Abhandlungen der zu London 1773 errichteten medicinischesn Gesellschaft. - v.1-4, 1789-1796. Altenburg: Richter, 1789-1796. Translated by Johann Dietrich Philipp Christian Ebeling. **TRANSLATION OF:** Memoirs of the medical society of London. **SOURCES:** G78a, K3648 **LOCATIONS:** CCF, ZDB   **SUBJECT:** Medicine - Societies - London - Translations.

1134   Merkwürdige Abhandlungen holländischer Aerzte. - v.1-2, 1794-1797. Leipzig: Georg August Grieshammer, 1794-1797. Translated by Daniel Collenbusch. **SOURCES:** G377, OCLC   **SUBJECT:** Medicine - Periodicals - Netherlands - Translations.

1135   Merkwürdige Himmelsbegebenheiten. - St.1-34, 1736-1740. Kiel: 1736-1740. **EDITOR:** Michael Adelbulner.   **SOURCES:** K3190 **LOCATIONS:** CCF **SUBJECT:** Astronomy - Periodicals - Germany.

1136   Merkwürdige Krankengeschichte und seletene practische Beobachtungen. - Bd.1, 1795. Halle: Renger, 1795. **TRANSLATION OF:** Acta societatis medicae Havniensis. **SOURCES:** GI3c **LOCATIONS:** ULS **SUBJECT:** Medicine - Societies - Copenhagen - Translations.

1137   A Meteorological register kept at Mansfield Woodhouse in Nottinghamshire. - Nottingham: Samuel Tupnian, 1794-1806.   **EDITOR:** Hayman Rooke. **SOURCES:** B7016, OCLC **LOCATIONS:** BUC **SUBJECT:** Meteorology - Periodicals - England.

1138   Meteorologische Ephemeriden / Akademie der Wissenschaften, Munich. - v.1-9, 1761-1789.   Munich: 1781-1789.   Also published in Neue philosophische Abhandlungen der baierischen Akademie der Wissenschaften. **EDITOR:** Franz Xaver Epp. **MICROFORMS:** RM(3).   **SOURCES:** G459c, K327, OCLC, S3109h   **LOCATIONS:** BUC, CCF, CCN, ULS, ZDB   **SUBJECT:** Meteorology - Societies - Munich.

1139   Mikroscopische Gemüths-und Augen-Ergötzung. - Nürnberg: Adam Wolfgang Winterschmidt, 1760-1765. Imprint varies. Another edition in Bayreuth 1761. **EDITOR:** Martin Frobenius Ledermüller.   **SOURCES:** K3231, OCLC **LOCATIONS:** BL **SUBJECT:** Microscopes and microscopy - Periodicals - Germany   **SUBJECT:** Natural history - Periodicals - Germany.

1140   Der Mineralog, oder, compendiöse Bibliothek alles Wissenswürdigen aus dem Gebiete der Minerologie. - v.1-5, 1792-1796. Gotha: 1792-1796. **SOURCES:** B3009, S2667   **SUBJECT:** Mineralogy - Periodicals - Germany.

1141   Mineralogische, chemische und alchymistische Briefe von Reisenden und andern Gelehrten an J.F. Henkel. - v.1-3, 1794-1795. Dresden: Walther, 1794-1795.

MICROFORMS: RM(3). **SOURCES:** B3012, G690, OCLC, S2514 **SUBJECT:** Mineralogy - Periodicals - Germany. **SUBJECT:** Chemistry - Periodicals - Germany. **SUBJECT:** Alchemy - Periodicals - Germany.

1142 Mineralogischer Briefwechsel. - v.1-2 (4 Heft each) 1779-1784. Giessen: Justus Friedrich Krieger, 1779-1784. **OTHER TITLE:** Mineralogische Briefe. **EDITOR:** Phillip Engel Klipstein. **SOURCES:** K3270 **LOCATIONS:** BL **SUBJECT:** Mineralogy - Periodicals - Germany.

1143 Mineralogische Belustigungen zum Behuf der Chymie und Naturgeschichte des Mineralreichs. - v.1-6, 1768-1771. Leipzig: Johann Friedrich Heineck; Copenhagen: Faber, 1768-1771. **EDITOR:** Johann Christoph Adelung. **SOURCES:** B3011, K3244, S2968 **LOCATIONS:** BL, CCF, ULS, ZDB **SUBJECT:** Mineralogy - Periodicals - Germany. **SUBJECT:** Chemistry - Periodicals - Germany. **SUBJECT:** Natural history - Periodicals - Germany.

1144 Minutes / Medical society of the county of Kings. - v.1-64, 1794-1885. Brooklyn: 1794-1885. **SOURCES:** G125 **LOCATIONS:** ULS **SUBJECT:** Medicine - Societies - Brooklyn.

1145 Minutes / Society for philosophical experiments and conversation. - v.1 (1794) 1795. London: 1795. **TRANSLATED AS:** Protocolle der Verhandlungen einer Privatgesellschaft in London über die neueren gegenstände der Chemie (S2714). **SOURCES:** G517, S453 **LOCATIONS:** BUC, ULS **SUBJECT:** Science - Societies - London.

1146 Miscellanea austriaca ad botanicam, chemiam et historiam naturalem spectantia. - v.1-2, 1778-1781. Vienna: Kraus, 1778-1781. **EDITOR:** Nikolaus Joseph Jacquin. **CONTINUED BY:** Collectanea austriaca ad botanicam, chemiam et historiam naturalem spectantium. **SOURCES:** B3048, G691, OCLC, S3605 **LOCATIONS:** BUC, ULS **SUBJECT:** Science - Periodicals - Austria.

1147 Miscellanea Berolinensia ad incrementum scientiarum ex scriptis Societatis regiae scientiarum. - v.1, 1710; v.2,1725; v.3,1727; v.4,1734; v.5,1737; v.6, 1740; v.7, 1743. Berlin: Johann Chrisopher Popenii, 1710-1743. **TRANSLATED AS:** Choix des mémoires et abrégé de histoire. Imprint varies. **CONTINUED BY:** Histoire avec les mémoires pour les même années. **CONTINUED BY:** Nouveaux mémoires de l'Académie royale des sciences et belles-lettres, Berlin. **CONTINUED BY:** Mémoires de l'Académie royale des sciences et belles lettres, Berlin. **CONTINUED BY:** Sammlung der deutschen Abhandlungen. A supplementary volume was issued in 1744. **INDEXES:** Kohnke, Otto, ed. Gesamtregister über die in den Schriften der Akademie von 1700-1899 erscheinen wissenschaftlichen Abhandlungen und Festreden. Berlin, 1900 (Ref.126). **MICROFORMS:** IDC(8), RM(3), RP(1). **SOURCES:** B3049, G487, K24, OCLC, S2385 **LOCATIONS:** BUC, CCF, CCN, ULS, ZDB **SUBJECT:** Science - Societies - Berlin.

1148 Miscellanea curiosa, being a collection of some of the principle phaenomena in nature. - v.1-2, 1705-1707. London: Jeffry Vale and John Senex, 1705-1707.

There were at least 2 other editions 1708 and 1726-27.   Imprint varies.
SOURCES: GII:4d **LOCATIONS:** BUC, CCF, ULS **SUBJECT:** Science
- Periodicals - England.

1149 Miscellanea curiosa mathematica, or, the literary correspondence of some eminent
      mathematicians in Great Britain and Ireland. - v.1-2 no. 5, 1745-1753. London:
      Edward Cave, 1745-1753. Quarterly. v.1 also as no.1-9. There were several
      editions. **EDITOR:** Francis Holliday. **SOURCES:** B3032, OCLC, S383
      LOCATIONS: BUC, ULS, ZDB **SUBJECT:** Mathematics - Periodicals -
      England - Popular.

1150 Miscellanea curiosa medico-physica Academiae naturae curiosorum, sive
      ephemeridum medico-physicarum germanicarum curiosarum. - Published in 3
      Decades each covering 10 yrs: Dec.1, pt.1, 1670-1780 in 7 v.; Dec.1, pt.2,
      1684-88, in 5v., Dec.2, 1683-1692 in 10v., Dec.3, 1691-1706 in 7v.(Ref.21:
      v.23:953). Leipzig: Jacob Trescher, 1670-1706. Publisher and description
      vary. French translation in Collection académique,v.4,6-7. **CONTINUED BY:**
      Ephemerides sive observationum medico-physicarum, Academiae Caesareae
      Leopoldinae naturae curiosoum. **CONTINUED BY:** Acta physico-medica,
      Academia Caesarea Leopoldina-Carolina naturae curiosorum. **CONTINUED
      BY:** Nova acta physico-medica academiae Caesareae Leopoldinae- Carolinae
      naturae curiosorum. Separately paged supplements accompany most volumes.
      INDEXES: Index rerum Decuraie 1 et 2. Nuremberg, 1695. Index generalis et
      absolutissimus decuriae tertiae. Frankfurt, 1713. **MICROFORMS:** RM(3),
      IDC(8), RP(1)   **SOURCES:** B3051, GII:7, K3176, OCLC, S2808
      LOCATIONS: BUC, CCF, CCN, ULS, ZDB **SUBJECT:** Science - Societies
      - Halle.

1151 Miscellanea curiosae, or, entertainments for the ingenious of both sexes. - v.1-2,
      1734-1735. York: Thomas Gent, 1734-1735. Quarterly. **SOURCES:** B8565,
      OCLC **LOCATIONS:** BUC, ULS **SUBJECT:** Mathematics - Periodicals -
      England - Popular.

1152 Miscellânea curioza e proveitoza, ou, compilaçâo, tirada das melhores obras das
      macoens estrangerias. - v.1-7, 1779-1785.   Lisbon: 1779-1785.
      LOCATIONS: BUC **SUBJECT:** General - Periodicals - Portugal.

1153 Miscellanea italica physico-mathematica. - v.1-4, 1691-1692. Parma: Hippolyti
      & Francisci Moriae de Rosatis, 1691-1692. **EDITOR:** Gaudenzio Roberti.
      SOURCES: OCLC **LOCATIONS:** CCN NUC **SUBJECT:** Science -
      Periodicals - Italy. **SUBJECT:** Mathematics - Periodicals - Italy.

1154 Miscellanea mathematica, consisting of a large collection of curious mathematical
      problems and their solutions. - No. 1-13, 1771-1775. London: G. Robinson and
      R. Baldwin, 1771-1775.   **OTHER TITLE:** Mathematical miscellany.
      EDITOR: Charles Hutton. **SOURCES:** OCLC **LOCATIONS:** BL, ULS
      SUBJECT: Mathematics - Periodicals - England.

1155 Miscellanea medico-chirurgica practica et forensia. - v.1-7, 1731-1737. Leipzig; Görlitz: Johann Gottlob Laurentius, 1731-1737. EDITOR: Gottlieb Budaes. SOURCES: K3522 LOCATIONS: BL, DSG, ZDB SUBJECT: Medical jurisprudence - Periodicals - Germany. SUBJECT: Medicine - Periodicals - Germany.

1156 Miscellanea physico-medica. - v.1, 1789. Halle: 1789. EDITOR: Johann Hermann Pfingsten. SOURCES: B3057, G379, S2725 SUBJECT: Medicine - Periodicals - Germany.

1157 Miscellanea physico-medico-mathematica, oder, angenehme curieuse und nützliche Nachrichten von physical- und medicinischen auch dahin gehörigen Kunst-und Literatur- Geschichten. - v.1-4 (4 issues each) 1727-1730. Erfurt: C.F. Jungnicol, 1727-1730. CONTINUATION OF: Sammlung von Natur- und Medicin. EDITOR: Andreas Elias Büchner. INDEXES: Index to 1727-1730. MICROFORMS: RP(1). SOURCES: B4197a, GII:7o, K33187, S2556 LOCATIONS: BUC, CCF, CCN, ULS, ZDB SUBJECT: Science - Periodicals - Germany.

1158 Miscellanea scientifica curiosa. - No. 1-23, 1766-1767. York; London: 1766-1767. EDITOR: J. Randall (Ref.12, no. 16). LOCATIONS: BUC, ULS SUBJECT: Science - Periodicals - England.

1159 Mode-, Fabriken-, und Gewerbzeitung. - No. 1-52, 1787-1788. Prague: 1787-1788. LOCATIONS: ZDB SUBJECT: Technology - Periodicals - Germany. SUBJECT: Commerce - Periodicals - Germany.

1160 Modern husbandman, or, the practice of farming. - No. 1-12, May 1741-April 1744. London: T.Osborne, 1741-1744. Reprinted in Dublin by George Faulkner in 1743 and in London in 8 v.by Browne and Davis in 1750. EDITOR: William Ellis. MICROFORMS: RP(1). (Goldsmiths' Kress Library). SOURCES: OCLC LOCATIONS: BUC, ULS SUBJECT: Agriculture - Periodicals - England.

1161 Monath-Schrift von nützlichen und neuen Erfahrungen aus dem Reiche der Scheide-Kunst und andern Wissenschaften. - St. 1-2, 1773-1774. Tübingen: 1773-1774. EDITOR: Jacob Andreas Weber. SOURCES: B7105, K3253 LOCATIONS: ZDB SUBJECT: Chemistry - Periodicals - Germany.

1162 Monatlich herausgegebene Insektenbelustigungen. - Pt.1(1746), 2(1749), 3(1755), 4,(1761) Nürnberg: Fleischmann, 1746-1761. New edition in 1792 in 5v. Reprinted in 1975 in Stuttgart (Verlag Müller und Schindler). OTHER TITLE: Insekten-belustigungen. EDITORS: August Johann Rösel von Rosenhof and Christian Friedrich Carl Kleemann (v.4). SOURCES: B3100, OCLC SUBJECT: Entomology - Periodicals - Germany.

1163 Monatliche Beiträge zur Naturkunde. - v.1-12, 1752-1765. Berlin: Verlag der Realschule, 1752-1765. CONTINUATION OF: Physicalische Briefe. EDITOR: Johann Daniel Denso. CONTINUED BY: Fortgesetzte Beiträge zur

Naturkunde. **CONTINUED BY:** Neue Monatliche Beiträge zur Naturkunde.
MICROFORMS: RM(3). **SOURCES:** B3644a, K3212, OCLC, S2352
LOCATIONS: BUC, CCN, ULS, ZDB   **SUBJECT:** Natural history -
Periodicals - Germany

1164 Monatliche Belustigungen im Reiche der Natur. - v.1, 1753. Hamburg: 1753.
A new ed. edited by Johann Dominik Schultze (Hamburg, Herald, 1790).
EDITOR: Nicolaus Georg Geve. **SOURCES:** K3214 **LOCATIONS:** NUC
SUBJECT: Natural history - Periodicals - Germany.

1165 Monatliche Correspondenz zur Beförderung der Erd- und Himmelskunde. -
v.1-28, 1800-1813. Gotha: Becker, 1800-1813. **CONTINUATION OF:**
Allgemeine geographische Ephemeriden. **EDITORS:** Franz Xaver von Zach.
CONTINUED BY: Correspondance astronomique, géographique,
hydrographique et statistique. **INDEXES:** Index compiled by G.G. Galle
published in 1848. **SOURCES:** B3083, K3383, S2669 **LOCATIONS:** BUC,
CCF, CCN, ULS, ZDB **SUBJECT:** Astronomy - Periodicals - Germany
SUBJECT: Geography - Periodicals - Germany.

1166 Monatliche Erzehlungen allerhand künstlicher und natürlicher Curiositäten. - St.
1-4, May-Aug. 1689. Leipzig: Johann Herbart Kloss, 1689. **SOURCES:**
K3910 **SUBJECT:** Technology - Periodicals - Germany - Reviews.

1167 Monatliche neu eröffnete Anmerckungen über alle Theile der Artzney-Kunst. -
Pt.1-4, 1680-1682. Hamburg: G. Schultz; Marburg: Caspar Forberger,
1680-1683 (Ref.99). Translated by Johann Lange. **TRANSLATION OF:**
Nouvelles déscouvertes sur toutes les parties de la medécine. **SOURCES:**
GI:4b **LOCATIONS:** DSG, ZDB **SUBJECT:** Medicine - Periodicals -
France - Translations.

1168 Monatsschrift der Akademie der Künste und mechanischen Wissenschaften zu
Berlin. - Bd.1-3, 1788-1789. Berlin: 1788-1789. **CONTINUED BY:** Annalen
der Akademie der Künste und mechanischen Wissenschaften. **SOURCES:**
K4091 **LOCATIONS:** BUC, CCF, ULS **SUBJECT:** Technology - Societies
- Berlin.

1169 Monthly magazine and British register. - v.1-60, 1796-1826, ns. v.1-26,
1826-1838, s.3 v.1-9, 1839- 1843. London: R. Phillips, 1796-1843. Monthly.
Subtitle varies. **EDITOR:** John Aikin. **INDEXES:** v.6 1798 includes an index.
MICROFORMS: RP(1), UM(1). **SOURCES:** OCLC **LOCATIONS:** BUC,
CCF, ULS **SUBJECT:** General - Periodicals - England.

1170 Monthly miscellany, or, memoirs for the curious. - v.1-4, 1707-1710. London:
J. Morphew, 1707-1710. Monthly. **OTHER TITLE:** Memoirs for the curious.
EDITOR: James Petiver. **MICROFORMS:** UM(1). **SOURCES:** B3175,
OCLC **LOCATIONS:** BUC, ULS **SUBJECT:** General - Periodicals -
England.

1171 Monthly observations for the preserving of health, with a long and confortable life
by Phylotheus Physiologus. - v.1, 1688. London: A. Sowle, 1688. EDITOR:
Thomas Tryon. SOURCES: OCLC,(Ref.17). LOCATIONS: NUC
SUBJECT: Medicine - Periodicals - England - Popular.

1172 Monthly review, or, literary journal. - v.1-81, 1749-1789; ser. 2, v.1-108,
1790-1825; ser. 3, v.1-ser. 4, v.45, 1826-1844; London: Ralph Griffiths,
1749-1844. Monthly. Subtitle and imprint vary. EDITOR: Ralph Griffiths.
INDEXES: Nangle, Benjamin Christie. Indexes of contributors and articles.
Oxford, Clarendon Press, 1934. MICROFORMS: RP(1), UM(1).
SOURCES: OCLC   LOCATIONS: BUC, CCF, CCN, ULS, ZDB
SUBJECT: General - Periodicals - England - Reviews.

1173 Monthly weather-paper. - Feb.-July 1711. LOCATIONS: BUC SUBJECT:
Meteorology - Periodicals - England.

1174 Museum der Heilkunde, herausgegeben von der Helvetischen Gesellschaft
correspondirender Aerzte und Wundärzte. - v.1-4, 1792-1795. Zürich: Orell,
Gessner und Füssli, 1792-1797. CONTINUATION OF: Annalen der
Staatsarzneikunde. EDITOR: Johann Heinrich Rahn. SOURCES: G109,
K3676 LOCATIONS: BUC, CCF, ULS, ZDB SUBJECT: Medicine -
Periodicals - Switzerland.

1175 Museum for sunheds og kundskabs elskere. - Nos. 1-24, 1788-1789.
Copenhagen: 1788-1789. CONTINUATION OF: Sundstidende.
CONTINUATION OF: Nye sundstidende. CONTINUATION OF: Sundhed
og underholdning. CONTINUATION OF: Hygaea og muserne. EDITOR:
Johann Clemens Tode. CONTINUED BY: Hertha. CONTINUED BY:
Sundhedsbog. CONTINUED BY: Medicinalblade. CONTINUED BY:
Sundhedsjournal. CONTINUED BY: Sundhedsraad. CONTINUED BY:
Sundhedsjournal et maanedskrift. CONTINUED BY: Nyeste sundhedsblade.
SOURCES: G186 SUBJECT: Medicine - Periodicals - Denmar

1176 Museum rusticum et commerciale, oder auserlesene Schriften des Ackerbau, die
Handlung, die Künste und Manufacturen betreffend. - Bd.1-Bd.10, 1764-1769.
Leipzig: Johann Friedrich Junius, 1764-1769. EDITORS: Christian August
Schultze and Johann Joachim Schwabe. TRANSLATION OF: Museum
rusticum et commerciale, or, select papers on agriculture. SOURCES: K2880
SUBJECT: Technology - Societies - London - Translations. SUBJECT:
Economics - Societies - London - Translations. SUBJECT: Agriculture -
Societies - London - Translations.

1177 Museum rusticum et commerciale, or, select papers on agriculture, commerce,
arts and manufactures. Society instituted at London for the encouragement of
arts, manufactures and commerce. - v.1-v.6, 1763-1766. London: R. Davis,
1763-1766. TRANSLATED AS: Museum et commerciale oder auserlesene
Schriften des Ackerbau, die Handlung, die Künste und Manufacturen betreffend.
Publisher varies. CONTINUED BY: Memoirs of agriculture and other
economical arts. CONTINUED BY: Transactions of the Society instituted at

London for the encouragment of the arts, manufactures and commerce.
MICROFORMS: RP(1) (Goldsmiths' Kress Library).   **SOURCES:** OCLC
LOCATIONS: BUC, ULS   **SUBJECT:** Technology - Societies - London.
SUBJECT: Economics - Societies - London.   **SUBJECT:** Agriculture -
Societies - London.

1178 Mythohermetisches Archiv, aus dem Französischen des Claviere de Plessis. - St.
1-2. 1780. Gotha: Carl Wilhelm Ettinger, 1780. Reprinted in 1781 with a
slightly changed title and the same imprint. **EDITOR:** Clavier Duplessis.
TRANSLATION OF: Archives mythohermétiques. **SOURCES:** B3214, GXX
LOCATIONS: BUC   **SUBJECT:** Alchemy - Periodicals - France.
SUBJECT: Astrology - Periodicals - France.

1179 Naalezing van den Artz of geneesheer. v.1-v.4, 1772-1775.   Amsterdam:
1772-1775. **CONTINUATION OF:** Arzt of geneesheer.   **SOURCES:** G30
LOCATIONS: ULS   **SUBJECT:** Medicine - Periodicals - Netherlands -
Popular.

1180 Nachricht von dem chirurgischen Klinikum in Göttingen. - Anzeige 1-8,
1797-1800. Göttingen: Barmier, 1797-1800.   **SOURCES:** K3711
LOCATIONS: ZDB **SUBJECT:** Surgery - Periodicals - Germany.

1181 Nachricht von dem Fortgange der naturforschenden Gesellschaft in Jena. - v.1-9,
1794-1802. Jena: Göpferdt, 1794-1802. Appeared in 1793 under the title:
Nachricht von der Gründing. **OTHER TITLE:** Nachricht von der Gründing
einer naturforschenden Gesellschaft zu Jena.   **SOURCES:** K3358, S2830a
LOCATIONS: BUC, CCF   **SUBJECT:** Natural history - Societies - Jena.

1182 Nachricht von der Anstalt für arme Kranke zu Altdorf in Nürnbergischen. - St.
1-17, 1787-1807. Altdorf; Nürnberg: Monath, 1787-1807. **OTHER TITLE:**
Bemerkungen über die Kenntniss und Kur einiger Krankheiten. **EDITOR:** G.
Hoffmann and others.   **SOURCES:** G77, K3623   **LOCATIONS:** ZDB
SUBJECT: Medicine - Periodicals - Germany.

1183 Nachricht von der Einrichtung des Instituti clinici in Erlangen. - No. 1-6,
1780-1785. Erlangen: Palm, 1780-1785.   **EDITOR:** Friedrich Wendt.
SOURCES: K3584 **SUBJECT:** Medicine - Societies - Erlangen.

1184 Nachrichten aus dem Blumenreiche, eine Quartalschrift. - St. 1-6, 1784-1789.
Dessau; Leipzig: S.L. Crusius, 1784-1789. Imprint varies. **EDITOR:** Ludwig
Christopher Schmaling. **SOURCES:** K3301, OCLC **LOCATIONS:** CCF,
ZDB **SUBJECT:** Botany - Periodicals - Germany.

1185 Nachrichten der Hamburgischen Gesellschaft zur Beförderung der Künste und
nützlichen Gewerbe. - Hamburg: 1790. **CONTINUED BY:** Verhandlungen und
Schriften der Hamburgischen Gesellschaft zur Beförderung der Künste und
nützlichen Gewerbe. **SOURCES:**(Ref.6,p.289). **LOCATIONS:** ZDB
SUBJECT: Economics - Societies - Hamburg.

1186   Nachrichten von dem Fortgange der Westphälischen naturforschenden Gesellschaft zu Brockhausen. - No. 1-2, 1798-1799. Düsseldorf: Dänzer, 1798-1799. **SOURCES:** K3373, S2534a   **SUBJECT:** Natural history - Societies - Brockhausen.

1187   Nachrichten von dem Leben und den Schriften jeztlebender teutscher Aerzte und Wundärzte, Thierärzte, Apotheker und Naturforscher. - v.1, 1799. Hildesheim: J.G. Gerstenberg, 1799. **EDITOR:** Johann Caspar Philipp Elwert. **SOURCES:** OCLC **LOCATIONS:** ULS **SUBJECT:** Medicine - Periodicals - Germany - Biography.

1188   Nachrichten von dem Zustande der Wissenschaften und Künste in den Kgl. dänischen Reichen und Ländern. - v.1-3 (St. 1-24) 1753-1757. Copenhagen; Leipzig: Johann B. Ackermann, 1753-1757. **EDITOR:** Anton Friedrich Büsching. **CONTINUED BY:** Fortgesetzte Nachrichten von dem Zustande der Wissenschaften und Künste in den Kgl. dänischen Reichen und Ländern. **MICROFORMS:** RM(3). **SOURCES:** B3217, K193, S628 **LOCATIONS:** BUC, ULS, ZDB **SUBJECT:** General - Periodicals - Denmark - Reviews.

1189   Nachrichten von einer hallischen Bibliothek. - v.1-8 (St. 1-48), 1748-1751. Halle: J.J. Gebauer, 1748-1751. Siegmund Jacob Baumgarten. **CONTINUED BY:** Nachrichten von merkwürdigen Büchern. **INDEXES:** v.12 of continued title is index to both series. **SOURCES:** K159 **LOCATIONS:** BUC, CCF, ULS, ZDB **SUBJECT:** General - Periodicals - Germany - Reviews.

1190   Nachrichten von gelehrten Sachen von dem Akademie zu Erfurt. - v.1-7, 1797-1802. Erfurt: 1797-1802. **CONTINUATION OF:** Erfurtische gelehrte Nachrichten. **CONTINUATION OF:** Erfurtische gelehrte Zeitungen. **CONTINUATION OF:** Erfurtische gelehrte Zeitung. **SOURCES:** B1623b, K454, S2547d **LOCATIONS:** ULS, ZDB **SUBJECT:** General - Periodicals - Germany - Reviews.

1191   Nachrichten von jetzt lebenden Aerzten und Naturforchern in und um Teutschland. - v.1-4, 1749-1773. Wolfenbüttel: Albrecht, 1749-1773. Imprint varies. **OTHER TITLE:** Nachrichten von den vornehmsten Lebenumständen und Schriften jetzlebender berühmten Aerzte und Naturforscher in und um Deutschland. **EDITORS:** Friedrich Börner and Ernst Gottfried Baldinger. **SOURCES:** K3530 **LOCATIONS:** BN, ZDB **SUBJECT:** Medicine - Periodicals - Germany - Biography.

1192   Nachrichten von merkwürdigen Büchern. - v.1-12 (St. 1-66) 1752-1758. Halle: J.J. Gebauer, 1752-1758. **CONTINUATION OF:** Nachrichten von einer hallischen Bibliothek. **EDITOR:** Siegmund Jacob Baumgarten. **INDEXES:** v.12 is an index to this title and its predecessor. **SOURCES:** K182 **LOCATIONS:** BUC, ULS, ZDB **SUBJECT:** General - Periodicals - Germany - Reviews.

1193   Nachrichten von Verbesserung der Landwirthschaft und des Gewerbes / Braunschweig-Lünebergische Landwirthschafts- Gesellschaft. - Bd.1-3,

1765-1778. Celle: Gselllius, 1765-1778. **CONTINUED BY:** Neue Abhandlungen und Nachrichten. **CONTINUED BY:** Annalen des Ackerbaues. **SOURCES:** K2894 **LOCATIONS:** ZDB **SUBJECT:** Agriculture - Societies - Celle.

1194 The Naturalists' journal. - London: Benjamin White, 1767-1775. A ruled notebook for entering meteorological observations (Ref.100). **EDITOR:** Daines Barrington. **SOURCES:** B3252, S388-9 **LOCATIONS:** BN **SUBJECT:** Meteorology - Periodicals - England.

1195 The Naturalist's miscellany. - v.1-24, 1789-1813. London: Elizabeth Nodder, 1789-1813. Text in English and Latin. **OTHER TITLE:** Vivarium natur. **EDITORS:** George Shaw and Frederick P. Nodder. **CONTINUED BY:** Zoological miscellany. **INDEXES:** Index in v.24. **SOURCES:** B3254, OCLC, S390 **LOCATIONS:** BUC, CCF, ULS **SUBJECT:** Natural history - Periodicals - England.

1196 The Naturalists' pocket magazine, or, compleat cabinet of the curiosities and beauties of nature. - v.1-7, 1798-1802. London: Harrison Cluse & Co., 1798-1802. A book in parts. **SOURCES:** B3256 **LOCATIONS:** BUC, ULS **SUBJECT:** Natural history - Periodicals - England.

1197 La Nature considérée sous des différents aspects, ou lettres sur les animaux, les végétaux et les minéraux. - v.1-47, 1771-1783, ns. v.1-9, 1787-1789. Paris: Durand, 1771-1789. **CONTINUATION OF:** Lettres périodique sur la méthode de s'enrichir promptement et conserver sa santé par la culture des vgetaux. **CONTINUATION OF:** Lettres périodiques, curieuses, utiles et intéresantes sur les avantages que la société peut retirer de a connaissance des animaux. **CONTINUATION OF:** Lettres hebdomadaires sur l'utilité minéraux dans la société civil. First ser. edited by Buc'hoz and the second by Bertholon and Boyer. **OTHER TITLE:** Journal d'histoire naturelle. **EDITOR:** Pierre-Joseph Buc'hoz, Alexis Boyer and Pierre Bertholon. **MICROFORMS:** RM(3). **SOURCES:** B2691a, S1466 **LOCATIONS:** BUC, CCF **SUBJECT:** Science - Periodicals - France. **SUBJECT:** Medicine - Periodicals - France.

1198 Der Naturforscher. - v.1-30, 1774-1804. Halle: Johann Jacob Gebauer, 1774-1804. **EDITORS:** Johann Ernst Immanuel Walch (1774-1779) and Johann Christian Daniel Schreber (1780-1804). **INDEXES:** Indexes to v.1-10 in 10, 11-20 in 20 and 21-30 in 30. See also Ref 53. **MICROFORMS:** RM(3). **SOURCES:** B3263, G695, K3256, OCLC, S2763 **LOCATIONS:** BUC, CCF, CCN, ULS, ZDB **SUBJECT:** Natural history - Periodicals - Germany.

1199 Der Naturforscher, eine physikalische Wochenschrift. - v.1-3 (No.1-78) 1747-1748 (Ref.101). Leipzig: Johann Gottlieb Crull, 1747-1748. Weekly. **EDITORS:** Christlob Mylius and Gotthold Ephraim Lessing. **SOURCES:** B3262, G694, K3198, S2976 **LOCATIONS:** BUC, CCF, ULS, ZDB **SUBJECT:** Natural history - Periodicals - Germany - Popular.

1200 Der Naturforscher, oder, Abhandlungen über ausgewählte Gegenstände aus dem Reiche der Natur. - Görlitz: Hermsdorff and Anton, 1795. **SOURCES:** B3264, G545, S2653 **LOCATIONS:** CCF, ZDB **SUBJECT:** Natural history - Periodicals - Germany.

1201 Natuurkundige verhandelingen van de Koninglijke Maatschappij der Wetenschappen te Haarlem. - v.1-24, 1799-1844. Amsterdam: Johannes Allart, 1799-1844. **CONTINUATION OF:** Verhandelingen uitgegeven door de Hollandsche maatschappij der wetenschappen te Haarlem. A translation in one vol. by D. von Halem was published in Leipzig in 1802 under the title: Naturhistorische Abhandlungen. **MICROFORMS:** RM(3). **SOURCES:** G462, S847 **LOCATIONS:** CCF, CCN, ULS, ZDB **SUBJECT:** Science - Societies - Haarlem.

1202 Natuur- en geneeskundig bibliotheek, bevattende den zaakelyken inhoud van alle nieuwe werken welke in de geneeskunde en natuurlyke historie buiten ons vaderland uitkomen. - v.1-11, 1765-1775. s'Gravenhage: Pieter van Cleef, 1765-1775. **EDITOR:** Eduard Sandifort. **CONTINUED BY:** Nieuwe natuur-en geneeskundige bibliotheek. **INDEXES:** v.11 contains an index to the series. **SOURCES:** G696, OCLC **LOCATIONS:** BUC, CCF, CCN, ULS **SUBJECT:** Medicine - Periodicals - Netherlands - Reviews **SUBJECT:** Natural history - Periodicals - Netherlands - Reviews

1203 Natuur en Konstkabinet, of, kabinet der natuurlyke historien, wetenschappen, konsten en handwerken. - v.1-9, 1719-1732. Amsterdam: Strik, 1719-1732. **OTHER TITLE:** Kabinet der natuurlyke historien, wetenschaappen, konsten en handwerken. **EDITOR:** Willem van Ranouw. **INDEXES:** Index to series in v.9. **SOURCES:** B3280, S758 **LOCATIONS:** BN, CCN, ULS, ZDB **SUBJECT:** Natural history - Periodicals - Netherlands.

1204 Natuur-scheikundige verhandelingen. - No. 1-4, 1799-1802. Amsterdam: 1799-1802. **CONTINUATION OF:** Recherches physico-chemiques. **EDITOR:** Johann Rudolph Deiman. **SOURCES:** B3285 **LOCATIONS:** CCF, CCN **SUBJECT:** Chemistry - Periodicals - Netherlands.

1205 Natuurkundige aanmerkingen, waarneemingen en onderviningen, getrokken uit de Philosophical transactions. - v.1-2, 1735. Amsterdam: 1735. Translated by Pieter Le Clercq. **TRANSLATION OF:** Philosophical transactions. **SOURCES:** S767 **SUBJECT:** Science - Societies - London - Translations.

1206 Natuurkundige verhandelingen of verzameling van stukken de natuurkunde, geneeskunde en natuurlyke historie betreffende, getrokken uit de geachte werken van engelsche, fransche en hoogjuitsche schryvers. - v.1, 1767. Amsterdam: 1767. **SOURCES:** B3284, S759 **LOCATIONS:** ULS **SUBJECT:** Science - Periodicals - Netherlands - Translations. **SUBJECT:** Medicine - Periodicals - Netherlands - Translations.

1207 Natuurkundige verhandelingen of verzameling van stukken de natuurkunde, geneeskunde, oeconomie, natuurlyke historieenz, betreffende. - v.1-5,

1772-1777. Amsterdam: Albert van der Kroe, 1772-1777. **SOURCES:** G697 **LOCATIONS:** BUC, CCN **SUBJECT:** Science - Periodicals - Netherlands - Translations. **SUBJECT:** Medicine - Periodicals - Netherlands - Translations.

1208 Nautical almanac and astronomical ephemeris for the meridien of the Royal observatory of Greenwich. - v.1-114, 1767-1880. London: Commissioners of Longitude, 1767-1880. Translated into French in the 18th century by the Académie de marine (Ref.102, p.392); 41 v. published in New York by Blunt, 1812-52; and 10 v.by Garnett in New Brunswick, 1804-1813. There were 2nd editions of some of the volumes. **SOURCES:** OCLC, S393, S4151, S4207 **LOCATIONS:** BUC, CCF, CCN, ZDB **SUBJECT:** Astronomy - Periodicals - England. **SUBJECT:** Naval art and science - Periodicals - England.

1209 The Naval chronicle. - v.1-40, 1799-1818. London: Bunney and Gold, 1799-1818. Imprint varies. **MICROFORMS:** PMC(1,5,6), RP(1). **SOURCES:** B3298 **LOCATIONS:** BUC, CCF, CCN, ULS **SUBJECT:** Naval art and science - Periodicals - England. **SUBJECT:** Travel - Periodicals - England.

1210 Neue Abhandlungen aus der Naturlehre, Haushaltungskunst und Mechanik. Bd.1-Bd.12, (1780-1790) 1784-1792. Leipzig: 1784-1792. **CONTINUATION OF:** Abhandlungen aus der Naturlehre, Haushaltungskunst und Mechanik. Translated by Abraham Gotthelf Kästner, Johann Dietrich Brandis and Heinrich Friedrich Link. **TRANSLATION OF:** Nya handlingar, Kungliga Svenska vetenskapsakademien. **SOURCES:** B14a, G477f, K3296 **LOCATIONS:** BUC, CCN, ULS, ZDB **SUBJECT:** Science - Societies - Stockholm - Translations.

1211 Neue Abhandlungen und Nachrichten / Königliche Grossbrittanische Churfürstliche Braunschweig-Lünebergische Landwirthschafts-Gesellschaft. - Bd.1-4, 1787-1794. Celle: Richter, 1787-1794. **CONTINUATION OF:** Nachrichten von Verbesserung der Landwirthschaft und des Gewerbes. **CONTINUATION OF:** Annalen der niersächsischen Landwirthschaft. 2nd ed. of v.1-2, Hannover, 1794. **CONTINUED BY:** Annalen des Ackerbaues. **SOURCES:** G464, K2953, S3336 **LOCATIONS:** BUC, ZDB **SUBJECT:** Agriculture - Societies - Celle.

1212 Neue alchymistische Bibliothek für Naturkündiger unseres Jahrhunderts. - v.1-2, 1770-1774. Frankfurt: Brönner, 1770-1774. **EDITOR:** Friedrich Joseph Wilhelm Schröder. **CONTINUED BY:** Neue Sammlung der Bibliothek für die höhere Naturwissenschaft. **SOURCES:** K3248 **LOCATIONS:** CCF, CCN, ZDB **SUBJECT:** Alchemy - Periodicals - Germany - Reviews.

1213 Neue allgemeine deutsche Bibliothek. - Bd.1-107, 1793-1806. Kiel: Carl Ernest Bohn; Stettin: Christoph Friedrich Nicolai, 1793-1806. **CONTINUATION OF:** Allgemeine deutsche Bibliothek. **EDITOR:** Christoph Friedrich Nicolai. Supplements published under the title Anhang. **INDEXES:** Indexes to Bd.1-28, 28-68, 69-194, and 105-107 are included in Anhänge. **MICROFORMS:**

GO(9). **SOURCES:** G588a, K444, OCLC **LOCATIONS:** BUC, CCF, CCN, ULS, ZDB **SUBJECT:** General - Periodicals - Germany - Reviews.

1214 Neue Anmerkungen über alle Theile der Natur-lehre, aus denen englischen Transactionen, denen Gedenkschriften der Akademie der Wissenschaften in Paris, und anderen mehre zusammengezogen und gesamelt. - v.1-2, 1753-1754. Copenhagen; Leipzig: Ackermann, 1753-1754. **TRANSLATION OF:** Philosophical transactions. **SOURCES:** K3213, S635 **LOCATIONS:** ZDB **SUBJECT:** Science - Societies - London - Translations. **SUBJECT:** Science - Societies - Paris - Translations.

1215 Neue Arzneien wider die medicinischen Vorurtheile. Bd.1-Bd.2, 1768-1769. Langensalza: 1768-1769. **CONTINUATION OF:** Arzneyen, eine physikalisch-medicinische Monatsschrift. **EDITOR:** Ernest Gottfried Baldinger. **SOURCES:** K3550 **SUBJECT:** Medicine - Periodicals - Germany - Popular.

1216 Neue Auszüge aus den besten ausländischen Wochen- und Monatschriften. - v.1-10, 1765-1769. Frankfurt: Scheper, 1765-1769. **EDITOR:** Christian Friedrich Schwan. **CONTINUED BY:** Bibliothek der ausländischen neuesten Litteratur in Auszügen aus den besten Wochen- und Monatschriften. **SOURCES:** B762, K4445, S2601 **LOCATIONS:** ZDB **SUBJECT:** General - Periodicals - Germany - Reviews.

1217 Neue Auszüge aus dissertationen für Wundärzte. - v.1-18, 1774-1783. Frankfurt; Leipzig: Adam Friedrich Böhme, 1774-1783. **EDITOR:** Friedrich Ausgust Weiz (Waitz). **CONTINUED BY:** Neue Lecture für deutsche Wundärzte in Auszügen. **SOURCES:** G303, K3564, OCLC **LOCATIONS:** NUC, ZDB **SUBJECT:** Surgery - Periodicals - Germany - Dissertations.

1218 Neue Beiträge zur Natur- und Arznei-Wissenschaft. - v.1-3, 1782-1786. Berlin: Mylius, 1782-1786. **EDITOR:** Christian Gottlieb Selle. **SOURCES:** B835, G605, S2359 **LOCATIONS:** CCF, CCN, ULS, ZDB **SUBJECT:** Natural history - Periodicals - Germany. **SUBJECT:** Medicine - Periodicals - Germany.

1219 Neue Beiträge zur Völker-und Länderkunde.Bd.1-Bd.3, 1790-1793. Leipzig: Gotthelf Kummer, 1790-1793. **CONTINUATION OF:** Beiträge zur Völker-und Länderkunde. **OTHER TITLE:** Auswahl kleiner Reisebeschreibungen und anderer statistischer und geographischer Nachrichten. **EDITORS:** Johann Reinhold Forster and Matthias Christian Sprengel. **CONTINUED BY:** Auswahl der besten ausländischen geographischen und statistischen Nachrichten. **SOURCES:** B760, G239a, K1232, S2898 **LOCATIONS:** BUC, CCF, CCN, ULS, ZDB **SUBJECT:** Geography - Periodicals - Germany.

1220 Neue Berliner Beiträge zur Landwirthschaftswissenschaft. - Bd.1-2 in 12 issues, 1791-1794. Berlin: Maurer, 1791-1794. **CONTINUATION OF:** Berliner Beiträge zur Landwirthschaftswissenschaft. **EDITOR:** G. Ludwig Grossmann.

SOURCES: K2977 **LOCATIONS:** CCN, ULS, ZDB **SUBJECT:** Agriculture - Periodicals - Germany.

1221 Neue Berlinische Monatsschrift. - Bd.1-Bd.26, 1799-1811. Berlin und Stettin: Friedrich Nicolai, 1799-1811. Monthly. **CONTINUATION OF:** Berlinische Monatsschrift. **CONTINUATION OF:** Berlinische Blätter. **EDITOR:** Johann Erich Biester. **INDEXES:** Indexed in Hocks (Ref.55). **MICROFORMS:** GO(9). **SOURCES:** G609, K4625 **LOCATIONS:** BUC, CCF, ULS, ZDB **SUBJECT:** General - Periodicals - Germany.

1222 Neue Denkwürdigkeiten für Aerzte und Geburtshelfer. - v.1 (in two parts) 1797-1799. Göttingen: Johann Georg Rosenbusch, 1797-1799. **CONTINUATION OF:** Denkwürdigkeiten für die Heilkunde und Geburtshülfe. **EDITOR:** Friedrich Benjamin Osiander. **SOURCES:** G140 **LOCATIONS:** CCF, ULS, ZDB **SUBJECT:** Medicine - Periodicals - Germany. **SUBJECT:** Obstetrics - Periodicals - Germany.

1223 Neue Entdeckungen und Beobachtungen aus der Physik, Naturgeschichte und Oeconomie. - v.1, 1791. Frankfurt: 1791. **EDITOR:** Bernhard Sebastien von Nau. **SOURCES:** B1581, S2602 **SUBJECT:** Science - Periodicals - Germany.

1224 Das Neue gelehrte Europa. - v.1-19, 1752-1773. Wolfenbüttel: Johann Christian Meisner, 1752-1773. **EDITOR:** Johann Christian Strodtmann. **SOURCES:** G699 **LOCATIONS:** BUC, ULS, ZDB **SUBJECT:** General - Periodicals - Germany - Reviews.

1225 Neue gerichtlich-medicinische Beobachtungen. - Bd.1, 1798. Königsberg: 1798. **CONTINUATION OF:** Gerichtlich-medicinische Beobachtungen. Published as a supplement to Metzger's Kurzgefassten System der gerichtlichen Arzneiwissenschaft. **EDITOR:** Johann Daniel Metzger. **SOURCES:** G381 **LOCATIONS:** ULS, ZDB **SUBJECT:** Medical jurisprudence - Periodicals - Germany.

1226 Neue gesellschaftliche Erzählungen für die Liebhaber der Naturlehre, Haushaltungs-wissenschaft, Arzneykunst und Sitten. - v.1-4, 1758-1762. **CONTINUATION OF:** Gesellschaftliche Erzählungen für die Liebhaber der Naturlehre, Haushaltungs-wissenschaft, Arzneykunst und Sitten. **EDITOR:** Johann Daniel Titius. **SOURCES:** B1905a, K3230, S1749 **LOCATIONS:** BUC, CCF, ULS, ZDB **SUBJECT:** Science - Periodicals - Germany.

1227 Neue Hallische gelehrte Zeitungen. - v.1-27, 1766-1792. Halle: Curt, 1766-1792. **OTHER TITLE:** Hallische neue gelehrte Zeitungen; Hallische gelehrte Zeitungen. **EDITORS:** Christian Adolph Klotz and others. **SOURCES:** B2022, K258 **LOCATIONS:** CCF, ZDB **SUBJECT:** General - Periodicals - Germany.

1228 Neue Handlungsbibliothek. - v.1 (St.1-4) 1799-1800. Ronneberg: Schumann; Leipzig: Barth, 1799-1800. **CONTINUATION OF:** Handlungsbibliothek. **EDITOR:** A. Schumann. **SOURCES:** K2786 **SUBJECT:** Commerce - Periodicals - Germany.

1229 Neue Lecture für deutsche Wundärzte in Auszügen. - v.1-2, 1785-1786. Leipzig; Mannheim: A.F. Böhme, 1785-1786. **CONTINUATION OF:** Neue Auszüge aus dissertationen für Wundärzte. **EDITOR:** Friedrich Ausgust Weiz (Waitz) **SOURCES:** G305 **LOCATIONS:** NUC **SUBJECT:** Surgery - Periodicals - Germany - Dissertations.

1230 Neue Litterarische Nachrichten für Aerzte Wundärzte und Naturforscher. - v.1 (St. 1-48) (1785-86) 1787, v.2 (St. 1-60) 1787, v.3 (Quart. 1-4) (1788-89) 1789. Halle: Johann Christian Hendel, 1786-1789. **OTHER TITLE:** Neue Hallische Nachrichten für Aerzte. **EDITOR:** Curt Sprengel. **SOURCES:** K3628 **LOCATIONS:** CCF, ULS **SUBJECT:** Medicine - Periodicals - Germany. **SUBJECT:** Natural history - Periodicals - Germany. **SUBJECT:** Surgery - Periodicals - Germany.

1231 Neue Litteratur und Beiträge zur Kenntniss der Naturgeschichte vorzüglich der Conchylien und Fossilien. - Leipzig: J.G. Müller, 1784-1787. **CONTINUATION OF:** Journal für die Liebhaber des Steinreichs. **CONTINUATION OF:** Für die Litteratur und Kenntniss der Naturgeschichte, sonderlich der Conchylien und der Steine. **EDITOR:** Johann Samuel Schröter. **SOURCES:** B247, G547, K3299, OCLC, S2980 **LOCATIONS:** BUC, CCF, CCN, ULS **SUBJECT:** Natural history - Periodicals - Germany. **SUBJECT:** Mollusks - Periodicals - Germany. **SUBJECT:** Mineralogy - Periodicals - Germany.

1232 Neue Mannigfaltigkeiten, eine gemeinnützige Wochenschrift. - v.1-4, 1774-1777. Berlin: Bosse, 1774-1777. Weekly. **CONTINUATION OF:** Mannigfaltigkeiten. Imprint varies. **EDITORS:** Friedrich Heinrich Wilhelm Martini. **CONTINUED BY:** Neueste Mannigfaltigkeiten. **CONTINUED BY:** Allerneueste Mannifgaltigkeiten. **SOURCES:** B2381, K5355, S2347 **LOCATIONS:** BUC, CCF, ULS, ZDB **SUBJECT:** General - Periodicals - Germany.

1233 Neue medicinische Bibliothek. - v.1-8 (St.1-45) 1754-1773. Göttingen: Abraham Vandenhoek, 1754-1773. **CONTINUATION OF:** Medicinische Bibliothek (Vogel). **EDITOR:** Rudolf Augustin Vogel. **CONTINUED BY:** Medicinisch-practische Bibliothek (Murray). **CONTINUED BY:** Medicinische Bibliothek (Blumenbach). **SOURCES:** G307, K3536, OCLC **LOCATIONS:** BUC, CCF, ULS **SUBJECT:** Medicine - Periodicals - Germany - Reviews.

1234 Neue medicinische Litteratur. - v.1-4 (4 St. each) 1787-1794. Leipzig: Carl Friedrich Schneider, 1787-1794. **CONTINUATION OF:** Medicinische Litteratur für practische Aerzte. **EDITOR:** Johann Christian Traugott Schlegel. **CONTINUED BY:** Übersicht der neuesten medicinischen Litteratur.

SOURCES: G301a, K3639   **LOCATIONS:** CCN, ULS, ZDB   **SUBJECT:** Medicine - Periodicals - Germany - Reviews.

1235 Neue medicinische und chirurgische Anmerkungen. - St. 1-2, 1769-1772. Berlin; Stralsund: Gottfried August Lange, 1769-1772. **CONTINUATION OF:** Sammlung medicinischer und chirurgischer Anmerkungen. **EDITOR:** Johann Friedrich Henkel. **SOURCES:** K3553,OCLC   **LOCATIONS:** BN **SUBJECT:** Medicine - Periodicals - Germany.   **SUBJECT:** Surgery - Periodicals - Germany.

1236 Neue medicinischshe und chirurgische Wahrnehmungen. - v.1, 1778. Altenburg: Richter, 1778.   **LOCATIONS:** ZDB   **SUBJECT:** Medicine - Periodicals - Germany.   **SUBJECT:** Surgery - Periodicals - Germany.

1237 Neue Monatliche Beiträge zur Naturkunde. - St.1-5, 1770. Schwerin: W. Bärensprung, 1770.   **CONTINUATION OF:** Physicalische Briefe. **CONTINUATION OF:** Monatliche Beiträge zur Naturkunde. **CONTINUATION OF:** Fortgesetzte Beiträge zur Naturkunde. **EDITOR:** Johann Daniel Denso.   **SOURCES:** K3246, S3217   **LOCATIONS:** ZDB **SUBJECT:** Natural history - Periodicals - Germany.

1238 Neue monatliche Beiträge zur Naturkunde (Dessau). - St. 1-5, 1782. Dessau: Buchhandlung der Gelehrten, 1782.   **SOURCES:** B3646a, K3286, S2495 **SUBJECT:** Natural history - Periodicals - Germany.

1239 Neue Nordische Beiträge zur physicalischen und geographischen Erd- und Völkerbeschreibung, Naturgeschichte und Oekonomie. - v.1-7, 1781-1796. St. Petersburg; Leipzig: Johann Zacharias Logan, 1781-1796. v.5-7 also as Neueste Beiträge. Includes many of the editor's own papers. **OTHER TITLE:** Neueste Nordische Beiträge; Nordische Beiträge zum Wachsturn der Naturkunde. **EDITOR:** Peter Simon Pallas.   **SOURCES:** B3341, G701, K3276, OCLC, S3729   **LOCATIONS:** BUC, CCF, CCN, ULS, ZDB   **SUBJECT:** Anthropology - Periodicals - Russia.   **SUBJECT:** Natural history - Periodicals - Russia.   **SUBJECT:** Geography - Periodicals - Russia.

1240 Neue ökonomische Beiträge zur Verbesserung der Landwirtschaft in Niedersachsen. - St. 1, 1800. Berlin: 1800.   **CONTINUATION OF:** Oekonomische Beiträge zur Verbesserung der Landwirtschaft in Niedersachsen. **EDITOR:** Johann Daniel Denso.   **SOURCES:** K3028   **SUBJECT:** Agriculture - Periodicals - Germany.

1241 Neue ökonomische Nachrichten der patriotischen Gesellschaft in Schlesien. - v.1-5, 1780-1784. Breslau: Wilhelm G. Korn, 1780-1784. **CONTINUATION OF:** Oekonomische Nachrichten der patriotischen Gesellschaft in Schlesien. **EDITOR:** Carl Friedrich Tschirner.   **SOURCES:** K2930   **LOCATIONS:** ZDB   **SUBJECT:** Economics - Societies - Selesia.

1242 Neue ökonomische Nachrichten (Leipzig). - v.1-6 (St. 1-60) 1764-1773. Leipzig: Johann Wendler, 1764-1773.   **CONTINUATION OF:** Oekonomische

Nachrichten (Leipzig). **EDITOR:** Peter von Hohenthal. **SOURCES:** K2881 **LOCATIONS:** BUC, CCF, CCN, ULS, ZDB **SUBJECT:** Economics - Periodicals - Germany.

1243   Neue philosophische Abhandlungen der baierischen Akademie der Wissenschaften, Munich. - Bd.1(1778); Bd.2(1780), Bd.3(1783), Bd.4(1786), Bd.5(1790), Bd.6(1794), Bd.7(1797) each in two parts 1) Abhandlungen, 2) Preisschriften. Munich: Paul Vötter, 1778-1797. **CONTINUATION OF:** Abhandlungen der Churfürstlich-baierischen Akademie der Wissenschaften. **CONTINUED BY:** Physikalische Abhandlungen der Königlich-baierischen Akademie der Wissenschaften. **INDEXES:** Indexed in Vesenyi (Ref.1). **MICROFORMS:** RM(3). **SOURCES:** G459a, K329, OCLC, S3109i **LOCATIONS:** CCF, CCN, ULS, ZDB **SUBJECT:** Science - Societies - Munich.

1244   Neue physicalische Belustigungen. - v.1-2, 1770-1771, v.3, 1773. Prague: F.A. Höchenberger, 1770-1773. **EDITOR:** Franz Zeno. **SOURCES:** B3639. K3247, S3505 **LOCATIONS:** CCF, ULS **SUBJECT:** Natural history - Periodicals - Germany.

1245   Neue Sammlung auserlesener Wahrnehmungen aus allen Theilen der Arzneywissenschaft. - v.1-10, 1766-1775. Strassburg: Johann Gottfried Bauer, 1766-1775. **CONTINUATION OF:** Sammlung auserlesener Wahrnehmungen aus allen Theilen der Arzney-wissenschaft, der Wund-Arzney- und der Apotheker- Kunst. **TRANSLATION OF:** Journal de médecine, chirurgie et de pharmacie. **SOURCES:** GIII:24b, K3547 **LOCATIONS:** ZDB **SUBJECT:** Medicine - Periodicals - France - Translations.

1246   Neue Sammlung der auserlesensten und neuesten Abhandlungen für Wundärzte aus verschiedene Sprachen übersetzt. St.1-St.24, 1782-1789. Leipzig: Weygand, 1782-1789. **CONTINUATION OF:** Sammlung auserlesensten und neuesten Abhandlungen für Wundärzte. Same title published in 7v. in Mannheim, 1790 (ZDB) **EDITOR:** Ernst Benjamin Gottlieb Hebenstreit. **CONTINUED BY:** Neueste Sammlung der auserlesensten und neuesten Abhandlungen für Wundärzte. **SOURCES:** G382, K3579 **LOCATIONS:** BUC, CCF, ULS, ZDB **SUBJECT:** Surgery - Periodicals - Germany.

1247   Neue Sammlung der Bibliothek für die höhere Naturwissenschaft und Chemie. - v.1-2, 1775-1776. Leipzig: F. Fluttner and J.C. Müller, 1775-1776. **CONTINUATION OF:** Neue alchymistische Bibliothek für Naturkündiger unseres Jahrhunderts. Imprint varies. A 2nd ed. appeared in Leipzig in 1779. **EDITOR:** Friedrich Joseph Wilhelm Schröder. **SOURCES:** K3262 **LOCATIONS:** ZDB **SUBJECT:** Alchemy - Periodicals - Germany - Reviews.

1248   Neue Sammlung kleiner Abhandlungen einiger Gelehrten in Schweden über Naturgeschichte. - v.1, 1774. Copenhagen: 1774. **SOURCES:** B4184, S636 **LOCATIONS:** ZDB **SUBJECT:** Natural history - Periodicals - Denmark.

1249   Neue Sammlung physisch-ökonomischer Schriften herausgegeben von der ökonomischen Gesellschaft des Kantons Bern. - Bd.1-3, 1779-1785. Bern:

Typographische Gesellschaft, 1779-1785. **CONTINUATION OF:** Sammlungen von landwirthschaftlichen Dingen der Schweitzerischen Gesellschaft in Bern. **CONTINUATION OF:** Abhandlungen und Beobachtungen durch die ökonomische Gesellschaft zu Bern gesammelt. **CONTINUED BY:** Neueste Sammlung von Abhandlungen und Beobachtungen von der ökonomischen Gesellschaft in Bern. **SOURCES:** K2927, S2141b   **LOCATIONS:** BUC, ULS, ZDB   **SUBJECT:** Economics - Societies - Bern.

1250   Neue Sammlung vermischter ökonomischer Schriften und Beendigung der Bienenbibliothek. - Heft 1-Heft 14, 1792-1798. Dresden: Walther, 1792-1798. **CONTINUATION OF:** Auserlesene Sammlung vermischter ökonomischer Schriften. **EDITOR:** Johann Riem. **CONTINUED BY:** Neufortgesetzte Sammlung ökonomischer und Bienenschriften. **CONTINUED BY:** Halbjahr-beiträge zur ökonomie und Naturgeschichte für Landwirthe und Bienenfreunde. **SOURCES:** K2987   **LOCATIONS:** BUC, CCF, NUC, ZDB **SUBJECT:** Economics - Periodicals - Germany.

1251   Neue Sammlung verschiedener Schriften der grösten Gelehrten in Schweden, für die Liebhaber der Arzneiwissenschaft, der Naturgeschichte, Chymie und Oeconomie. - v.1, 1775. Copenhagen: 1775. **SOURCES:** B4192, S637 **LOCATIONS:** ZDB   **SUBJECT:** Natural history - Periodicals - Sweden - Translations. **SUBJECT:** Medicine - Periodicals - Denmark.

1252   Neue Sammlung von Versuchen und Abhandlungen der naturforschenden Gesellschaft in Danzig. - v.1, 1778. Danzig: Wedel, 1778. **CONTINUATION OF:** Versuche und Abhandlungen der naturforschenden Gesellschaft zu Danzig. **SOURCES:** G482a, K3469, S2485,2   **LOCATIONS:** BUC, CCF, CCN, ULS, ZDB   **SUBJECT:** Natural history - Societies - Danzig.

1253   Neue Schriften der Gesellschaft naturforschender Freunde zu Berlin. - Bd.1-Bd.4, 1795-1803.   Berlin: Reiner, 1795-1803.   **CONTINUATION OF:** Beschäftigungen der Berlinischen Gesellschaft naturforschender Freunde. **CONTINUATION OF:** Schriften der Berlinischen Gesellschaft Naturforschender Freunde. **INDEXES:** Indexed in Geuss (Ref.53, v.3). **MICROFORMS:** IDC(8).   **SOURCES:** B909b, K3364, OCLC **LOCATIONS:** BUC,CCN,ULS   **SUBJECT:** Natural history - Societies - Berlin.

1254   Neue Schriften / Gesellschaft der naturforschenden Freunde, Düsseldorf. - v.1-2, 1798-1805.   Düsseldorf: 1798-1805.   **OTHER TITLE:** Neue Schriften, Gesellschaft der naturforschenden Freunde, Westphalen. **SOURCES:** G470, K3374, S2533   **LOCATIONS:** BUC, ULS   **SUBJECT:** Natural history - Societies - Düsseldorf.

1255   Neue Untersuchungen über den thierischen Magnetismus. - Tübingen: Cotta, 1789. **CONTINUATION OF:** Untersuchungen den thierischen Magnetismus. **EDITOR:** Eberhard Gmelin.   **CONTINUED BY:** Materialen für die Anthropologie.   **SOURCES:** G237a   **SUBJECT:** Animal magnetism - Periodicals - Germany.

1256 Neue Versuche nützlicher Sammlungen zu der Natur- und Kunst-Geschichte, sonderlich von Ober-Sachsen. - v.1-4 (no. 1-48) 1747-1765. Schneeberg: Carl Wilhelm Fulden, 1747-1765. **EDITOR:** Christian Gottlob Grundig. **SOURCES:** B4673, K3200, S3214 **LOCATIONS:** BL, CCF, ZDB **SUBJECT:** Natural history - Periodicals - Germany.

1257 Neue Versuche und Bemerkungen aus der Arztneykunst und übrigen Gelehrsamkeit / Gesellschaft zu Edinburg. - v.1-3, 1755-1775. Altenburg: 1755-1775. **CONTINUATION OF:** Medicinischen Versuche und Bemerckungen, Gesellschaft in Edinburg. Translated by Abraham Gotthelf Kästner. **TRANSLATION OF:** Essays and observations, physical and literary. **SOURCES:** K3538 **LOCATIONS:** ZDB **SUBJECT:** Medicine - Societies - Edinburgh - Translations.

1258 Neue Wahrheiten zum Vortheil der Naturkunde und des gesellschaftlichen Lebens der Menschen. - St.1-12, 1754-1758. Leipzig: Bernhard Christoph Breitkopf, 1754-1758. **EDITOR:** Johann Heinrich Gottlob von Justi. **CONTINUED BY:** Fortgesetzte Bemühungen zum Vortheil der Naturkunde. **SOURCES:** K2867 **LOCATIONS:** ZDB **SUBJECT:** Natural history - Periodicals - Germany. **SUBJECT:** Economics - Periodicals - Germany.

1259 Neue Würzburger gelehrte Anzeigen. - v.1-4, 1799-1802. Würzburg: Köl, 1799-1802. **CONTINUATION OF:** Würzburger gelehrte Anzeigen. **CONTINUATION OF:** Würzburger wöchentliche Anzeigen von gelehrten und andern gemeinnützigen Gegenständen. **SOURCES:** B4808b, K458, S3333.3 **LOCATIONS:** ZDB **SUBJECT:** General - Periodicals - Germany - Reviews.

1260 Neue Zeitung für Kaufleute, Fabrikanten und Manufacturisten. - v.1-3, 1800-1802. Weimar: 1800-1802. Weekly. **CONTINUATION OF:** Handlungszeitung. **OTHER TITLE:** Neue Zeitung, oder Wochentliche Nachrichten für Handel. **EDITOR:** Johann Adolph Hildt. **SOURCES:** K2792 **LOCATIONS:** CCF, ULS, ZDB **SUBJECT:** Commerce - Periodicals - Germany.

1261 Neuer Unterricht für Wundärzte, von einer Gesellschaft von Wundärzten. - v.1-2, 1785-1787. Halle: Hemmerde und Schwetschke, 1785-1787. **SOURCES:** G66 **SUBJECT:** Surgery - Periodicals - Germany.

1262 Neuere Abhandlungen der Königlichen böhmischen Gesellschaft der Wissenschaften zu Prag. - Bd.1-3, 1790-1798. Prague: Franz Geszabek, 1790-1798. **CONTINUATION OF:** Abhandlungen einer Privatgesellschaft in Böhmen zur Aufnahme der Mathematik, der vaterländischen Geschichte und der Naturgeschichte. **CONTINUATION OF:** Abhandlungen der königlichen böhmischen Gesellschaft der Wissenschaften zu Prag. **EDITOR:** Ignaz Edler von Born. **INDEXES:** Repertorium sämtlicher Schriften der königlichen böhmischen Gesellschaft der Wissenschaften vom Jahre 1769 bis 1868 zusammengestellt von Wilhelm Rudolph Weitenwever. Prague, 1860. **INDEXES:** Generalregister zu den Schriften den Königlichen böhmischen Gesellschaft der Wissenschaften, 1784-1884 zusammengestellt von Georg

Wegner. Prague, 1884. **SOURCES:** G463b, K429, S3492 **LOCATIONS:** BUC, CCF, CCN, ULS, ZDB **SUBJECT:** Science - Societies - Prague.

Neuerer Forstmagazine, see no. 1306.

1263 Neues allgemeines Journal für die Handlung, oder, gemeinnützige Aufsätze, Versuche und Nachrichten für Kaufleute. - v.1, 1788-1789. Frankfurt: Andreae, 1788-1789. **CONTINUATION OF:** Allgemeines Journal für die Handlung, oder gemeinnützige Aufsätze, Versuche und Nachrichten für Kaufleute. **EDITOR:** Johann Christian Schedel. **LOCATIONS:** ZDB **SUBJECT:** Commerce - Periodicals - Germany.

1264 Neues Archiv der practischen Arzneykunst für Aerzte, Wundärzte und Apotheker. - v.1-3, 1788-1795. Leipzig: Weygand, 1788-1795. **CONTINUATION OF:** Archiv der practischen Arneykunst für Aerzte, Wundärzte und Apotheker. **EDITOR:** Philipp Friedrich Theodor Meckel. **SOURCES:** G65a, K3643 **LOCATIONS:** BUC, CCF, ULS, ZDB **SUBJECT:** Medicine - Periodicals - Germany. **SUBJECT:** Surgery - Periodicals - Germany. **SUBJECT:** Pharmacy - Periodicals - Germany.

1265 Neues Archiv der Schwärmerei und Aufklärung. Heft 1-Heft 4, 1797-1798. Altona: Bechtold, 1797-1798. **CONTINUATION OF:** Archiv der Schwärmerei und Aufklärung. **EDITOR:** Friedrich Wilhelm von Schütz. **SOURCES:** K4612 **LOCATIONS:** ZDB **SUBJECT:** General - Periodicals - Germany.

1266 Neues Archiv für die allgemeine Heilkunde. St.1, 1793. Leipzig: Fleischer, 1793. **CONTINUATION OF:** Archiv für die allgemeine Heilkunde. **EDITOR:** August Friedrich Hecker. **SOURCES:** K3679 **LOCATIONS:** ZDB **SUBJECT:** Medicine - Periodicals - Germany.

1267 Neues Archiv für die Geburtshülfe, Frauenzimmer und Kinderkrankheiten mit Hinsicht auf die Physiologie, Diätetik und Chirurgie. - Bd.1-2 (each with 4 issues), Bd.3 (St. 1), 1798-1804. Jena: Wolfgang Stahl, 1798-1804. **CONTINUATION OF:** Archiv für die Geburtshülfe, Frauenzimmer und neugeborner Kinder-Krankheiten. **EDITOR:** Johann Christian Starke. **SOURCES:** K3716, OCLC **LOCATIONS:** BUC, CCF, ULS, ZDB **SUBJECT:** Obstetrics - Periodicals - Germany. **SUBJECT:** Pediatrics - Periodicals - Germany.

1268 Neues bergmännisches Journal. - Bd.1-4 (each with 6 issues). Freiberg: Craz, 1795-1817. **CONTINUATION OF:** Bergmännisches Journal. **EDITORS:** Alexander Wilhelm Köhler and Christian August Siegfried Hoffmann (Bd.3 by Hoffmann alone). **SOURCES:** B864a, K3361, S2625 **LOCATIONS:** CCF, ULS, ZDB **SUBJECT:** Mineral industries - Periodicals - Germany.

1269 Neues Bremisches Magazin zur Ausbreitung der Wissenschaften, Künste und Tugend. - Bd.1-3 (3 issues each). Bd.4 (St.1) 1766-1772. Bremen: G.L. Förster, 1766-1772. **CONTINUATION OF:** Bremisches Magazin zur Ausbreitung der Wissenschaften, Künste und Tugend. **SOURCES:** B1066a,

K5312   **LOCATIONS:** BUC, CCF, ULS, ZDB   **SUBJECT:** Science - Periodicals - Germany.

1270 Neues chemisches Archiv.- Bd.1-Bd.10, 1784-1791. Leipzig: J. G. Müller, 1784-1791. **CONTINUATION OF:** Chemisches Archiv. **EDITOR:** Lorenz Florenz Friedrich von Crell. **CONTINUED BY:** Neuestes chemisches Archiv. **INDEXES:** Indexed by Engelhardt (Ref.51). **SOURCES:** B1257a, G624a, K3298, OCLC, S2912 **LOCATIONS:** BUC, CCF, CCN, ULS, ZDB **SUBJECT:** Chemistry - Periodicals - Germany.

1271 Neues Deutsches Museum. - v.1-4, 1789-1791. Leipzig: G.J. Göschen, 1789-1791. Monthly. **EDITORS:** Heinrich Christian Boie and Christian Wilhelm Dohm.   **CONTINUATION OF:** Deutsches Museum. **MICROFORMS:** GO(9), RM(3).   **SOURCES:** K4575, OCLC, S2920 **LOCATIONS:** BUC, CCN, ULS, ZDB **SUBJECT:** General - Periodicals - Germany - Reviews.

1272 Neues Forst Archiv.- v.1-13, 1796-1807. Ulm: Stettin, 1796-1807. **CONTINUATION OF:** Forst Archiv zur Erweiterung der Forst und Jagd Wissenschaft. Also under the title Forst Archiv v.18-30. **EDITORS:** Wilhelm Gottfried Moser and Christopher Wilhelm Jacob Gatterer. **CONTINUED BY:** Annalen der Forst und Jagdwissenschaft.   **SOURCES:** B6270, K2957 **LOCATIONS:** BUC, CCF, CCN, ULS, ZDB   **SUBJECT:** Forestry management - Periodicals - Germany.

1273 Neues geographisches Magazin. - Bd.1-Bd.4(St.1-St.12), 1785-1789. Halle: Waisenhaus, 1785-1789. **CONTINUATION OF:** Geographisches Magazin. **EDITOR:** Johann Ernst Fabri.   **CONTINUED BY:** Historische und geographische Monatsschrift. **CONTINUED BY:** Beiträge zur Geographie, Geschichte und Staatenkunde. **CONTINUED BY:** Magazin für die Geographie, Staatenkunde und Geschichte. **SOURCES:** B1892a, K1167 **LOCATIONS:** ZDB **SUBJECT:** Geography - Periodicals - Germany.

1274 Neues hamburgisches Magazin, oder, Fortsetzung gesammleter Schriften zum Unterricht und Vergnügen aus der Naturforschung der allgemeine Stadt- und Landöconomie und den angenehmen Wissenschaften überhaupt.- St.1-120, 1767-1781. Hamburg; Leipzig: Holle, 1767-1781.   **CONTINUATION OF:** Hamburgisches Magazin. **MICROFORMS:** RM(3).   **SOURCES:** B2025a, G657a, K3242, S2764 **LOCATIONS:** BUC, CCF, CCN, ULS, ZDB **SUBJECT:** Science - Periodicals - Germany.   **SUBJECT:** Natural history - Periodicals - Germany.

1275 Neues Hannoverisches Magazin. - v.1-34, 1791-1824. Hannover: Schlüter, 1791-1824.   **CONTINUATION OF:** Hannoverische gelehrte Anzeigen. **CONTINUATION OF:** Nützliche Sammlungen. **CONTINUATION OF:** Hannoverische Beiträge zum Nutzen und Vergnügen. **CONTINUATION OF:** Hannoverisches Magazin.   Beilage published as no.1-23, 1791-1813. **SOURCES:** B4185d, G659a, K439, S2782.2 **LOCATIONS:** BUC, CCF, CCN, ULS, ZDB **SUBJECT:** General - Periodicals - Germany - Reviews.

1276 Neues Journal aller Journale, oder sciagraphische Übersicht der vorzüglichsten fremden und einheimischen Zeitschriften. - Heft 1-12, 1790. Hamburg: Chaidron, 1790. **CONTINUATION OF:** Journal aller Journale. **EDITOR:** Heinrich Wilhelm Lawätz. **SOURCES:** G664a, K4578 **SUBJECT:** General - Periodicals - Germany - Reviews.

1277 Neues Journal der Erfindungen, Theorien un Widersprüche in der Natur-und Arzneiwissenschaft. - v.1-5, 1798-1909. Gotha: Justus Perthes, 1798-1809. **CONTINUATION OF:** Journal der Erfindungen. August Friedrich Hecker. **CONTINUED BY:** Neuestes Journal der Erfindungen. **SOURCES:** B2405, G110, K3675, OCLC, S2665 **LOCATIONS:** BUC, CCF, CCN, ULS, ZDB **SUBJECT:** Science - Periodicals - Germany. **SUBJECT:** Technology - Periodicals - Germany.

1278 Neues Journal der Physik. - Bd.1-4 (each with 4 issues). Leipzig; Halle: Johann Abrose Barth, 1795-1797. **CONTINUATION OF:** Journal der Physik. **EDITOR:** Friedrich Albrecht Gren. **CONTINUED BY:** Annalen der Physik. **INDEXES:** Vollständiges Register über Gren's Neues Journal der Physik. Leipzig, J. A. Barth, 1800. **MICROFORMS:** RM(3). **SOURCES:** B2410a, G669a, K3362, OCLC, S2718 **LOCATIONS:** BUC, CCF, CCN, ULS, ZDB **SUBJECT:** Science - Periodicals - Germany.

1279 Neues Magazin für Aerzte. - v.1-12, 1779-1800. Leipzig: Friedrich Gotthaes Jacobäer, 1779-1800. **CONTINUATION OF:** Magazin vor Aerzte. **EDITOR:** Ernst Gottfried Baldinger. **INDEXES:** Index 1-5 in v.5, 1-10 in 1790, 11-20 in 1800 (Ref.21, 25:2). **SOURCES:** G36a, K3581 **LOCATIONS:** BUC, CCF, ULS, ZDB **SUBJECT:** Medicine - Periodicals - Germany.

1280 Neues Magazin für die Botanik in ihrem ganzen Umfange. - Bd.1, 1791-1794. Zürich: Ziegler, 1794. **EDITOR:** Johann Jakob Römer. **MICROFORMS:** RM(3). **SOURCES:** B2778a, G564a, K3357, S2208 **LOCATIONS:** BUC, CCF, CCN, ULS, ZDB **SUBJECT:** Botany - Periodicals - Switzerland.

1281 Neues Magazin für die gerichtliche Arzneikunde und medicinische Polizei. - v.1-2 (4 issues each) 1785-1788. Stendal: Christian Franzen und Johann Christian Grosse, 1785-1788. **CONTINUATION OF:** Magazin für die gerichtliche Arzneikunde und medicinische Polizei. **EDITOR:** Johann Theodor Pyl. **CONTINUED BY:** Repertorium für die öffentliche und gerichtliche Arzneiwissenschaft. **SOURCES:** G67, K3622 **LOCATIONS:** BUC, CCF, CCN, ULS, ZDB **SUBJECT:** Medical jurisprudence - Periodicals - Germany.

1282 Neues Magazin für die Liebhaber der Entomologie. - v.1-3, 1782-1787. Zürich; Winterthur: Heinrich Steiner, 1782-1787. Quarterly. **CONTINUATION OF:** Magazin für die Liebhaber der Entomologie. **EDITORS:** Johann Caspar Fuessli and Johann Jacob Römer. **CONTINUED BY:** Neuestes Magazin für die Liebhaber der Entomologie. **MICROFORMS:** IDC(8). **SOURCES:** B2784a, G582a, K3288, OCLC, S2201 **LOCATIONS:** BUC, CCF, ULS, ZDB **SUBJECT:** Entomology - Periodicals - Switzerland.

1283 Neues Magazin für die neuere Geschichte, Erd- und Völkerkunde. - v.1, 1790. Leipzig: Jacobäer, 1790. **CONTINUATION OF:** Magazin für die neue Historie und Geographie. **EDITOR:** Friedrich Gottlieb Canzler. **SOURCES:** K1240 **LOCATIONS:** CCN, ULS, ZDB **SUBJECT:** Geography - Periodicals - Germany. **SUBJECT:** Anthropology - Periodicals - Germany.

1284 Neues medicinisches Archiv für denkende Leser allen Ständen zur Belehrung und Unterhaltung. - No.1-3, 1793-1796. Mannheim; Eberfeld: Schönian, 1793-1796. **EDITOR:** Gerhard Wilhelm von Eicken. **SOURCES:** G114, K3680 **LOCATIONS:** ULS **SUBJECT:** Medicine - Periodicals - Germany - Popular.

1285 Neues medicinisches und physisches Journal. - v.1-3, 1797-1802. Marburg: Akademisches Buchhandlung, 1797-1802. **CONTINUATION OF:** Medicinisches Journal v.1-3 also as v.10-12. **OTHER TITLE:** Physisch-medicinisches Journal. **EDITOR:** Ernst Gottfried Baldinger. **SOURCES:** B2907, K3709, S3082 **LOCATIONS:** BUC, CCF, ULS, ZDB **SUBJECT:** Medicine - Periodicals - Germany.

1286 Neues medicinisches Wochenblatt für Aerzte, Wundärzte und Apotheker und Freunde der Naturwissenschaft, unter Aufsicht der medicinischen Facultät zu Giessen. - v.1 (nos.1-52) 1789. Giessen; Frankfurt: 1789. **CONTINUATION OF:** Medicinisches Wochenblatt für Aerzte, Wundärzte und Apotheker. **EDITOR:** Georg Thom. **CONTINUED BY:** Frankfurter medicinische Annalen für Aerzte, Wundärzte, Apotheker. **CONTINUED BY:** Medicinisches Wochenblatt oder forgesetzte medicinische Annalen. **CONTINUED BY:** Medicinischer Rathgeber für Aerzte und Wundärzte. **CONTINUED BY:** Medicinisches Repertorium für Gegenstände aus allen Fächern der Arzneiwissenschaft. **SOURCES:** G42a, K3657 **LOCATIONS:** BUC, ULS **SUBJECT:** Medicine - Periodicals - Germany. **SUBJECT:** Surgery - Periodicals - Germany. **SUBJECT:** Pharmacy - Periodicals - Germany.

1287 Neues polytechnisches Magazin, oder, die neuesten Entdeckungen der Naturlehre, Chemie, der Land- und Hauswirtschaft und der nützlichsten Künste und Gewerbe. - v.1-2, 1798-1799. Winterthur: 1798-1799. **SOURCES:** B3691, G704, K3919, S2189 **SUBJECT:** Economics - Periodicals - Germany. **SUBJECT:** Science - Periodicals - Germany.

1288 Neues Schwedisches Magazin kleiner Abhandlungen welche in die Natur- und Haushaltungskunde einschlagen. - v.1, 1783. Nürnberg: Felssecker, 1783. **CONTINUATION OF:** Schwedisches Magazin. **EDITOR:** Johann Christian Daniel Schreber. **SOURCES:** B4232a, K3294, S3169 **LOCATIONS:** ZDB **SUBJECT:** Natural history - Periodicals - Germany - Popular. **SUBJECT:** Home economics - Periodicals - Germany.

1289 Neues Taschenbuch für Ärzte und Nichtärzte. - Bd.1, 1797. Jena: 1787. **CONTINUATION OF:** Almanach für Ärzte und Nichtärzte. **EDITOR:** Christian Gottfried Gruner. **SOURCES:** G137, K3712 **LOCATIONS:** BUC, ULS, ZDB **SUBJECT:** Medicine - Periodicals - Germany - Popular.

1290 Neues technologisches Magazin. - v.1, 1793. Heidelberg: Braun, 1793. **CONTINUATION OF:** Technologisches Magazin. **EDITOR:** Christoph Wilhelm Jacob Gatterer. **SOURCES:** K2991 **LOCATIONS:** BUC, ULS, ZDB **SUBJECT:** Technology - Periodicals - Germany.

1291 Neues ungrisches Magazin, oder, Beiträge zur ungrischen Geschichte, Geographie, Naturwissenschaft und der dahin einschlagenden Litteratur. - v.1-2 (no. 1-6) 1795-1798. Pressburg: 1795-1798. **CONTINUATION OF:** Ungrisches Magazin. **EDITOR:** Carl Gottlieb von Windisch. **SOURCES:** K1291, OCLC **LOCATIONS:** BUC, CCF   **SUBJECT:** Science - Periodicals - Austria - Popular.

1292 Neues Westphälisches Magazin zur Geographie, Historie und Statistik. - v.1-3 (no. 1-12) 1789-1794. Bielefeld: Meyer, 1789-1794. **CONTINUATION OF:** Westphälisches Magazin zur Geographie, Historie und Statistik. Imprint varies. **EDITOR:** Peter Florenz Weddigen. **SOURCES:** K1218 **LOCATIONS:** BUC, CCF, ZDB **SUBJECT:** Geography - Periodicals - Germany.

1293 Neues Wittenbergisches Wochenblatt, eine Sammlung von Aufsätzen und Wahrnehmungen über die Witterungen, die Haushaltungskunst, das Gewerbe etc. - v.1-12, 1793-1814. Leipzig; Wittenberg: Zimmermann, 1793-1814. **CONTINUATION OF:** Wittenbergisches Wochenblatt. Carl Heinrich Ludwig Pölitz. **SOURCES:** B4783a, K2992, S3324 **LOCATIONS:** CCF, ULS **SUBJECT:** Meteorology - Periodicals - Germany. **SUBJECT:** Economics - Periodicals - Germany.

1294 Neueste Annalen der Französischen Arzneikunde und Wundarzneikunst. - v.1-3, 1791-1800. Leipzig: A.F.B. Böhme, 1791-1800. **EDITOR:** Christoph Wilhelm Hufeland. **SOURCES:** G308, OCLC **LOCATIONS:** BUC, CCF, CCN, ULS, ZDB **SUBJECT:** Medicine - Periodicals - France - Translations.

1295 Neueste Mannigfaltigkeiten, eine gemeinnützige Wochenschrift. - v.1-4, 1778-1780. Berlin: 1778-1780. Weekly. **CONTINUATION OF:** Mannigfaltigkeiten. **CONTINUATION OF:** Neues Mannigfaltigkeiten. **EDITORS:** Friedrich Heinrich Wilhelm Martini and Johann Friedrich Wilhelm Otto. **CONTINUED BY:** Allerneueste Mannifgaltigkeiten. **SOURCES:** B2831b, K5509, S2347 **LOCATIONS:** BUC, CCF, ULS, ZDB **SUBJECT:** General - Periodicals - Germany.

1296 Neueste Sammlung der auserlesensten und neuesten Abhandlungen für Wundärzte. St.1-St.7, 1790-1794. Leipzig: 1790-1794. **CONTINUATION OF:** Sammlung auserlesensten und neuesten Abhandlungen für Wundärzte. **CONTINUATION OF:** Neue Sammlung der auserlesensten und neuesten Abhandlungen für Wundärzte. **EDITORS:** Ernst Benjamin Gottlieb Hebenstreit, Carl Gottlob Kühn and Christian Gotthold Eschenbach. **SOURCES:** G383, K3667 **LOCATIONS:** BUC, CCF, ULS, ZDB **SUBJECT:** Surgery - Periodicals - Germany.

1297 Neueste Sammlung von Abhandlungen und Beobachtungen von der ökonomischen Gesellschaft in Bern. - Bd.1, 1796. Bern: Adler, 1796. **CONTINUATION OF:** Sammlungen von landwirthschaftlichen Dingen der Schweitzerischen Gesellschaft in Bern. **CONTINUATION OF:** Abhandlungen und Beobachtungen durch die ökonomische Gesellschaft zu Bern gesammelt. **CONTINUATION OF:** Neue Sammlung physisch-ökonomischer Schriften herausgegeben von der ökonomischen Gesellschaft des Kantons Bern. **SOURCES:** K3011 **LOCATIONS:** ULS **SUBJECT:** Economics - Societies - Bern.

1298 Das Neueste und Nützliche der Chemie, und hauptsächlich für Kaufleute, Fabrikanten, Künstler und Handwerker. - v.1-24, 1798-1831. Nürnberg: 1798-1831. Monthly. **LOCATIONS:** CCF **SUBJECT:** Chemistry - Periodicals - Germany. **SUBJECT:** Economics - Periodicals - Germany.

1299 Die neuesten Endeckungen in der Chemie. - Theil 1-Theil 12, 1781-1786. Leipzig: Weygand, 1781-1786. **CONTINUATION OF:** Chemische Journal für die Freunde der Naturlehre. **EDITOR:** Lorenz Florenz Friedrich von Crell. **CONTINUED BY:** Auswahl aller eigenthümlicher Abhandlungen und Beobachtungen in der Chemie. **INDEXES:** Indexed by Engelhardt (Ref.51). **MICROFORMS:** RM(3). **SOURCES:** B1258a, G705, K3278, OCLC, S2982 **LOCATIONS:** BUC, CCF, ULS, ZDB **SUBJECT:** Chemistry - Periodicals - Germany.

1300 Die Neuesten Entdeckungen und Erläuterungen aus der Arzneikunde systematisch dargestelt. - v.1-5, 1798-1804. Berlin: Felisch, 1798-1804. Imprint varies. **EDITOR:** Friedrich Ludwig Augustin. **SOURCES:** K3719 **LOCATIONS:** BUC, CCF, ZDB **SUBJECT:** Medicine - Periodicals - Germany.

1301 Die neuesten Erfahrungen Britischer Ärzte über die Wirkungen der Lustseuche. - 1 v., 1797. Breslau: 1797. Translated by Friedrich Gotthelf Friese. **TRANSLATION OF:** Reports principally concerning the effects of nitrous acid in the venereal disease. **SOURCES:** G344a **SUBJECT:** Medicine - Periodicals - Germany - Translations.

1302 Neuestes chemisches Archiv.- Bd.1, 1798. Weimar: Hoffmann, 1798. **CONTINUATION OF:** Chemisches Archiv. **CONTINUATION OF:** Neues chemisches Archiv. **EDITOR:** Lorenz Florenz Friedrich von Crell. **INDEXES:** Indexed by Engelhardt (Ref.51). **SOURCES:** B1257b, G624b, S2912.3 **LOCATIONS:** BUC, CCF, ULS, ZDB **SUBJECT:** Chemistry - Periodicals - Germany.

1303 Neuestes Magazin für die Liebhaber der Entomolgie. - Heft. 1-5, 1791-1794. Stralsund: C.L. Struck, 1791-1794. **CONTINUATION OF:** Magazin für die Liebhaber der Entomologie. **CONTINUATION OF:** Neues Magazin für die Liebhaber der Entomolgie. **EDITOR:** David Heinrich Schneider. **SOURCES:** B2784b, G582b, K3345, OCLC **LOCATIONS:** BUC, CCF, ULS, ZDB **SUBJECT:** Entomology - Periodicals - Germany.

1304 Neufortgesetzte Sammlung ökonomischer und Bienenschriften. - Bd.1-Bd.5, 1799-1803. Leipizg: J. G. Müller, 1799-1803. **CONTINUATION OF:** Auserlesene Sammlung vermischter ökonomischer Schriften. **CONTINUATION OF:** Neue Sammlung vermischter ökonomischer Schriften. **OTHER TITLE:** Neue fortgesetzte Sammlung vermischte ökonomische Schriften. **EDITOR:** Johann Riem. **CONTINUED BY:** Halbjahr-beiträge zur Ökonomie und Naturgeschichte für Landwirthe und Bienenfreunde. **LOCATIONS:** CCN, NUC, ZDB   **SUBJECT:** Economics - Periodicals - Germany.

1305 Neujahrsblatt der Gesellschaft auf dem schwarzen Garten. - Heft 1-42, 1786-1832. Zürich: 1786-1832. Annual. None issued from 1799 to 1803. Addressed to the children of Zürich. **OTHER TITLE:** Neujahrsgeschenk, Gesellschaft auf dem schwarzen Garten; Neujahrsstuck, Gesellschaft auf dem schwarzen Garten. **LOCATIONS:** BUC   **SUBJECT:** Science - Societies - Zürich - Popular.

1306 Neueres Forstmagazin, eine Sammlung zerstreuter Forstschriften. - No.1-5, Aug. 1776-Feb./Mar. 1777. Frankfurt: Johann Jacob Friedrich Straube, 1776-1777. Imprint varies. **EDITOR:** Matthias Joseph Anton Franzmadhes. **SOURCES:** K2917 **LOCATIONS:** CCF, LS, ZDB   **SUBJECT:** Forestry management - Periodicals - Germany.

1307 The new asiatic miscellany. v.1, 1789. Calcutta: 1789. **CONTINUATION OF:** Asiatic miscellany. Also printed in Paris by Langlés. **EDITOR:** Francis Gladwin. **MICROFORMS:**IDC(8), RP(1). **LOCATIONS:** BUC, ULS **SUBJECT:** General - Periodicals - India.

1308 The New London medical journal. - v.1-2, 1792-1793. London: E. Hodson, 1792-1793. **SOURCES:** G106, OCLC   **LOCATIONS:** BUC, ULS **SUBJECT:** Medicine - Periodicals - England.

1309 Nieuwe algemeen magazijn van wetenschap, konst en smaak, behelzende: I. Wysbergeerte en zedekunde; II. Natuurkunde en natuurlyke historie, III. Historiekunde; IV. Beschaafde letteren, fraaije kunsten, en mengelwerk. - Deel 1, (stuk 1-5)-deel 5, (stuk 2), 1792-1799. Amsterdam: D'erven P. Meijer en G. Warnars, 1792-1799. **CONTINUATION OF:** Algemeen magazijn van wetenschap, konst en smaak. Each volume is in two parts with separate title pages. Imprint varies slightly. **CONTINUED BY:** Vaderlandsch magazijn van wetenschap, kunst en smaak. **MICROFORMS:** RM(3). **SOURCES:** B80a, G706, OCLC, S721.2 **LOCATIONS:** CCN, ULS   **SUBJECT:** General - Periodicals - Netherlands.

1310 Nieuwe algemeene konst-en letter-bode. - Deel 1-14, 1794-1800. Haarlem: A. Loosjes, 1794-1800. Weekly. **CONTINUATION OF:** Algemeene konst-en letter-bode. Imprint varies. **INDEXES:** Algemeene register op de 25 deelen (1788-1899). Haarlem, 1803. **MICROFORMS:** IDC(8). **SOURCES:** B81 **LOCATIONS:** BUC, CCF, CCN, ULS   **SUBJECT:** General - Periodicals - Netherlands.

1311 Nieuwe chemische en phijsische oefeningen. - St.1-2, 1797. Amsterdam: 1797. **CONTINUATION OF:** Chemische en phijsische oefeningen voor de beminnaars der schei en naturkunde. **SOURCES:** B1256a, S728 **SUBJECT:** Science - Periodicals - Netherlands.

1312 Nieuwe genees- natuur- en huishoudkundige jaarboeken / Genootschap van genees-en natuurkundigen, Dordrecht. - v.1-5, 1782-1784. Dordrecht: 1792-1784. **CONTINUATION OF:** Genees- natuur- en huishoudkundige jaarboeken. **CONTINUED BY:** Algemeene genees- natuur- en huishoudkundige jaarboeken. **SOURCES:** G649a **LOCATIONS:** BUC, CCN, ULS **SUBJECT:** Medicine - Societies - Dordrecht. **SUBJECT:** Natural history - Societies - Dordrecht. **SUBJECT:** Home economics - Societies - Dordrecht.

1313 Nieuwe natuur-en geneeskundige bibliotheek. - v.1, 1774-1775. Amsterdam: Petrus Conradi, 1774-1775. **CONTINUATION OF:** Natuur-en geneeskindige bibliothek. **EDITOR:** Bartholomaeus Teissier (Ref.75, no.17). **SOURCES:** G696a **LOCATIONS:** BUC **SUBJECT:** Medicine - Periodicals - Netherlands - Reviews. **SUBJECT:** Natural history - Periodicals - Netherlands

1314 Nieuwe Nederlandsche historische en astronomische almanach. - s'Gravenhage: 1732-1808. **LOCATIONS:** CCN **SUBJECT:** Astronomy - Periodicals - Netherlands.

1315 Nieuwe scheikundige bibliotheek. - v.1-3, 1799-1802. Amsterdam: 1799-1802. **CONTINUATION OF:** Scheikundige bibliotheek. **SOURCES:** B4213a, G707, S801.2 **LOCATIONS:** CCF **SUBJECT:** Chemistry - Periodicals - Netherlands - Reviews.

1316 Nieuwe vaderlandsche bibliotheek van wetenschap, kunst en smaak. - Amsterdam: 1797-1811. **CONTINUATION OF:** Vaderlandsche bibliotheek van wetenschap, kunst en smaak. **SOURCES:** B4648a, S782.2 **LOCATIONS:** CCF, CCN, ZDB **SUBJECT:** General - Periodicals - Netherlands - Reviews.

1317 Nieuwe verhandelingen van het Bataafsch Genootschap der Proefondervindelijke wiisbegeerte te Rotterdam. - Deel 1-12, 1800-1865. Amsterdam: Johannes Allart, 1800-1865. **CONTINUATION OF:** Verhandelingen van het Bataafsch Genootschap der Proefondervindelijke wiisbegeerte te Rotterdam. Published in Rotterdam, 1810-1865. Publisher varies. **MICROFORMS:** RM(3). **SOURCES:** OCLC, S898a **LOCATIONS:** BUC, CCF, CCN, ULS, ZDB **SUBJECT:** Science - Societies - Rotterdam.

1318 Nordische Beiträge zum Wachsthum der Naturkunde und der Wissenschaften, wie auch der nützlichen und schönen Künste überhaupt.- v.1(no.1-3)v.2(no.1) 1756-1758. Altona: David Iversen, 1756-1758. **EDITOR:** Peter Simon Pallas. **SOURCES:** B3340, G708, K3222, S2241 **LOCATIONS:** BUC, CCF, ZDB **SUBJECT:** Science - Periodicals - Germany.

1319 Nordisches Archiv für Natur- und Arzneiwissenschaft. - v.1-4, 1799-1805. Copenhagen: Friedrich Brummer, 1799-1805. Imprint varies. **EDITORS:** Christian Heinrich Pfaff and Paul Scheel. **CONTINUED BY:** Neues Nordisches Archiv. **SOURCES:** G165, K3724 **LOCATIONS:** BUC, CCF, ULS, ZDB **SUBJECT:** Natural history - Periodicals - Denmark. **SUBJECT:** Medicine - Periodicals - Denmark.

1320 Notice de l'almanach sous-verre des associés contenant des decouvertes, inventions ou expériences nouvellement faites dans les sciences, les métiers, l'industrie. - No.1-44, 1768-1810. Paris: 1768-1810. Supplements: Calendrier de l'almanach sous-verre. **LOCATIONS:** BUC, CCF **SUBJECT:** Science - Periodicals - France.

1321 Notice générale des inventions et découvertes / Lycée des arts, Paris. - No.1-9, Mar. 30,1795-18 June,1796. Paris: Lycée des arts, 1795-1796. **SOURCES:** OCLC **LOCATIONS:** CCF, ULS **SUBJECT:** Technology - Societies - Paris.

1322 Nouveaux mémoires de l'Académie de Dijon pour la partie des sciences et des arts. - T.1-8, 1782-1786. Dijon: Causse, 1782-1786. **CONTINUATION OF:** Mémoires de l'Académie de Dijon. **INDEXES:** Table générale et particulieres des travaux contenues dans les mémoires (1769-1913). Dijon, 1915. **SOURCES:** S1137c **LOCATIONS:** BUC, CCF, CCN, ULS **SUBJECT:** Science - Societies - Dijon.

1323 Nouveaux mémoires de l'Académie royale des sciences et belles lettres, avec l'histoire, Berlin. - T.1(1770)1772,-t.17(1786) 1788. Berlin: Christien Frederic Voss, 1772-1788. **TRANSLATED AS:** Abhandlungen der Königlichen Preussischen Academie der Wissenschaften in Berlin. **CONTINUATION OF:** Miscellanea Berolinensis. **CONTINUATION OF:** Histoire avec les mémoires, Académie royale des sciences et belles lettres, Berlin. Publisher varies. **CONTINUED BY:** Mémoires de l'Académie royale des sciences et belles lettres, Berlin. **CONTINUED BY:** Sammlung der deutschen Abhandlungen. **INDEXES:** Kohnke, Otto, ed. Gesamtregister über die in den Schriften der Akademie von 1700-1899 erscheinen wissenschaftlichen Abhandlungen und Festreden. Berlin, 1900,(Ref.126). **MICROFORMS:** RM(3), SMS(1), IDC(8), RP(1). **SOURCES:** G487c, K294, OCLC, S2262a **LOCATIONS:** BUC, CCF, CCN, ULS, ZDB **SUBJECT:** Science - Societies - Berlin.

1324 Nouvelle collection des mémoires sur différents parties interessantes des sciences. - v.1-3, 1786. Paris: 1786. **CONTINUATION OF:** Mémoires sur différentes parties des sciences et des arts. **LOCATIONS:** BUC **SUBJECT:** Science - Periodicals - France.

1325 Nouvelle table des articles contenue dans les volumes de l'Académie royale des sciences de Paris, depuis 1666 jusqu'en 1770, dans ceux des Arts et Métieres, publiés par cette Académie, et dans la Collection Académique. - T.1-4, 1775-1776. Paris: Ruault, 1775-1776. Edited by Jean Baptiste François

Rozier. **SOURCES:** GII:17f, S1276i **LOCATIONS:** BUC, ULS
SUBJECT: Science - Societies - Paris - Indexes.

1326 Nouvelles de la republique des lettres. - v.1-40, 1684-1718. Amsterdam: Henry
Desbordes, 1684-1718. Suspended May 1687-Dec. 1698 and Jan. 1711-Dec.
1715. Reprinted 1966 by Slatkine Reprints. Analyzed in Ref.13, p.107-196.
EDITORS: Pierre Bayle, Daniel de Larroque, Jean Barrin and Jacques Bernard.
MICROFORMS: RP(1). **SOURCES:** OCLC **LOCATIONS:** BUC, CCF,
CCN, ULS, ZDB **SUBJECT:** General - Periodicals - Netherlands - Reviews.

1327 Nouvelles de la republique des lettres et des arts. - v.1-8, 1779-1788. Paris:
1779-1788. Bimonthly. **EDITOR:** Pahin de Champlain de la Blancherie.
CONTINUED BY: Salon de la correspondence pour les sciences et les arts.
Published by Salon de la correspondances as a supplement Dec. 1786-Aug. 1787
(Ref.91, p.51). **MICROFORMS:** ACR(1). **LOCATIONS:** BUC, CCF
SUBJECT: General - Periodicals - France - Reviews.

1328 Les Nouvelles découvertes sur totutes les parties de la medécine. - v.1-4,
1679-1682. Paris: L. d'Houry, 1679-1682. Imprint varies (Ref.98). OTHER
TITLE: Temple d'Esculape (v.2); Journal des nouvelles découvertes
concernant les sciences et les arts qui font partie de la medécine (v.3);
Nouveauxtes journalieres concernant les sciences et les arts qui sont partie de
la medécine (v.4). **EDITOR:** Nicolas de Blegny. **SOURCES:** GI:4
LOCATIONS: BUC, ULS, ZDB **SUBJECT:** Medicine - Periodicals - France.

1329 Nouvelles instructives, bibliographiques, historiques et critiques de médecine et
de chirurgie. - v.1-7, 1785-1791. Paris: Bureau des annales, 1785-1791. v.5-7
under the title: Nouvelles ou annales de l'art de querir. OTHER TITLE:
Nouvelles, ou annales de l'art de querir. **EDITOR:** N. Retz. **SOURCES:**
G68 **LOCATIONS:** CCF, ULS **SUBJECT:** Medicine - Periodicals - France.
SUBJECT: Surgery - Periodicals - France.

1330 Nova Acta Academia Electoralis Moguntinae scientiarum utilium quae Erfurt est.
- v.1-4, 1799-1806. Erfurt: Beyer et Maring, 1799-1806. **CONTINUATION
OF:** Acta Academia Electoralis Moguntinae Scientiarum utilium quae Erfordiae
est. **OTHER TITLE:** Abhandlungen der Kurfürstlich Mainzer Akademie
nützlicher Wissenschaften. **INDEXES:** Graesel, Arnim. Repertorium zu den
Acta und Nova Acta der Akademie. Halle, 1894. **SOURCES:** G439c, K457,
S2546 **LOCATIONS:** BUC, CCF, CCN, ULS, ZDB **SUBJECT:** Science
- Societies - Erfurt.

1331 Nova acta Academiae scientiarum imperialis Petropolitanae. - v.1-13 (1783-1795)
1787-1795. St. Petersburg: Typis Academiae Scientiarum, 1787-1795.
CONTINUATION OF: Commentarii Academiae scientiarum imperialis
Petropolitanae. **CONTINUATION OF:** Novi Commentarii Academiae
scientiarum imperialis Petropolitanae. **CONTINUATION OF:** Acta Academia
scientiarum imperialis Petropolitana. **CONTINUED BY:** Mémoires de
l'Académie impériale des sciences, St. Petersburg. **INDEXES:** Cumulative
index (Ref.115) **MICROFORMS:** IDC(8), RM(3). **SOURCES:** G422d,

OCLC, S3706 **LOCATIONS:** BUC, CCF, CCN, ULS, ZDB **SUBJECT:**
Science - Societies - St. Petersburg.

1332 Nova acta eruditorum. - T.1-8, 1732-1782. Leipzig: 1732-1782.
CONTINUATION OF: Acta eruditorum. Only 2 vols. appeared between 1764
and 1767 (Ref.2). Johnson Reprint. **EDITORS:** Friedrich Otto Mencke
1735-1754, Karl Andrea Bel, 1754-1757. Ad nova acta eruditorum quae
Lipsiae publicantur supplementa. Leipzig, 1735-1757, v.1-8.
MICROFORMS: GO(9), RP(1).   **SOURCES:** B25, G12d, OCLC
LOCATIONS: BUC, CCF, CCN, ULS, ZDB   **SUBJECT:** General -
Periodicals - Germany - Reviews.

1333 Nova acta Helvetica, physico-mathematico-botanico-medica, Societas
physico-medica Basiliensis. - Basle: Johann Schweighäuser, 1787. v.1, 1787.
CONTINUATION OF: Acta Helvetica, physico-mathematico-botanico-medica.
MICROFORMS: RM(3). **SOURCES:** B26a, G508, K3315, OCLC, S2128b
LOCATIONS: BUC, CCF, CCN, ULS, ZDB **SUBJECT:** Science - Societies
- Basle.

1334 Nova acta physico-medica academiae Caesareae Leopoldinae- Carolinae naturae
curiosorum. - T.1(1757), 2(1761), 3(1767), 4(1770), 5(1773), 6(1778), 7(1783),
8(1791) Nürnberg: Wolfgang Schwarzkopf, 1757-1791. **TRANSLATED AS:**
Medicinische Beobachtungen. **CONTINUATION OF:** Miscellanea curiosa
Academia naturae curiosorum. **CONTINUATION OF:** Ephemerides sive
observationum medico-physicarum, Academiae Caesareae Leopoldinae naturae
curiosoum. **CONTINUATION OF:** Acta physico-medica, Academia Caesarea
Leopoldina-Carolina naturae curiosorum. **CONTINUED BY:** Verhandlungen
der Kaiserlich-Leopoldinisch-Carolinische Academie der Naturforscher.
MICROFORMS: IDC(8), RP(1).   **LOCATIONS:** BUC, CCF, CCN, ULS
SUBJECT: Science - Societies - Halle.

1335 Nova litteraria circuli Franconici, oder, Fränkische Gelehrten- historie. - St.1-2,
1725. Nürnberg: 1725. **EDITOR:** E. Fr. Just Heinrich. **CONTINUED BY:**
Fränkische Acta erudita et curiosa.   **CONTINUED BY:** Nützliche und
auserlesene Arbeiten der Gelehrten im Reich. **SOURCES:** K73
LOCATIONS: ZDB **SUBJECT:** General - Periodicals - Germany - Reviews.

1336 Novi commentarii Academiae scientiarum imperialis Petropolitanae. - v.1-20
(1747-1776) 1750-1776.   St. Petersburg: Typis Academiae Scientiarum,
1750-1776.   **TRANSLATED AS:** Physikalische und medicinische
Abhandlungen der Kayserliche Akademie der Wissenschaften in St. Petersburg.
CONTINUATION OF: Commentarii Academia scientiarum imperialis
Petropolitanae.   **CONTINUED BY:** Acta Academia scientiarum imperialis
Petropolitanae.   **CONTINUED BY:** Nova acta Academiae scientiarum
imperialis Petropolitanae.   **INDEXES:** Cumulative index (Ref.115).
MICROFORMS: IDC(8).   **SOURCES:** G442a, OCLC, S3706
LOCATIONS: BUC, CCF, CCN, ULS, ZDB **SUBJECT:** Science - Societies
- St. Petersburg.

1337 Novi commentarii societatis regiae scientarum Gottingensis. - T.1-8, (1769-1777)
1771-1778.   Göttingen: Dieterich, 1771-1778.   **CONTINUATION OF:**
Commentationes regiae societatis Gottingensis.   **CONTINUATION OF:**
Commentarii Societatis regiae scientiarum gottingensis.   Kraus Reprint.
EDITOR: Christian Gottlob Heyne.   **CONTINUED BY:** Commentationes
societatis regiae scientiarum Gottingensis.   **CONTINUED BY:** Deutsche
Schriften der königlichen Societät der Wissenschaften zu Göttingen. **INDEXES:**
Index to the society's publications from 1751-1808 in v.16 of its Commentarii
for 1804/1808.   Also indexed in Vezenyi (Ref.1) **MICROFORMS:** RM(3).
SOURCES: K281, OCLC, S2694f **LOCATIONS:** BUC, CCF, CCN, ULS,
ZDB **SUBJECT:** General - Societies - Göttingen.

1338 Novoe prodol'zenie trudov / Imperatorskoi vol'noe ekonomicheskago obtchestva.
- No. 1-61, 1795-1809.   St. Petersburg: 1795-1809.   **CONTINUED BY:**
Novoe prodol'zenie trudov.   **LOCATIONS:** BUC, ZDB   **SUBJECT:**
Economics - Societies - St. Petersburg.

1339 Nuova raccolta d'opuscoli scientifici e filologici. - v.1-42, 1755-1787.   Venice:
Simone Occhi, 1755-1787.   **CONTINUATION OF:** Raccolta d'opuscoli
scientifici e filologici.   **EDITORS:** Angelo Calgerà and Fortunato Mandalli.
SOURCES: G727a, OCLC **LOCATIONS:** BN, ULS **SUBJECT:** Science
- Periodicals - Italy.

1340 Nuovo giornale della più recente letteratura medico chirurgica d'Europa. - v.1-12,
1791-1797.   Milan: Gaetano Motta, 1791-1797.   **EDITORS:** Crispi, Frank and
others.   **SOURCES:** G309, OCLC   **LOCATIONS:** BUC, CCF, ULS
SUBJECT: Medicine - Periodicals - Italy.   **SUBJECT:** Surgery - Periodicals
- Italy.

1341 Nuovo giornale di medicina. - v.1, 1781. Venice: 1781.   **CONTINUATION OF:**
Giornale di medicina.   **EDITOR:** Francesco Vitalio.   **SOURCES:** G44
LOCATIONS: BUC **SUBJECT:** Medicine - Periodicals - Italy.

1342 Nuovo giornale d'Italie, spettante alle science naturali e principalmente
all'agricoltura, alle arti ed al commercio. - v.1-7, 1777-1788, ns. v.1-6,
1790-1795.   Venice: 1777-1795.   **CONTINUATION OF:** Giornale d'Italie,
spettante alle science naturali e principalmente all'agricoltura, alle arti ed al
commercio.   **SOURCES:** B1948a, G635a, S2094.2 **LOCATIONS:** BUC,
CCF, ULS **SUBJECT:** Science - Periodicals - Italy.

1343 Nützliche Nachrichten und Abhandlungen, das Oekonomie und Cammerzwesen
betreffend Monatsschrift. - v.1-3 (St. 1-12) 1767-1769. Vienna: Johann Georg
Trattner, 1767-1769. Monthly.   **OTHER TITLE:** Wienerische Nachrichten
und Abhandlungen aus dem Oeconomie und Cameralwesen.   **EDITOR:** Johann
Georg Wolf.   **SOURCES:** K2888 **LOCATIONS:** BUC, ZDB **SUBJECT:**
Economics - Periodicals - Austria.

1344 Nützliche Sammlungen. - v.1-4, 1755-1758.   Hannover: Schlüter, 1755-1758.
CONTINUATION OF: Hannoverische gelehrte Anzeigen.   **OTHER TITLE:**

Hannoverische Anzeigen von allerhand Sachen. **CONTINUED BY:**
Hannoverische Beiträge zum Nutzen und Vergnügen. **CONTINUED BY:**
Hannoverisches Magazin. **CONTINUED BY:** Neues Hannoverisches Magazin.
SOURCES: B4185a, K206, G711, S2786 **LOCATIONS:** BUC, CCF, ULS,
ZDB **SUBJECT:** General - Periodicals - Germany - Reviews.

1345 Nützliche und auserlesene Arbeiten der Gelehrten im Reich. - St.1-7, 1733-1736.
Nürnberg: Endter, 1733-1736. **CONTINUATION OF:** Nova litteraria circuli
Franconici. **EDITOR:** E. Fr. Just Heinrich. **SOURCES:** K95
LOCATIONS: BUC, CCF, ULS, ZDB **SUBJECT:** General - Periodicals -
Germany - Reviews.

1346 Nützlicher und getreuer Unterricht für den Land-und Bauersmann. - Jahrg.3-10,
1772-1779. Stuttgart: Johann Benedict Mezler, 1772-1779. **CONTINUATION
OF:** Allgemeiner öconomischer oder Landwirthschafts-calender auf das Jahr.
EDITOR: Balthasar Sprenger. **CONTINUED BY:** Ökonomische Beiträge und
Bemerkungen zur Landwirthschaft. **MICROFORMS:** RP(1) (Goldsmiths'
Kress Library). **SOURCES:** K2908 **LOCATIONS:** CCF, ZDB
SUBJECT: Agriculture - Periodicals - Germany - Popular. **SUBJECT:**
Economics - Periodicals - Germany - Popular. **SUBJECT:** Medicine -
Periodicals - Germany - Popular.

1347 Ny journal uti hushallningen. - Stockholm: Carlbohm, 1790-1813.
CONTINUATION OF: Hushallnings-journal. **SOURCES:** (Ref.54:292).
SUBJECT: Economics - Periodicals - Sweden.

1348 Nya handlingar / Kungliga Svenska vetenskapsakademien. Stockholm: Johan
Georg Lange, 1780-1812. Quarterly. **TRANSLATED AS:** Neue
Abhandlungen aus der Naturlehre, Haushaltungskunst und Mechanik.
CONTINUATION OF: Handlingar, Kungliga Svenska Vetenskapsakademien.
Publisher varies. **INDEXES:** Indexes to v.1-33 (1780-1812) published
1798-1812. **MICROFORMS:** RM(3). **SOURCES:** G477e, OCLC, S693
LOCATIONS: BUC, CCF, CCN, ULS **SUBJECT:** Science - Societies -
Stockholm.

1349 Nye samling af det Kongelige Danske videnskabers skrifter. - v.1-5, 1781-1799.
Copenhagen: Lingreens, 1781-1799. **TRANSLATED AS:** Physikalische,
chemische, naturhistorische und mathematische Abhandlungen, aus den neuen
Sammlungen der Schriften der königlichen dänischen Gesellschaft der
Wissenschaften. **CONTINUATION OF:** Skrifter, som udi det Kjöbenhavnske
selskab af laerdoms ogvidenskabs elstere ere fremlagte og oplaeste. 1798 also
has the title: Nyeste Samling (Ref.21,v.25:604a). **CONTINUED BY:**
Kongelige Danske videnskabers selskabs skrifter. **INDEXES:** Fortegnelse over
de Kongelige danske Videnkabernes selskabs publicationer, 1742-1930. Ed.
Asger Lomholt. Copenhagen, 1930. **MICROFORMS:** RM(3). **SOURCES:**
G475c, OCLC, S615c **LOCATIONS:** BUC, CCF, ULS **SUBJECT:** Science
- Societies - Copenhagen.

1350 Nye Samling det Kongelige Norske Videnkabers Selskab. - v.1-2, 1784-1788. Copenhagen: 1784-1788. **CONTINUATION OF:** Skrifter det Trondhjemske Selskab. **CONTINUATION OF:** Skrifter der Kongelige Norske videnskabers Selskab. **CONTINUED BY:** Nyeste Samling der Kongelige Norske Videnskabers Selskab. **SOURCES:** G476c, OCLC, S641.2 **LOCATIONS:** BUC, CCF, ULS, ZDB **SUBJECT:** Science - Societies - Trondhjem.

1351 Nye sundhedstidende. - v.1-2, 1782-1783. Copenhagen: Johann Rudolph Thiele, 1782-1783. **CONTINUATION OF:** Sundstidende. **CONTINUATION OF:** Sundhed og underholdning. **EDITOR:** Johann Clemens Tode. **CONTINUED BY:** Sundhedsblade. **CONTINUED BY:** Hygaea og muserne. **CONTINUED BY:** Museum for sunheds og kundskabs elskere. **CONTINUED BY:** Hertha. **CONTINUED BY:** Sundhedsbog. **CONTINUED BY:** Medicinalblade. **CONTINUED BY:** Sundhedsjournal. **CONTINUED BY:** Sundhedsraad. **CONTINUED BY:** Sundhedsjournal et maanedskrift. **CONTINUED BY:** Nyeste sundhedsblade. **SOURCES:** G179 **LOCATIONS:** DSG **SUBJECT:** Medicine - Periodicals - Denmark.

1352 Nyeste Samling der Kongelige Norske Videnskabers Selskab. - v.1, 1798. Copenhagen: 1798. **CONTINUATION OF:** Skrifter det Trondhjemske Selskab. **CONTINUATION OF:** Skrifter der Kongelige Norske Videnskabers Selskab. **CONTINUATION OF:** Nye Samling det Kongelige Norske Videnskabers Selskab. **SOURCES:** G476d, OCLC, S641.3 **LOCATIONS:** BUC, CCF, ULS, ZDB **SUBJECT:** Science - Societies - Trondhjem.

1353 Nyeste sundhedstidende, et ugeblad auf blandet inhold. - Nos. 1-3, 1799-1800. Copenhagen: 1799-1800. **CONTINUATION OF:** Sundstidende. **CONTINUATION OF:** Nye sundstidende. **CONTINUATION OF:** Sundhed og underholdning. **CONTINUATION OF:** Hygaea og muserne. **CONTINUATION OF:** Museum for sunheds og kundskabs elskere. **CONTINUATION OF:** Hertha. **CONTINUATION OF:** Sundhedsbog. **CONTINUATION OF:** Medicinalblad. **CONTINUATION OF:** Sundhedsjournal. **CONTINUATION OF:** Sundhedsraad. **CONTINUATION OF:** Sundhedjournal et maanedskrift. **EDITOR:** Johann Clemens Tode. **SOURCES:** G207 **LOCATIONS:** ULS **SUBJECT:** Medicine - Periodicals - Denmark.

1354 Ober-Sächsische Berg-Akademie, in welcher die Bergwercks-Wissenschaften nach ihren Grundwahrheiten untersucht werden. - v.1-3, 1746. Dresden; Leipzig: Hekel, 1746. **EDITOR:** Carl Friedrich Zimmermann. **SOURCES:** K3194 **SUBJECT:** Mineral industries - Periodicals - Germany.

1355 Oberdeutsche Beiträge zur Naturlehre und Oekonomie. - v.1, 1787. Salzburg: Mayr, 1787. **EDITOR:** Carl Ehrenbert Moll. **CONTINUED BY:** Abhandlunmgen einer Privatgesellschaft von Naturforscher und Oeconomen in Oberdeutschland. **SOURCES:** B3387, K3316, S3530 **LOCATIONS:** ZDB **SUBJECT:** Natural history - Societies - Salzburg. **SUBJECT:** Economics - Societies - Salzburg.

1356 Obererzgebürgisches Journal, oder, Sammlung von allerhand in die hiesige Natur-Wissenschaft überhaupt einschlagenden merckwürdigen Abhandlungen. - No.1-12, 1748-1751. Freiberg; Leipzig: Theodor Gottlob Reinhold, 1748-1753. There was a 3rd ed. (Freiburg, 1750-1756) which also had the title: Sammlung kleine ungedruckter Ober-ertzgebügischer Schriften. **EDITOR:** Johann Christian Themel. **SOURCES:** B3389, K3202, S2245 **LOCATIONS:** BUC, CCF   **SUBJECT:** Mineral industries - Periodicals - Germany.

1357 Observaciones astronómicas hechas en Cadiz, en el observatorio real de la Compana de Cavalleros, Guardias-marinas. - Cadiz: 1776- . Title varies. **EDITOR:** Vicente Tofino de San Miguel.   **SOURCES:** B3394 **LOCATIONS:** BUC, ULS   **SUBJECT:** Astronomy - Societies - Spain.

1358 Observaciones sobra la fisica, historia natural y artes utiles. - No. 1-14, 1788. Mexico: J.F. Rangel, 1788. No.13-14 undated. **EDITORS:** José Antonio de Alzate y Ramìrez.   **SOURCES:** OCLC **LOCATIONS:** ULS   **SUBJECT:** Science - Periodicals - Mexico.

1359 Observationes anatomicae selectiores collegii privatii Amstelodami. - v.1-2, 1667-1673. Amsterdam: Caspar Commelin, 1667-1673.   Another edition (Reading, 1938) and a facsimile reprint (Nieukoop, B. de Graaf, 1975). **MICROFORMS:** RP(1).   **SOURCES:** GI:2, OCLC   **LOCATIONS:** BUC, DSG   **SUBJECT:** Anatomy - Societies - Amsterdam.

1360 Observationes chymico-physico-medicae curiosae, mensibus singulis continuandae. - Frankfurt; Leipzig: Georg Heinrich Müller, 1697-1698.   Imprint varies. **EDITOR:** Georg Ernst Stahl.   **SOURCES:** K3177, OCLC   **LOCATIONS:** BL   **SUBJECT:** Science - Periodicals - Germany.

Observationes siderum habitae Pisis, see no. 1371.

1361 Observations curieuses sur toutes les parties de la physique, extraites et recueillies de meilleurs auteurs. - v.1-4 (1719-1730) 1771. Paris: A. Cailleau, 1719-1771. Imprint varies.   **EDITORS:** Guillaume Hyacinthe Bougeant and Nicolas Grozelier.   **SOURCES:** OCLC, S1476   **SUBJECT:** Science - Periodicals - France.

1362 Observations et mémoires sur la physique, sur l'histoire naturelle et sur les arts et métiers. - v.1-43, 1773-1793 and supplement v.44. Paris: Hotel de Thou, 1773-1793. Monthly. **CONTINUATION OF:** Observations périodique sur la physique, l'histoire naturelle.   **CONTINUATION OF:** Introduction aux observations sur la physique, sur l'histoire naturelle et sur les arts. **OTHER TITLE:** Observation sur l'histoire naturelle, sur la physique et sur la peinture; Observations sur la physique, sur l'histoire naturelle et des arts. **EDITORS:** Jean Baptiste François Rozier, Jean André Mongez, and Jean Claude de la Methérie.   **CONTINUED BY:** Journal de physique, de chimie, d'histoire naturelle et des arts.   **INDEXES:** Index to 1778-1786 in v.29 of Journal de physique. **MICROFORMS:** IDC(8), RM(3).   **SOURCES:** B2201a, G714,

OCLC, S1607 **LOCATIONS:** BUC, CCF, CCN, ULS, ZDB   **SUBJECT:** Science - Periodicals - France.

1363 Observations et recherches des médecins de Londres, sur les objets les plus importants de médecine et de chirurgie. - v.1-2 (1757-1784) 1810. Paris: 1810. Translated by Louis Caullet de Veaumorel. **TRANSLATION OF:** Medical observations and inquiries. **SOURCES:** OCLC **LOCATIONS:** CCF, ZDB **SUBJECT:** Medicine - Periodicals - England - Translations.

1364 Observations faites dans le département des hopitaux-civils. - v.1-3, 1785-1787. Paris: 1785-1787. Reprinted from Journal de médecine. **OTHER TITLE:** Journal de médecine, chirurgie et de pharmacie. **LOCATIONS:** DSG **SUBJECT:** Medicine - Periodicals - France.

1365 Observations médicales. - v.1, 1798.Paris; Avignon: 1798. **CONTINUATION OF:** Essais de médecine et d'histoire naturelle. **EDITORS:** Waton and Guérin. **SOURCES:** G148 **SUBJECT:** Medicine - Periodicals - France.

1366 Observations périodique sur la physique, l'histoire naturelle et les arts. - v.1-3, 1756-1757. Paris: 1756-1757. **CONTINUATION OF:** Observations sur l'histoire naturelle, sur la physique et sur la peinture. Publisher varies. **EDITORS:** Jacques Gautier d'Agoty (v.1, 1756) and François Vincent Toussaint (1757). **CONTINUED BY:** Observations périodique sur l'histoire naturelle, la physique et les arts. **CONTINUED BY:** Observations et mémoires sur la physique, sur l'histoire naturelle. **SOURCES:** B3147b, G715, OCLC, S1481 **LOCATIONS:** BUC, CCF, ULS **SUBJECT:** Science - Periodicals - France.

1367 Observations périodique sur l'histoire naturelle, la physique, les arts. - v.1, 1786. Paris: 1786. **CONTINUATION OF:** Observations sur l'histoire naturelle, sur la physique et sur la peinture. **CONTINUATION OF:** Observations périodique sur la physique, l'histoire naturelle et les arts. **CONTINUED BY:** Observations et mémoires sur la physique, sur l'histoire naturelle. **MICROFORMS:** RM(3). **SOURCES:** B3417c, S1479 **LOCATIONS:** BUC, CCF, ULS **SUBJECT:** Science - Periodicals - France.

1368 Observations physique dediées au Roy. - v.1-4, 1750-1753. Paris: 1750-1753. 2nd edition 1753. **SOURCES:** OCLC **LOCATIONS:** CCF **SUBJECT:** Science - Periodicals - France.

1369 Observations sur l'histoire naturelle, sur la physique et sur la peinture. - v.1-6, 1752-1755. Paris: Delaguette, 1752-1755. **EDITOR:** Jacques Gautier d'Agoty. **CONTINUED BY:** Observations périodique sur la physique, l'histoire naturelle et les arts. **CONTINUED BY:** Observations périodique sur l'histoire naturelle, la physique et les arts. **CONTINUED BY:** Observations et mémoires sur la physique, sur l'histoire naturelle. **MICROFORMS:** RM(3). **SOURCES:** B3417, G713, S1477 **LOCATIONS:** BUC, CCF, ULS, ZDB **SUBJECT:** Science - Periodicals - France.

1370 Observationum chymico-physico-medicarum. - No. 1-6, July 1697-April 1698.
Frankfurt; Leipzig: 1697-1698. **EDITOR:** Georg Ernst Stahl. **LOCATIONS:**
BUC **SUBJECT:** Science - Periodicals - Germany.

1371 Observationes siderium habitae Pisis. - Pisa: Augustinus Pizzornius, 1769-1786.
Imprints may vary. **EDITORS:** Tommaso Perelli, R.C. Petri Leopoldi and
Joseph Slop von Cadenberg. **SOURCES:** B3400, OCLC **LOCATIONS:**
BL, CCF **SUBJECT:** Astronomy - Periodicals - Italy.

1372 Oeconomisch-botanisches Gartenjournal. - v.1-6 (each 2 issues) 1795-1803.
Eisenach: Bärecke, 1795-1803. Publisher varies. **EDITOR:** Friedrich Gottlob
Dietrich. **CONTINUED BY:** Neues botanisches Gartenjournal. **SOURCES:**
B3426, K3005, S2538 **LOCATIONS:** BUC, CCF, ULS, ZDB **SUBJECT:**
Gardening - Periodicals - Germany.

1373 Oeconomisch-physicalsche Abhandlungen. - v.1-20, 1747-1763. Leipzig: Carl
Ludwig Jacobi, 1747-1763. **EDITOR:** Peter von Hohenthal. **SOURCES:**
B3427, K3195, S2985 **LOCATIONS:** BUC, CCF, CCN, ULS, ZDB
SUBJECT: Science - Periodicals - Germany. **SUBJECT:** Economics -
Periodicals - Germany.

1374 Oeconomisch-veterinärische Hefte. - No.1-8, 1799-1802. Leipzig: 1799-1802.
EDITORS: Johann Riem and Gottlob Siegmund Reutter. **SOURCES:** K3024
SUBJECT: Veterinary medicine - Periodicals - Germany.

1375 Oeconomische Abhandlungen von der Verbesserung des Ackerbaues, Vermehrung
des Fleisses und Anwuchs des Volkes, Beförderung der Handwerker, Fabriken
und Manufacturen und zur Aufnahme des Handels. - Vienna: Trattner, 1788.
EDITOR: J. Wiegand. **SOURCES:** K2956 **SUBJECT:** Economics -
Periodicals - Vienna.

1376 Oeconomische Hefte, oder, Sammlung von Nachrichten, Erfahrungen und
Beobachtungen für den Stadt-und Landwirth. - v.1-31, 1792-1808. Leipzig:
Brockhaus, 1791-1808. Publisher varies. **EDITORS:** Friedrich Gottlieb
Leonhardi and Johann Christian Schedel. **CONTINUED BY:** Archiv der
deutschen Landwirtschaft. **INDEXES:** Two index vols.were issued.
MICROFORMS: BHP(1). **SOURCES:** B3429, K1986 **LOCATIONS:**
BUC, CCF, ULS, ZDB **SUBJECT:** Economics - Periodicals - Germany.

1377 Oeconomische Nützllichkeiten, Vortheile und Wahrheiten für Naturkunde,
Landwirthschaft und Haushaltungen. - v.1-4, 1790-1792. Göttingen: Dietrich,
1790-1792. **EDITOR:** Georg Heinrich Piepenbring. **SOURCES:** K2613
SUBJECT: Economics - Periodicals - Germany.

1378 Oeconomische Schriften und Verhandlungen der K.K. Gesellschaft der Ackerbaues
und der Künste in Steyermark. - Salzburg: 1788. **SOURCES:** (Ref.6,p.289)
SUBJECT: Agriculture - Societies - Steyermark.

1379 Oeconomische Weisheit und Thorheit, oder, Journal von und für Oeconomen, Cameralisten, Hausmutter, Gartenliebhaber und Freunde der Stadt und Landwirthschaftskunde. - No.1-6, 1789-1794. Erfurt: Kaiser, 1789-1794. **SOURCES:** K2966 **LOCATIONS:** CCF **SUBJECT:** Economics - Periodicals - Germany.

1380 Oeconomisches Archiv zum nützlichen Gebrauch für Künstler. - Coburg: 1800. **SOURCES:** B3431 **SUBJECT:** Technology - Periodicals - Germany.

1381 Oeconomisches und cameralistischer Taschenbuch. - Leipzig: Baumgärtner, 1793. **EDITOR:** M.T.G. Leonhardi. **LOCATIONS:** BUC **SUBJECT:** Economics - Periodicals - Germany.

1382 Oeconomishes Portfeuille, zur Ausbreitung nützlicher Kenntnisse und Erfahrungen aus allen Theilen der Oeconomie. - v.1-3 (3 issues each) 4 (no.1) 1786-1789. Lübeck: Christian Gottfried Donatiue, 1786-1789. **EDITOR:** Johann Heinrich Pratje. **SOURCES:** K2951 **LOCATIONS:** CCF, ZDB **SUBJECT:** Economics - Periodicals - Germany.

1383 Oeconomiske Annaler. - Copenhagen: 1797-1810. **CONTINUED BY:** Nye oeconomiscke annaler. **LOCATIONS:** BUC **SUBJECT:** Economics - Periodicals - Denmark.

1384 Oeconomiske, physiske og mechaniske afhandlingar / Kungliga Svenska Vetenskapsakademien, Stockholm. v.1-v.8 (1739-1746) 1757-1765. Copenhagen: Friedrich Christian Pelt, 1757-1765. **TRANSLATION OF:** Handlingar, Kungliga Svenska Vetenskapsakademien. **SOURCES:** G477b, S656 **LOCATIONS:** NUC **SUBJECT:** Science - Societies - Stockholm.

1385 Oekonomische Beiträge zur Verbesserung der Landwirtschaft in Niedersachsen. - St.1-2, 1793-1797. Lübeck: Niemann, 1793-1797. **EDITOR:** Johann Daniel Denso. **CONTINUED BY:** Neue ökonomische Beiträge zur Verbesserung der Landwirtschaft in Niedersachsen. **SOURCES:** K2990 **SUBJECT:** Agriculture - Periodicals - Germany.

1386 Oekonomische Nachrichten. - v.1-15, 1750-1763. Leipzig: Johann Wendler, 1750-1763. **EDITOR:** Peter von Hohenthal. **CONTINUED BY:** Neue ökonomische Nachrichten. **SOURCES:** K2863 **LOCATIONS:** BUC, CCF, CCN, ULS, ZDB **SUBJECT:** Economics - Periodicals - Germany.

1387 Oekonomische Nachrichten der patriotischen Gesellschaft in Schlesien. - v.1-7, 1773-1779. Breslau: W.G. Korn, 1773-1779. **EDITORS:** Emmanuel Carl Heinrich Börner and Carl Friedrich Tschirner. **CONTINUED BY:** Neue ökonomische Nachrichten der patriotischen Gesellschaft in Schlesien. **SOURCES:** K2910 **LOCATIONS:** BN, CCF, ZDB **SUBJECT:** Economics - Societies - Selesia.

1388 Ökonomische Beiträge und Bemerkungen zur Landwirthschaft. - Stuttgart: Johann Benedict Metzler, 1780-1793. **CONTINUATION OF:** Allgemeiner ökonomischer oder Landwirthschafts-kalender. **CONTINUATION OF:** Nützlicher und getreuer Unterricht für den Land-und Bauersmann. **EDITOR:** Balthasar Sprenger. **MICROFORMS:** RP(1) (Goldsmiths' Kress Library). **SOURCES:** K2929 **LOCATIONS:** CCF, ZDB **SUBJECT:** Agriculture - Periodicals - Germany - Popular. **SUBJECT:** Economics - Periodicals - Germany - Popular. **SUBJECT:** Medicine - Periodicals - Germany - Popular.

1389 Opuscoli scelti sulle scienze e sulle arti, tratti, e degli atti della accademia e dalle altre collezione filosofiche e letterarie e dalle opere più recenti. - v.1-23, 1778-1803. Milan: Guiseppe Marelli: Guiseppe Galezzi, 1776-1803. **CONTINUATION OF:** Scelta opuscoli interessanti. **EDITORS:** Carlo Amoretti, Francesco Soave and others. **CONTINUED BY:** Nuova scelta d'opuscoli interessanti. **INDEXES:** v.17:402-448 contains author index to the first 17 v. and of the Sceli d'opuscoli nuove edizione (Milan, 1781-1783). v.22 contains author index to the last 5 v. of the Opusculi scelti. **SOURCES:** B4211a, OCLC, S1903 **LOCATIONS:** BN, BUC, ULS **SUBJECT:** Science - Periodicals - Italy.

1390 Opuscules et rapports de la Société libre d'émulation de la Seine-infèrieure. - v.1-3, 1800-1802. Rouen, 1800-1802. **CONTINUATION OF:** Mémoires de la Société du commerce et de l'industrie de la Seine-Infèriere. Continued to 1876 under various titles. **OTHER TITLE:** Rapports des travaux de la Socité d'émulation de Rouen. **SOURCES:** S1646b **LOCATIONS:** NUC **SUBJECT:** Economics - Societies - Seine-infèriere.

1391 Oriental collection, consisting of original essays and dissertations, translations and miscellaneous papers illustrating the history and antiquities, the arts, sciences, and literature of Asia. - v.1-3, 1797-1800. London: Cooper and Graham, 1797-1800. Subtitle varies. **EDITOR:** William Ouseley. **SOURCES:** G717, OCLC **LOCATIONS:** BUC, ULS, ZDB **SUBJECT:** Asia - Periodicals - England.

1392 Oriental repertory, tracts on the history, industries, etc. of India, China and Indo-China. - v.1-2, 1791-1797. London: G. Bigg, 1791-1797. Reprint in 2 vol. 1793-1801. **EDITOR:** Alexander Dalrymple. **SOURCES:** B3491, OCLC, S405 **LOCATIONS:** BUC, ULS **SUBJECT:** Asia - Periodicals - England.

1393 Osservazioni spettanti alla fisica, alla storia naturale ed alle arti. - Venice: 1776. An unauthorized translation of Rozier's Journal (Ref.p.103,p.435). **TRANSLATION OF:** Introduction aux observations sur la physique. **SOURCES:** S2098 **SUBJECT:** Science - Periodicals - France - Translations.

1394 Parisian chirurgical journal. - v.1-v.2, 1793-1794. London: T.Boosey and R. Cheesewright, 1793-1794. Translated by Robert Gosling. **TRANSLATION OF:** Journal de chirurgie. **SOURCES:** G97b, OCLC **LOCATIONS:** BUC, ULS **SUBJECT:** Surgery - Periodicals - France - Translations.

1395 Der Patriotische Medicus (Hamburg). - v.1 (St. 1-38) v.2 (St. 39-88) 1765-1768. Hamburg: Beneke, 1765-1768. A 2nd ed. was published in Hamburg, 1768-1769. **EDITOR:** Anton Heins. **SOURCES:** K3544 **SUBJECT:** Medicine - Periodicals - Germany - Popular.

1396 Der Patriotische Medicus (Nürnberg). - No.1-59, Nov.6, 1724-Dec. 30, 1726. Nürnberg: J.C. Rissner, 1724-1727. Weekly. Imprint varies. **SOURCES:** GIII:6, K3518 **LOCATIONS:** ULS, ZDB **SUBJECT:** Medicine - Periodicals - Germany - Popular.

1397 Philosophical collections, containing an account of such physical, anatomical, chymical, mechanical, astronomical, optical, or other mathematical and philsophical experiments and observations as have lately come to the publishers hands. - No. 1-7, 1679-1782. London: John Martyn, 1679-1682. **CONTINUATION OF:** Philosophical transactions. Imprint varies. Published during the suspension of the Philosophical transactions. Kraus and Johnson reprints in 1965. **EDITOR:** Robert Hooke. **CONTINUED BY:** Philosophical transactions. **MICROFORMS:** RM(3), RP(1). **SOURCES:** B3602, GII:4b, OCLC, S434f **LOCATIONS:** BUC, CCF, CCN, ULS, ZDB **SUBJECT:** Science - Societies - London.

1398 The philosophical history and memoirs of the Royal academy of sciences at Paris from 1699 to 1720, or, an abridgement of all the papers relating to natural philosophy which have been published by the members of that illustrious society. - v.1-5, 1742. London: J & P. Knapton, 1742. Translated and abridged by John Martyn. **TRANSLATION OF:** Histoire de l'Académie royale des sciences, Paris. **MICROFORMS:** RM(3). **SOURCES:** OCLC, S171a **LOCATIONS:** BUC, ULS **SUBJECT:** Science - Societies - Paris - Translations.

1399 Philosophical magazine, comprehending the various branches of science, the liberal and fine arts, agriculture, manufactures and commerce. - v.1-42, 1798-1813. London: Taylor and Francis, 1798-1813. Imprint and subtitle vary. **EDITOR:** Alexander Tilloch. **CONTINUED BY:** London and Edinburgh philosophical magazine and journal of science. **CONTINUED BY:** Journal of natural philosophy, chemistry and the arts. **MICROFORMS:** PMC(1,5,6), RP(1). **SOURCES:** B3603, G718, OCLC, S412.1 **LOCATIONS:** BUC, CCF, CCN, ULS, ZDB **SUBJECT:** Science - Periodicals - England. **SUBJECT:** Technology - Periodicals - England.

1400 Philosophical transactions / Royal Society of London. - v.1- , 1665- . London: John Martyn and others, 1665- . **TRANSLATED AS:** Transactions philosophique. Appeared in several reprints, translations and abridgments in several languages (Ref.104). Reprints by Johnson Reprint Co. and Kraus Reprints. Translated into French in Collection académique v.2,4,6-7. **EDITORS:** Henry Oldenburg and others. **MICROFORMS:** RM(3), DA(1), IDC(8), MIM(1,5,6), PMC(1,5,8). **SOURCES:** GII:4, OCLC, S434 **LOCATIONS:** BUC, CCF, CCN, ULS, ZDB **SUBJECT:** Science - Societies - England.

1401 Der Philosophische Arzt. - St.1-4, 1775-1777. Frankfurt: 1775-1777. Imprint varies. There were at least 3 other editions before 1800. **EDITOR:** Melchoir Adam Weickard. **SOURCES:** G35, K3571, OCLC **LOCATIONS:** BUC, ULS **SUBJECT:** Medicine - Periodicals - Germany.

1402 Philosophische und historische Abhandlungen / K. Gessellschaft der Wissenschaften in Edinburg. - v.1, 1789. Göttingen: 1789. Translated by Johann Gottlieb Bühle. **TRANSLATION OF:** Transactions, Royal Society of Edinburgh. **SOURCES:** G502a, S2679 **LOCATIONS:** CCF **SUBJECT:** Science - Societies - Edinburgh - Translations.

1403 Philosophisches Archiv.- v.1-2, 1792-1795. Berlin: Reimer, 1792-1795. **CONTINUATION OF:** Philosophisches Magazin. **EDITOR:** Johann August Eberhard. **SOURCES:** B3605a, K568 **LOCATIONS:** CCF, ULS, ZDB **SUBJECT:** Philosophy - Periodicals - Germany.

1404 Philosophisches Journal, von einer Gesellschaft deutscher Gelehrten. - v.1-10, 1795-1800. Neustrelitz: Michaelis, 1795-1800. Imprint varies. **EDITORS:** Friedrich Immanuel Niethammer and Johann Gottlob Fichte. **MICROFORMS:** IDC(8). **SOURCES:** B3606, G719, S3139 **LOCATIONS:** BUC, CCF, ULS, ZDB **SUBJECT:** Philosophy - Periodicals - Germany.

1405 Philosophisches Magazin. - v.1-4, 1788-1791. Berlin: J. Gebauer, 1788-1791. Reprinted in 1968 by Culture et Civilization, Brussels. **EDITOR:** Johann August Eberhard. **CONTINUED BY:** Philosophisches Archiv. **MICROFORMS:** s: IDC(8). **SOURCES:** B3605, K559, OCLC **LOCATIONS:** BUC, CCF, CCN, ULS, ZDB **SUBJECT:** Philosophy - Periodicals - Germany.

1406 Physical and mathematical memoirs extracted from the registers of the Royal academy of sciences at Paris to be continued monthly. - No. 1, 1692. London: Randal Taylor, 1692. Only one issue seems to have been appeared. **TRANSLATION OF:** Mémoires de mathematiques et de physique (1692). **MICROFORMS:** RM(3). **LOCATIONS:** BUC **SUBJECT:** Science - Societies - Paris - Translations.

1407 The Physical post, or, the Doctors weekly visit to the practitioners as well as to the sick and well. - No. 1-3, 1715? London: 1715? Selected from Davies' Athenae Brittanicae. **OTHER TITLE:** Doctor's harangue at the wells, or, the Physical post. **EDITOR:** Myles Davies. **LOCATIONS:** NUC **SUBJECT:** Medicine - Periodicals - England - Popular.

1408 Physicalisch-chemisches Magazin für Aerzte, Chemisten und Künstler. - v.1-2, 1780. Berlin: Arnold Wever, 1780. **EDITOR:** Jacob Andreas Weber. **SOURCES:** B3642, G720, K3271, S2342 **LOCATIONS:** CCF, ZDB **SUBJECT:** Science - Periodicals - Germany. **SUBJECT:** Medicine - Periodicals - Germany. **SUBJECT:** Chemistry - Periodicals - Germany.

1409 Physicalisch-öconomische Monats-und Quartalschrift. - v.1 (St. 1-12), 2 (St. 1-4), 1787-1788. Dresden; Leipzig: Breitkopf, 1787-1788. CONTINUATION OF: Physicalische Zeitung. CONTINUATION OF: Physicalisch-öconomische Zeitung. EDITOR: Johann Riem. SOURCES: B3648a, K2955, S2732 SUBJECT: Economics - Periodicals - Germany.

1410 Physicalisch-öconomische Zeitung, eine Monatsschrift. - St.1-12, 1785-1786. Dresden; Leipzig: Breitkopf, 1785-1786. Monthly. CONTINUATION OF: Physicalische Zeitung. OTHER TITLE: Physicalisch-öconomische Monatsschrift. EDITOR: Johann Carl Christian Löwe. CONTINUED BY: Physicalisch-öconomische Monats-und Quartalschrift. SOURCES: B3648, K2948, S2732 LOCATIONS: CCF SUBJECT: Economics - Periodicals - Germany.

1411 Physikalisch-ökonomische Bienbibliothek, oder Sammlung auserlesener Abhandlungen und Bienwahrnehmungen. Breslau: Gottlieb Löwe, 1776-1787. v.1-3,1726-1737. EDITOR: Johann Riem. SOURCES: OCLC SUBJECT: Bee culture - Periodicals - Germany.

1412 Physicalisch-ökonomische Auszüge aus den neuesten und besten Schriften. - v.1-10, 1758-1770. Stuttgart: Metzler, 1758-1770. CONTINUATION OF: Physicalisch-ökonomische Wochenschrift. EDITOR: Johann Ernst Friedrich Bernhard. CONTINUED BY: Etwas für Alle. SOURCES: B3655, K3227, S3257 LOCATIONS: ZDB SUBJECT: Science - Periodicals - Germany. SUBJECT: Economics - Periodicals - Germany.

1413 Physicalisch-ökonomische Bibliothek, worin von den neuesten Büchern, welche die Naturgeschichte, Naturlehre und die Land-und Stadtwirthschaft betreffen. - v.1-23 (St. 1-92) 1770-1806. Göttingen: Vandenhoek und Ruprecht, 1770-1806. Imprint varies. EDITOR: Johann Beckmann. SOURCES: B3656, G722, K3249, OCLC, S2691 LOCATIONS: BUC, CCF, CCN, ULS, ZDB SUBJECT: Science - Periodicals - Germany - Reviews. SUBJECT: Economics - Periodicals - Germany - Reviews.

1414 Physicalisch-ökonomische Wochenschrift, welche als eine Realzeitung das nützliche und neueste aus der Natur-und Haushaltungswissenschaft enthält. - Quartal 1-6, 1753-1766. Stuttgart: Metzler, 1753-1766. OTHER TITLE: Physicalisch-ökonomische Realzeitung. EDITOR: Johann Ernst Friedrich Bernhard. CONTINUED BY: Physicalisch-ökonomische Auszüge. CONTINUED BY: Etwas für Alle. SOURCES: B3658, K3221, S3259 LOCATIONS: CCF, ZDB SUBJECT: Science - Periodicals - Germany. SUBJECT: Economics - Periodicals - Germany.

1415 Physicalisch-technologisches Magazin. - St. 1, 1793. Berlin: 1793. EDITOR: Friedrich Franz Daniel Wadzeck. SOURCES: K3350 SUBJECT: Technology - Periodicals - Germany.

1416 Physicalische Arbeiten der einträchtigen Freunde in Wien. - v.1 (no.1-4)-v.2(1-3) 1783-1788. Vienna: Christian Friedrich Wappler, 1783-1788. EDITOR: Ignaz

Edler von Born. **SOURCES:** B3644, K3292, OCLC, S3631 **LOCATIONS:** BUC, CCF, CCN, ULS, ZDB **SUBJECT:** Science - Periodicals - Austria.

1417 Physicalische Belustigungen. - v.1 (no. 1-10) 1751, v.2 (11-20) 1752, v.3 (20-30) 1755-1757 (Ref.101, p.112). Berlin: Christian Friedrich Voss, 1751-1757. v.2-3 paged continuously. **EDITORS:** Christlob Mylius and Abraham Gotthelf Kästner. **CONTINUED BY:** Neue physicalische Belustigungen. **SOURCES:** B3638, OCLC, S2363 **LOCATIONS:** BUC, CCF, CCN, ULS, ZDB **SUBJECT:** Science - Periodicals - Germany - Popular.

1418 Physicalische Bibliothek. - v.1 (St. 1-8)-v.2 (St. 1-2) 1754-1757. Rostock; Wismar: Berger und Bödner, 1754-1757. **EDITOR:** Johann Daniel Desno. **SOURCES:** B3645, K3218, S3204 **LOCATIONS:** BL, CCN, ZDB **SUBJECT:** Science - Periodicals - Germany - Reviews.

1419 Physicalische Bibliothek, oder, Nachricht von den neuesten Büchern, die in die Naturkunde einschlagen. - v.1-4 (4 issues each) 1775-1779. Göttingen: Johann Christian Dieterich, 1775-1779. **EDITOR:** Johann Christian Polykarp Erxleben. **SOURCES:** K3260 **LOCATIONS:** BUC, ZDB **SUBJECT:** Science - Periodicals - Germany - Reviews.

1420 Physicalische Briefe. - No.1-12, 1750-1751. Dessau: Kunkel, 1750-1751. Imprint varies. **EDITOR:** Johann Daniel Denso. **CONTINUED BY:** Monatliche Beiträge zur Naturkunde. **CONTINUED BY:** Fortgesetzte Beiträge zur Naturkunde. **CONTINUED BY:** Neue monatliche Beiträge zur Naturkunde. **SOURCES:** B3646, K3207, S3227 **LOCATIONS:** BN **SUBJECT:** Natural history - Periodicals - Germany.

1421 Physicalische Brieftasche. - Vienna: 1781. **SOURCES:** K3277 **SUBJECT:** Science - Periodicals - Austria.

1422 Physicalische, chemische, anatomische und medizinische Abhandlungen aus den Denkschriften / Königliche Akademie der Wissenschaften, Berlin. - Bd.1 (1746-57) 1764. Berlin: Lange, 1764. A translation of Johann Theodor Eller's papers from the Academy's Histoire, by Carl Abraham Gerhard. **OTHER TITLE:** Physisch - chymisch - medicinische Abhandlungen. **TRANSLATION OF:** Histoire de l'Académie royale des sciences et des belles-lettres de Berlin. **SOURCES:** B3640, G487i **SUBJECT:** Science - Societies - Berlin - Translations.

1423 Physicalische und medicinische Abhandlungen der königlichen Academie der Wissenschaften zu Berlin. - Bd.1 (Miscellanea Bd.1-3, 1710-1727), Bd.2 (Miscellanea Bd.4-7, 1734-1743), Bd.3 (Histoire 1745-50), Bd.4 (Histoire 1751-56), 1781-1786. Gotha: Ettinger, 1781-1786. Translated from the Latin and French by Johann Ludwig Conrad Mümler. **TRANSLATION OF:** Miscellanea Berolinensis. **TRANSLATION OF:** Histoire de l'Académie royale des sciences et des belles-lettres de Berlin. **SOURCES:** B3649, G476j, K3587, S2656 **LOCATIONS:** BUC, ULS, ZDB **SUBJECT:** Science - Societies - Berlin - Translations.

1424 Der Physicalische und öconomische Patriot, oder, Bemerkungen und Nachrichten aus der Natur-histoire, der allgemeinen Haushaltungskunst und der Handlungskunst. - v.1-3, 1756-1758. Hamburg: Georg Christian Grund, 1756-1758. EDITOR: Johann August Unzer. SOURCES: K2869, S2766 LOCATIONS: BUC, CCF, ZDB SUBJECT: Science - Periodicals - Germany - Popular. SUBJECT: Economics - Periodicals - Germany - Popular.

1425 Physicalischer Almanach. - v.1-2, 1786. Vienna: 1786. SOURCES: B3651, S3630 SUBJECT: Science - Periodicals - Austria.

1426 Physicalisches Tagebuch für Freunde der Natur. - v.1-4 (St. 1-14) 1784-1788. Salzburg: Akademische Waisenhausbuchhandlung, 1784-1788. EDITOR: Lorenz Hübner. SOURCES: G721, K3303, S3531 LOCATIONS: BUC, ULS, ZDB SUBJECT: Natural history - Periodicals - Austria.

1427 Physicalisches Taschenbuch für Freunde der Naturlehre und Künstler. - v.1, 1786. Göttingen: Johann Christian Dietrich, 1786. EDITOR: Johann Georg Tralles. SOURCES: B3641, K3314, S2692 LOCATIONS: BUC SUBJECT: Science - Periodicals - Germany.

1428 Physicalische Zeitung. - St.1-48, 1784. Halle: Trampe, 1784. EDITOR: Johann Carl Christian Löwe. CONTINUED BY: Physicalisch-öconomische Zeitung. CONTINUED BY: Physicalische-ökonomische Monats-und Quartalschrift. SOURCES: B3684, K2945, S2732 LOCATIONS: ZDB SUBJECT: Science - Periodicals - Germany.

1429 Physicalsk, ökonomisk og medico-chirurgisk Bibliothek for Danmark og Norge. - v.1-12, 1794-1797. Copenhagen: C.L. Buchs and Sebastien Papp, 1794-1797. Monthly. OTHER TITLE: Fysicalsk, ökonomisk og medico-chirurgisk Bibliothek for Danmark og Norge. EDITOR: Ole Hieronymus Mynster (Ref.42). CONTINUED BY: Bibliothek for physik, medicin og oeconomie. CONTINUED BY: Nyt Bibiliothek for physik, medicin og oeconomie. SOURCES: B3660, G723, S3660 LOCATIONS: CCF, ULS SUBJECT: Science - Periodicals - Denmark.

1430 Physicalske aarbog. - v.4-7, 1786-1793. Copenhagen: 1786-1793. CONTINUATION OF: Christianias physicalske aarbog. Volume 6 also bears the title Physicalske tillaeg till den physicalske aarbog. OTHER TITLE: Physicalske tillaeg till den physicalske aarbog. EDITOR: Christian Ernst Wildberg Schulze. SOURCES: B3661, S644 SUBJECT: Science - Periodicals - Denmark.

1431 Den Physicalske aarbog. - v.1, 1800. Fredericia: 1800. EDITOR: Nicolai Bötcher. SOURCES: B3661, S601 SUBJECT: Science - Periodicals - Denmark.

1432 Physikalische, chemische, naturhistorische und mathematische Abhandlungen, aus den neuen Sammlungen der Schriften der königlichen Dänischen Gesellschaft der Wissenschaften. - v.1-2, 1798-1803. Copenhagen: Brummer, 1798-1803.

Translated by Paul Scheel and Carl Ferdinand Degen. **TRANSLATION OF:** Nye samling af det Kongelige Danske videnskabers skrifter. **SOURCES:** B3647, G475d, S613b   **SUBJECT:** Science - Societies - Copenhagen - Translations.

1433   Physikalische und medicinische Abhandlungen der Kayserliche Akademie der Wissenschaften in St. Petersburg. - Bd.1-3, 1782-1785. Riga: Johann Friedrich Kartknoch, 1782-1785. Translated by Johann Ludwig Conrad Mümler. **TRANSLATION OF:** Novi commentarii Academiae scientiarum imperialis Petropolitanae. **SOURCES:** G442b, K3284, S3699   **LOCATIONS:** CCF, ULS, ZDB   **SUBJECT:** Science - Societies - St. Petersburg - Translations.

1434   Physikalische und philosophische Abhandlungen der Gesellschaft der Wissenschaften zu Manchester. Bd.1-Bd.2, 1788. Leipzig: Weidmann, 1788. Translated by Ernst Benjamin Gottlieb Hebenstreit. **TRANSLATION OF:** Memoirs of the literary and philosophical society of Manchester. **SOURCES:** B3650, G480a, K3322   **LOCATIONS:** CCF   **SUBJECT:** Science - Societies - Manchester - Translations.

1435   Der Physiker, oder, compendiöse Bibliothek der Wissenswürdige aus dem Gebiete der Naturlehre. - No.1, 1795. Eisenach; Halle: 1795. **EDITOR:** Christian Carl André.   **SOURCES:** G548   **SUBJECT:** Natural history - Periodicals - Germany.

1436   Physiologisches und anthropologisches Magazin. - St.1-3, 1796. Jena: 1796. **SOURCES:** K3704   **SUBJECT:** Physiology - Periodicals - Germany. **SUBJECT:** Anthropology - Periodicals - Germany.

1437   Physisch-medicinisches Journal. - v.1-6, 1800-1802. Leipzig: Sommer, 1800-1802. Monthly. Translated by Carl Gottlob Kühn. **TRANSLATION OF:** Medical and physical journal. **CONTINUED BY:** Repertorium der neuesten Erfahrungen englischer Gelehrten. **SOURCES:** B3663, G116a, K3737, S2992 **LOCATIONS:** BUC, CCF, ULS, ZDB   **SUBJECT:** Medicine - Periodicals - England - Translations.

1438   Physische Abhandlungen / Königliche Akademie der Wissenschaften in Paris. - Bd.1-13, 1748-1759. Breslau: Johann Jakob Korn, 1748-1759. Translated by Wolfgang Balthasar Adolf von Steinwehr from the publications of the Académie des Science, Paris from 1692-1741. **OTHER TITLE:** Physikalisch-chemische Abhandlungen. **SOURCES:** GII:17k, K3201, S2453a **LOCATIONS:** BUC, CCN, ZDB   **SUBJECT:** Science - Societies - Paris - Translations.

1439   Der Poetische Medicus, oder, Sammlung auserlesener Medicin-und physicalischer Gedancken, Verse, Sprichwörter. - St. 1-3, 1730? Berlin; Leipzig: 1730? **SOURCES:** K3520   **SUBJECT:** Medicine - Periodicals - Germany - Popular.

1440   Politisk og physisk magazin, meest af udenlandsk laesning. - v.1-2, 1793. Copenhagen: 1793.   **SOURCES:** S646   **SUBJECT:** Science - Periodicals - Denmark - Popular.

1441 Der Pommersche und Neumärksche Wirth. - v.1-3 (St.1-46) 1778. Stettin: Johann
      Franz Struck, 1778. v.2-3 as Zuverlässige Nachrichten. OTHER TITLE:
      Zuverlässige Nachrichten für den wichtigsten Landes-und
      Wirthschafts-Verbesserungen. SOURCES: K2921 SUBJECT: Economics
      - Periodicals - Germany.

1442 Portefeuille für Gegenstände der Chemie und Pharmacie. - St.1, 1784. Hamburg:
      Hoffmann, 1784. Imprint varies. SOURCES: B3714, K3302, S2767
      SUBJECT: Chemistry - Periodicals - Germany. SUBJECT: Pharmacy -
      Periodicals - Germany.

1443 Le Pour et contre, ouvrage périodique d'un gout nouveau dans lequel on
      s'explique librement sur tout ce qui peut intéresser la curiosité du public en
      matière des sciences, d'arts, de livres, d'auteurs, etc. - v.1-20, 1733-1740.
      Paris: Didot, 1733-1740. Another edition was published at La Haye. Slatkine
      reprint, Geneva, 1967. EDITORS: Antoine François Prévost d'Exiles, Pierre
      François Guyot Desfontaines, Charles Hughes Le Febvre de Saint Mare.
      INDEXES: Index v.1-10 in 10. Indexed in Sgard, Jean. Le 'Pour et contre' de
      Prévost.   Introduction, table et index, Paris, A.G. Nizet, 1969.
      MICROFORMS: ACR(1).   SOURCES: B3729, OCLC, S1490
      LOCATIONS: BUC, CCF, ULS, ZDB SUBJECT: General - Periodicals -
      France.

1444 The Practical husbandman and planter, or, Observations on the ancient and
      modern husbandry, planting and gardening. - v.1-2 (no.1-6) 1733-1734.
      London: Stephen Switzer, 1733-1734.   EDITOR: Stephen Switzer.
      MICROFORMS: RP(1). (Goldsmiths' Kress Library). SOURCES: OCLC
      LOCATIONS: BUC, ULS SUBJECT: Agriculture - Periodicals - England.

1445 Practische Abhandlungen aus der Schriften der Königlichen medicinischen Societát
      zu Paris. - v.1-2, 1796-1797. Berlin: 1796-1797. Translated by Hermann
      Wilhelm Lindemann. TRANSLATION OF: Histoire de la Société royale de
      médecine, Paris. SOURCES: G41b SUBJECT: Medicine - Societies - Paris
      - Translations.

1446 Practische Annalen vom Militär-Lazareth in Cassel. - v.1 (St. 1-4), 2 (St. 1-3),
      1794-1809. Cassel: Krieger, 1794-1809. Imprint varies. OTHER TITLE:
      Practisches medicinisches Archiv.   EDITOR: Theodor John Piderit.
      SOURCES: K3686 SUBJECT: Medicine, Military - Periodicals - Germany.

1447 Practische Beiträge für Freunde der Oeconomie, Cameralwissenschaft,
      Arzneikunde und Scheidekunst. - v.1, 1790. Leipzig: 1790.   SOURCES:
      G726, K2969 SUBJECT: Economics - Periodicals - Germany. SUBJECT:
      Medicine - Periodicals - Germany.   SUBJECT: Chemistry - Periodicals -
      Germany.

1448 Practische Beiträge zur Geschichte der Kinderpocken und Kuhpocken. - Heft 1,
      1800. Leipzig: 1800.   EDITOR: Joseph Eyerel.   SOURCES: G222

SUBJECT: Smallpox - Periodicals - Germany.   **SUBJECT:** Medicine - Periodicals - Germany.

1449 Practische Bemerkungen zur Forstwissenschaft. - St. 1-3, 1783-1785. Frankfurt: Varrentrapp, 1783-1785. **EDITOR:** Franz Damian Friedrich Müllenkampf. **SOURCES:** K2935   **SUBJECT:** Forestry management - Periodicals - Germany.

1450 Practische Handlungs- und Industrie- Journal. - Prague: Hladky, 1793. **EDITORS:** Johann Wenzel Kumerle and Joseph Lange. **SOURCES:** K2764 **SUBJECT:** Economics - Periodicals - Czechoslovakia.

1451 Der Practische Landarzt, eine Wochenschrift. - v.1-2, 1773-1774. Mitau: J.F. Staffenhagen; Leipzig: W. Nauck, 1773-1774. Weekly. **EDITOR:** Peter Ernst Wilde. **SOURCES:** K3562 **LOCATIONS:** ULS **SUBJECT:** Medicine - Periodicals - Russia - Popular.

1452 Praesciptiones Lynceae Academiae. - v.1, 1624. Interamnae: Thomas Gueireis, 1614. Drafted 1604-05 but not published until 1624, (Ref.105). An edition in 1745 is noted. **SOURCES:** GII:1 **LOCATIONS:** BL **SUBJECT:** Science - Societies - Rome.

1453 Précis analytique des travaux de l'académie des sciences, belles-lettres et arts de Rouen, depuis sa fondation en 1744 jusqu'a l'epoque de sa restauration, le 29 Juin 1803, précedé de l'histoire de l'académie. - T.1(1744-1754) 1814; t.2(1751-1760) 1816; t.3(1761-1770) 1817; t.4(1771-1780) 1819; t.5(1781-1793) 1821. Rouen: 1814-1821. **SOURCES:** G452, OCLC **LOCATIONS:** BUC, CCF, CCN, ULS, ZDB **SUBJECT:** General - Societies - Rouen.

1454 Preisschriften der Königlichen Akademie der Wissenschaften. - v.1-5, 1796-1840. Berlin: 1796-1840. **INDEXES:** Kohnke, Otto, ed. Gesamtregister über die in den Schriften der Akademie von 1700-1899 erscheinen wissenschaftlichen Abhandlungen und Festreden. Berlin, 1900 (Ref.126). **SOURCES:** S2370i **SUBJECT:** General - Societies - Berlin - Prize essays.

1455 Preissschriften und Abhandlungen der Kaiserlichen freien ökonomischen Gesellschaft zu St. Petersburg. - St. Petersburg: J.D. Gerstenberg, 1796. Theil 1, 1796. **CONTINUATION OF:** Trudy vol'nago ekonomicheskoi obchestvo, St. Petersburg. **LOCATIONS:** BUC, NUC **SUBJECT:** Economics - Societies - St. Petersburg - Prize essays.

1456 Premiums by the Society established at London for the encouragement of arts, manufactures and commerce. - London: 1759-1782. Issued annually until included in the first volume of the society's Transactions in 1783. **LOCATIONS:** NUC **SUBJECT:** Technology - Societies - London. **SUBJECT:** Economics - Societies - London.   **SUBJECT:** Agriculture - Societies - London.

1457 Premiums offered by the Dublin Society. - Dublin: Sleaton, 1768-1805. Broadsides. Issues noted for 1768, 1781, 1787, 1799, 1801 and 1805 (Ref.106). SOURCES: (Ref.106). SUBJECT: Science - Societies - Dublin.

1458 Premiums offered by the Society of Agriculture at Manchester. - Manchester: 1772-1781. MICROFORMS: RP(1) (Goldsmiths' Kress Library). SUBJECT: Agriculture - Societies - Manchester.

1459 Prijsverhandelingen bekroond door het genootschap ter bevordering van natuurkunde, genees-en heelkunde te Amsterdam. - v.1-6, 1791-1807. Amsterdam: J.B. Elive, 1791-1807. Society name varies. CONTINUED BY: Nieuwe prijsverhandelingen bekroond door het genootschap ter bevordering van natuurkunde, genees-en heelkunde te Amsterdam. LOCATIONS: BUC, CCF, CCN, ULS SUBJECT: Natural history - Societies - Amsterdam - Prize essays SUBJECT: Medicine - Societies - Amsterdam - Prize essays.

1460 Primitiae physico-medicae, ab iis qui in Polonia et extra eam medicinam faciunt collectae. - v.1-3, 1750-1753. Zullichow; Leipzig: Fromann, 1750-1753. EDITORS: Ernst Jeremias Neifeld (Ref.107) and Gottlob Ephraim Hermann. SOURCES: G391, K3532 LOCATIONS: BUC SUBJECT: Medicine - Periodicals - Poland.

1461 Prix proposés par l'Académie royale des sciences, Paris. - Paris: Imprimerie Royale, 1719-1791. Published as separate pamphlets announcing the subject(s) proposed for the year. Title varies. LOCATIONS: BUC SUBJECT: Science - Societies - Paris - Prize essays.

1462 Prize essays and transactions of the Highland society of Scotland, to which is prefixed an account of the institution and principal proceedings of the society. - v.1-6, 1799-1824. Edinburgh: Constable, 1799-1824. Imprint varies. EDITOR: Henry Mackenzie. SOURCES: OCLC LOCATIONS: BUC, CCF, ULS SUBJECT: Agriculture - Societies - Edinburgh - Prize essays.

1463 Proceedings of a society for promoting innoculation, and preventing the natural small-pox in Chester. - Chester: 1778-1782. In Haygarth, John. An inquiry how to prevent the small- pox, Chester, 1785, p.147-208. SOURCES: (Ref.17,no.18). SUBJECT: Smallpox - Periodicals - England.

1464 Proceedings / Royal Dublin society. - v.1-109, 1764-1803. Dublin: 1764-1803. INDEXES: General index to the proceedings of the Royal Dublin Society. Dublin, Graisberry, 1826. General index to the printed proceedings of the Dublin society in 50 volumes. Dublin, Griasberry and Campbell, 1814 (Ref.106). SOURCES: S86b LOCATIONS: BUC SUBJECT: Science - Societies - Dublin. SUBJECT: Agriculture - Societies - Dublin.

1465 Procès-verbal de la séances publique de la Société d'agriculture, du commerce et des arts de l'arrondisement de Boulogne- sur-mer. - v.1-17, 27 Apr. 1799-24 Sept.1834. Boulogne-sur-mer: 1798-1834. CONTINUED BY: Mémoires, Société d'agriculture, du commerce et des arts de l'arrondissment de

Boulogne-sur-mer. **LOCATIONS:** BUC, CCF, ULS **SUBJECT:** Agriculture - Societies - Boulogne. **SUBJECT:** Economics - Societies - Boulogne.

1466 Procés verbal de la séances publiques / Société des sciences et des arts de Potiers. - v.1-6, 1797-1803. Potiers: 1797-1803. **CONTINUED BY:** Athénée de Potiers. **LOCATIONS:** CCF, ULS **SUBJECT:** General - Societies - Potiers.

1467 Procès verbal de travaux / Société libre d'émulation du commerce et de la industrie de la Seine inferieure. - **INDEXES:** Table génerale du bulletin publié par la Société 1797 à 1899. Rouen, E. Cogniard, 1900. **SOURCES:** (Ref.139). **SUBJECT:** Economics - Societies - Seine Inferieure. **SUBJECT:** Agriculture - Societies - Seine inferieure.

1468 Procès verbal et journal / Académie des sciences, lettres et beaux arts de Marseilles. - v.1-5 (1796-1805) 1803-1814. Marseilles: 1796-1805. **CONTINUATION OF:** Recueil Académie des sciences, lettres et beaux arts de Marseilles. **CONTINUED BY:** Mémoires, Académie des sciences, lettres et beaux arts de Marseilles. **SOURCES:** G454a **SUBJECT:** General - Societies - Marseilles.

1469 Procés-verbaux de l'Académie royale d'architecture. - v.1-10 (1691-1793) 1911-1929. Paris: Eduard Champion, 1911-1929. Edited by Henry Limonnier. **INDEXES:** Index v.1-9 in v.10. **SOURCES:** OCLC **LOCATIONS:** BUC, CCF, CCN, ULS, ZDB **SUBJECT:** Architecture - Societies - Paris.

1470 Procés-verbaux des séances / Academia scientiarum imperialis Petropolitana. - v.1-4, 1725-1803. St. Peterburg: 1725-1803. **OTHER TITLE:** Protokoly zas'panii conferentsia. **SOURCES:** G442f **LOCATIONS:** BUC, CCF, CCN, ULS, ZDB **SUBJECT:** Science - Societies - St. Petersburg.

1471 Procès-verbaux des séances tenues depuis la fondation de l'Institut jusqu'au mois d'aout, 1835/ Institut nationale des sciences lettres et arts de France. - T.1 (1795-1799)-10(1832-1835), 1910-1922. Hendaye: Observatoire d'Abbadia, 1910-1922. **SOURCES:** OCLC **LOCATIONS:** CCN, ZDB **SUBJECT:** Science - Societies - Paris.

1472 Prodol'zenie trudov / Imperatorskoi vol'noe ekonomicheskago obtchestva. - No. 1-31, 1765-1794. St. Petersburg: 1765-1794. There were 2nd and 3rd editions of some of the volumes. **CONTINUED BY:** Novoe prodol'zenie trudov. **LOCATIONS:** BUC, ZDB **SUBJECT:** Economics - Societies - St. Petersburg.

1473 Prodomus praevertens continuata Acta medica Havniensia. Copenhagen: 1753. v.1,1753. **CONTINUATION OF:** Acta medica et philosophica (Ref.42:205). **SOURCES:** (Ref.42,p.205). **SUBJECT:** Medicine - Periodicals - Denmark.

1474 Programma / Hollandsche maatschappij der wetenschappen, Haarlem. - no. 1-121, 1754-1876. Haarlem: 1754-1876. Also published in its Verhandlingen

1754-1793, then as separates and in French, 1773-1875. **SOURCES: S849a**
LOCATIONS: BUC SUBJECT: Science - Societies - Haarlem.

1475 Programma van het genootschap ter bevordering der heelkunde te Amsterdam. -
Amsterdam: 1790-1828. **LOCATIONS: CCN SUBJECT:** Medicine -
Societies - Amsterdam.

1476 Programmas da Academia Real das sciencias de Lisboa. - Folheto 1-57,
1780-1854. Lisbon: 1780-1854. **LOCATIONS: NUC SUBJECT:** Science
- Societies - Lisbon.

1477 Le Progrès de la médecine, contenant un recueil de tout ce qui s'observe de
singulier par rapport à sa pratique avec un jugement de tous les ouvrages qui ont
rapport à la théorie de cette science. - v.1-4, 1695-1709. Paris: Laurent
d'Houry, 1695-1709. **EDITOR:** Claude Brunet. **MICROFORMS:** RP(1).
SOURCES: GI:10, OCLC **LOCATIONS:** BUC, CCF, ULS **SUBJECT:**
Medicine - Periodicals - France.

1478 Protokoli / Vetenskaps-societeten i Upsala. - v.1 (1732-1784) 1950, v.2
(1786-1803) 1960. Upsala: Lindequisst, 1950-1960. **LOCATIONS:** ZDB
SUBJECT: Science - Societies - Upsala.

1479 Provinzialblätter oder Sammlungen zur Geschichte, Naturkunde, Moral und
anderen Wissenschaften. - Bd.1-2(St.1-6)1781-1783. Leipzig und Dessau:
Buchhandlung der Gelehrten, 1781-1783. **EDITOR:** Karl Gottlob Anton.
CONTINUED BY: Lasusitzisches Wochenblatt zu Ausbreitung nützlicher
Kenntnisse aus der Natur-, Haushaltungs-, Staats- und Völker- Kunde der Ober-
und Niederlausitz. **CONTINUED BY:** Lausitzische Monatsschrift,
Oberlausitzische Gesellschaft der Wissenschaften, Görlitz. **SOURCES:** K355
LOCATIONS: CCF, ULS, ZDB **SUBJECT:** General - Periodicals -
Germany.

1480 Prüfung des Brown'schen Systems der Heilkunde, durch Erfahrungen am
Krankenbette. - St. 1-4, 1797-1799. Weimar: 1797-1799. **EDITOR:** Adalbert
Friedrich Marcus. **SOURCES:** G256, OCLC **LOCATIONS:** DSG
SUBJECT: Brunonianism - Periodicals - Germany. **SUBJECT:** Medicine -
Periodicals - Germany.

1481 Psychologisches Magazin. - v.1-3, 1796-1798. Jena: Cröker, 1796-1798.
CONTINUATION OF: Gnothi sauton. **EDITOR:** Carl Christian Erhard
Schmid. **CONTINUED BY:** Zeitschrift für die spekulative Physik.
SOURCES: B7612, G252, K580 **LOCATIONS:** ULS, ZDB **SUBJECT:**
Psychology - Periodicals - Germany.

1482 Psychologisches Magazin (Altenburg). - v.1-3, 1796-1798. Altenburg:
1796-1798. **EDITOR:** J.G. Heynig. **SOURCES:** G251, K579, OCLC, S2232
LOCATIONS: CCF, ULS **SUBJECT:** Psychology - Periodicals - Germany.

1483 Raccolta di memorie delle publiche Accademia di agricoltura, arti e commercio
      dello Stato Veneto. - No. 1-18, 1789-1798. Venice: Giovanni Antonio Perlini,
      1789-1798. LOCATIONS: ULS   SUBJECT: Agriculture - Societies -
      Venice. SUBJECT: Economics - Societies - Venice.

1484 Raccolta di vari opuscoli publicati sin' ora interno all' uso delle lucertole, per la
      guariggione de' cancri, ed altri mali. - v.1, 1785.   Naples: 1785.
      TRANSLATED AS: Ueber den Nutzen der Eideschen in Krebschäden.
      EDITOR: Jacques Christophe Valmont de Bomare. SOURCES: G399, OCLC
      LOCATIONS: BL   SUBJECT: Medicine - Periodicals - Italy.

1485 Raccolta d'opuscoli fiscico-medici. - v.1-6, 1775-1782.   Florence: Moücke,
      1775-1782. EDITOR: Giovanni Ludovici Targioni.   SOURCES: B3841,
      G396, OCLC, S1844 LOCATIONS: CCF SUBJECT: Science - Periodicals
      - Italy. SUBJECT: Medicine - Periodicals - Italy.

1486 Raccolta d'opuscoli medico-practici. - v.1-7, 1773-1783. Florence: 1773-1783.
      EDITOR: Giovanni Ludovico Targioni. SOURCES: G397 LOCATIONS:
      ULS   SUBJECT: Medicine - Periodicals - Italy.

1487 Raccolta d'opuscoli medico-teorico-pratiche ed anatomici, tratti da' foglj medici
      d'oltremonti e d'Italia. - v.1-3, 1762-1764.   Parma: Filippo Carmignani,
      1762-1764. SOURCES: G398 LOCATIONS: DSG SUBJECT: Medicine
      - Periodicals - Italy.

1488 Raccolta d'opuscoli scientifici e filologici. - v.1-58, 1728-1757.   Venice:
      Cristoforo Zane, 1728-1757. EDITOR: Angelo Calgerà. CONTINUED BY:
      Nuovo raccolta d'opuscoli scientifico e filologici. SOURCES: B3842, G727a,
      OCLC, S2100   LOCATIONS: CCN, ULS, ZDB   SUBJECT: Science -
      Periodicals - Italy.

1489 Raccolta d'opuscoli scientifici e letterari degli autori Italiani. - v.1-25, 1779-1796.
      Ferrara: Guiseppi Rinaldo, 1779-1796. OTHER TITLE: Raccolta ferranese
      di opuscoli scientifici e letterari. SOURCES: B3840, S1819 LOCATIONS:
      BUC, CCF, ZDB   SUBJECT: Science - Periodicals - Italy - Reviews
      SUBJECT: General - Periodicals - Italy - Reviews.

1489a Rapport des mémoires préséntés à la Société libre d'agriculture, commerce et arts
      du department du Doubs. v.1-8, 1799-1808. Besançon: 1799-1808. OTHER
      TITLE: Rapport générale des travaux de la Société libre d'agriculture,
      commerce et arts du department du Doubs. Extraits des memoires de la Société
      libre d'agriculture, commerce et arts du department du Doubs. LOCATIONS:
      BL, CCF, ULS   SUBJECT: Agriculture - Societies - Besançon. SUBJECT:
      Economics - Societies - Besançon.

1490 Rapport généraux des travaux de la Société Philomathique de Paris. - T.1-4
      (1788/92-1800) 1793-1800. Paris: Ballard, 1793-1800. OTHER TITLE:
      Rapports généraux des travaux de la Socété Philomathique de Paris.

SOURCES: G512a, S1594e   **LOCATIONS:** BUC, CCF, CCN, ULS, ZDB
SUBJECT: Science - Societies - Paris.

1491   Rapports / Société d'agriculture, sciences et arts, Meaux. - No. 1-12, 1798-1813.
Meaux: 1798-1813.   **SOURCES:** S1217   **SUBJECT:** Agriculture - Societies
- Meaux.   **SUBJECT:** Economics - Societies - Meaux.

1492   Der Rathgeber für alle Stände in Angelegenheiten, welche die Gesundheit, den
Vermögens- und Erwerbstand und den Lebensgenuss betreffen. - v.1-3,
1799-1803. Gotha: Becker, 1799-1803. Monthly. **CONTINUATION OF:**
Wochenblatt des aufrichtigen Volksarztes. Imprint varies. **EDITOR:** Daniel
Collenbusch.   **SOURCES:** G208, K3732   **SUBJECT:** Medicine - Periodicals
- Germany - Popular.   **SUBJECT:** Economics - Periodicals - Germany -
Popular.

1493   Recherches asiatiques, ou, mémoires de la Société établie à Bengale pour faire des
recherches sur l'histoire et les antiquitées, les arts, les sciences et la litérature
de l'Asie. - v.1-2, 1805. Paris: Marcel, Treutel et Würtz, 1805. Translated by
A. Labeaume and reviewed and augmented with notes from the oriental part by
M. Langlés and for the sciences by Cuvier, Delambre, Lamarck and Olivier.
TRANSLATION OF: Asiatic researches or transactions of the Society instituted
in Bengal for inquiring into the history and antiquities, the arts, sciences and
literature of Asia.   **SOURCES:** G458b   **LOCATIONS:** BUC, CCF, CCN,
ULS, ZDB   **SUBJECT:** General - Societies - Calcutta - Translations.

1494   Recherches de mathematique et de physique. - v.1-2, 1703. Paris: Florentin
Delaulne et Jean Goubert, 1703.   Reprinted in 3 v.1713 (Ref.20:211-3).
OTHER TITLE: Recherches de physique et de mathematique.   **EDITOR:**
Antoine Parent.   **LOCATIONS:** BUC,NUC   **SUBJECT:** Science - Periodicals
- France.   **SUBJECT:** Mathematics - Periodicals - France.

1495   Recherches physico-chimques. - No. 1-3, 1792-1794. Amsterdam: L. van Hurst,
1792-1794.   Translation in Ref.67.   **EDITOR:** Johan Rudolph Deiman.
CONTINUED BY: Natuur-scheikundige verhandelingen. **SOURCES:** B3872,
G728,   **LOCATIONS:** CCN, ULS   **SUBJECT:** Chemistry - Periodicals -
Netherlands.

1496   Records of the New Hampshire medical society. - v.1 (1791-1854) 1911.
Concord, N.H.: Rumford, 1911.   **CONTINUED BY:** Transactions, New
Hampshire medical society.   **LOCATIONS:** CCF, ULS   **SUBJECT:**
Medicine - Societies - New Hampshire

1497   Recreations in agriculture, natural history, arts and miscellaneous literature. -
v.1-6, 1799-1802. London: T.Bensley, 1799-1802. v.5-6 also as ser.2,v.1-2.
EDITOR: James Anderson.   **MICROFORMS:** RP(1) (Goldsmiths' Kress
Library).   **SOURCES:** B7649, OCLC   **LOCATIONS:** BUC, ULS
SUBJECT: Science - Periodicals - England - Popular. **SUBJECT:** Agriculture
- Periodicals - England - Popular.

1498 Recreations physiques, économiques et chimiques. - v.1-2, 1774. Paris: Monory, 1774. Translated with additions by Antoine Augustin Parmentier. **TRANSLATION OF:** Chymische Nebenstunden. **SOURCES:** OCLC **LOCATIONS:** BL   **SUBJECT:** Chemistry - Periodicals - Germany - Translations.

1499 Recueil / Académie de Montauban. - Montauban: 1742-1750. **SOURCES:** G444 **SUBJECT:** General - Societies - Montauban.

1500 Recueil / Académie des sciences, lettres et beaux-arts de Marseilles. - Ser. 1 (no. 1-31) 1727-1767, ser. 2 (no. 1-12) 1768-1786.   Marseilles: 1727-1786. **OTHER TITLE:** Recueil de plusieurs pièces de poésie presentée à l'Académie de belles-lettres de Marseilles. **CONTINUED BY:** Procès verbal et journal, Académie des sciences lettres et beaux arts de Marseilles. **CONTINUED BY:** Mémoires, Académie des sciences, lettres et beaux arts de Marseilles. **SOURCES:** G454, S1207 **LOCATIONS:** BUC, CCF **SUBJECT:** General - Societies - Marseilles.

1501 Recueil / Académie royale des sciences, arts et belles- lettres de Caen. - Caen: 1731-1816. **SOURCES:** G445 **SUBJECT:** General - Societies - Caen.

1502 Recueil contenant les délibérations de la Société royale d'agriculture de la géneralité de Paris. - v.1, 1761. Paris: d'Houry, 1761. **MICROFORMS:** RP(1). (Goldsmiths' Kress Library). **LOCATIONS:** CCF   **SUBJECT:** Agriculture - Societies - Paris.

1503 Recueil de mémoires concernant l'économie rurale. - T.1-2, 1760-1761. Zürich: Heydegger, 1760-1761.   Published simultaneously in French and German. **OTHER TITLE:** Sammlungen von landwirthschaftlichen Dingen der Schweitzerischen Gesellschaft in Bern.   **EDITOR:** Elie Bertrand (Ref.29,p.152). **CONTINUED BY:** Mémoires et observations recueillies par la Société oeconomique de Bern. **MICROFORMS:** RP(1) (Goldsmiths' Kress Library).   **LOCATIONS:** BUC, CCF, CCN, ULS, ZDB   **SUBJECT:** Agriculture - Societies - Bern. **SUBJECT:** Economics - Societies - Bern.

1504 Recueil des mémoires, ou, collection des pièces académiques, concernant la médecine, l'anatomie et la chirurgie, la chymie, la physique expérimentelle, la botanique et l'histoire naturelle tirées des meilleurs sources. - v.1-16, 1754-1787.   Dijon: Desventes, 1754-1787.   Various editions are cited (Ref.21,v23:1503a). Imprints vary. Also published under the title: Collection académique, partie française, including abstracts of the publications of the Académie des Sciences, Paris, and: Collection académique, partie étrangères including translations of proceedings of various scientific societies and periodicals (Ref.26,p.212-15). OTHER TITLE: Collection académique, partie française; Collection académique, partie étrangères. Edited by Jean Berryat and others.   **INDEXES:** Indexed in Rozier's Nouvelle table des articles contenue dans les Collection Académique. **MICROFORMS:** RM(3). **SOURCES:** B3884, G498, OCLC, S1141 **LOCATIONS:** BUC, CCF, CCN,

ULS   **SUBJECT:** Science - Periodicals - Europe - Translations.   **SUBJECT:** Science - Societies - Europe - Translations.

1505   Recueil de pièces en prose et en vers / Académie des belles- lettres, sciences et arts, La Rochelle. - v.1-3, 1747-1763.   Paris: Thibout, 1747-1763. **SOURCES:** S1634c   **LOCATIONS:** BUC, CCF   **SUBJECT:** General - Societies - La Rochelle.

1506   Recueil de plusiers traitez de mathematiques / Académie royale des sciences, Paris. - Paris: Imprimerie Royale, 1671-1677.   A series of six monographs on various subjects published 1671- 1677.   **LOCATIONS:** BUC, NUC **SUBJECT:** Science - Societies - Paris.   **SUBJECT:** Mathematics - Societies - Paris.

1507   Recueil de programmes et de listes des prix proposés et distribués par la Société royale de médeicine de Paris.   - Paris: 1778-1791 (Ref.14,p.162). **LOCATIONS:** BN   **SUBJECT:** Medicine - Societies - Paris - Prize essays.

1508   Recueil des actes, ou, mémoires et observations / Société de santé de Lyon. - v.1-2 (1792-1797) 1798-(1797-1801) 1802. Lyon: Bruyset, 1798-1802. Imprint varies.   **OTHER TITLE:** Mémoires et observations sur la chirurgie, la médecine, l'histoire naturelle, Société de Santé, Lyon; Journal de la Société de médecine de Lyon.   **EDITORS:** Pitt, Petit and Martin.   **SOURCES:** G155, 170, OCLC   **LOCATIONS:** BUC, CCF, ULS   **SUBJECT:** Medicine - Societies - Lyon.

1509   Recueil des délibérations et des mémoires du Bureau de Tours / Société d'agriculture de la géneralité du Maine. - LeMans: 1761.   Also published for 1762-1783? (Ref.32,p.288)   **SOURCES:** S1203a   **SUBJECT:** Agriculture - Societies - LeMans.

1510   Recueil des déliberations et des mémoires de la Société royale d'agriculture de la Généralité de Tours. - v.1, 1763.   Tours: F. Lambert, 1763.   **SOURCES:** S1687a   **LOCATIONS:** CCF, ULS   **SUBJECT:** Agriculture - Societies - Tours.   **SUBJECT:** Science - Societies - Tours.

1511   Recueil des discours prononcez dans les conférences académique / Académie des sciences, inscriptions et belles- lettres, arts, Toulouse. - v.1, 1692. Toulouse: 1692.   The society was founded in 1640 and was known up to 1704 as the Société des Lanternistes.   **SOURCES:** S1679e   **LOCATIONS:** NUC **SUBJECT:** General - Societies - Toulouse.

1512   Recueil des dissertations qui ont remporté le prix / Académie royale des science, belles-lettres et arts de Bordeaux. - Bordeaux: Pierre Brion, 1715-1741.   A collection of separately published prize essays gathered together.   **OTHER TITLE:** Recueil des dissertations couronées, Académie royale des sciences, belles-lettres et arts, Bordeaux.   **INDEXES:** Table historique des travaux et publications (1712-1875). Bordeaux. 1877.   **SOURCES:** G447, S1061a

LOCATIONS: BUC, CCF, CCN, ZDB   **SUBJECT:** General - Societies - Bordeaux - Prize essays.

1513 Deleted.

1514 Recueil des experiences et des découvertes faites ces derniers temps dans la physique, la médecine, la chirurgie, d'histoire naturelle, la chimie, l'économie domestique, l'agriculture, le commerce, etc. - Gothenberg: 1781. Monthly. **EDITOR:** Clas Alströmer. **SOURCES:** (Ref.110). **SUBJECT:** Medicine - Periodicals - Sweden.   **SUBJECT:** Economics - Periodicals - Sweden. **SUBJECT:** Science - Periodicals - Sweden.

1515 Recueil des lettres, mémoires et autres pieces / Académie des sciences et belle lettres, Béziers. - Parts 1-5, 1728-1736. Béziers: Estienne Barbut, 1728-1736. **SOURCES:** G449, S1056 **SUBJECT:** General - Societies - Béziers.

1516 Recueil des mémoires couronnés par la Société royale d'agriculture de Limoges. - Lyon: Perisse, 1771. **SOURCES:** (Ref.140).   **SUBJECT:** Agriculture - Societies - Limoges - Prize essays.   **SUBJECT:** Science - Societies - Limoges - Prize essays.

1517 Recueil des mémoires et conférences qui ont esté présentées à Monseigneur le Dauphin. - No.1- 12, Feb. 1672-June 1672. Paris: Fréderic Leonard, 1672. Monthly. Published as supplement to the Journal des Sçavans during the period it was suspended. There were several editions in quarto in Paris. Reprinted in Amsterdam and Brussels in 1682. **OTHER TITLE:** Mémoires concernant les arts et les sciences, présentez à Monseigneur le Dauphin; Journal des Sçavans; Mémoires sur les arts et les sciences.   **EDITOR:** Jean Baptiste Denis. **CONTINUED BY:** Conférence sur les sciences presentée à Monseigneur le Dauphin.   **SOURCES:** GII:11, OCLC   **LOCATIONS:** BUC, CCF, CCN, DSGS   **SUBJECT:** Science - Periodicals - France.

1518 Recueil des mémoires les plus intéressant de chimie et d'histoire naturelle / Kungliga Vetenskapsakademien, Stockholm. T.1-t.2, (1720-1760)1764. Paris: P.F.Didot, 1764.   Translated by Augustin Roux and Baron d'Holbach. **TRANSLATION OF:** Handlingar, Kungliga Svenska Vetenskapsakademien. **SOURCES:** B3886, OCLC, S1501 **LOCATIONS:** ULS **SUBJECT:** Science - Societies - Stockholm - Translations.

1519 Recueil des mémoires relatifs aux établissement d'humanité. - v.1-38, 1799-1804. Paris: Agasse, 1799-1804. Translated from English.   **SOURCES:** G400 **LOCATIONS:** BUC, DSG **SUBJECT:** Social sciences - Periodicals - England - Translations.

1520 Recueil des ouvrages lus aux séances / Lycée de Toulouse. - No. 1-5, 1798-1801. Toulouse: Benichet, 1798-1801. **SOURCES:** S1681a, OCLC   **SUBJECT:** General - Societies - Toulouse.

1521 Recueil des ouvrages présentés a l'Académie des belles- lettres et arts de Marseilles pour le prix. - 1727, 1774, 1777. Marseilles: François Brebion, 1727-1777. Imprint varies. **LOCATIONS:** BUC **SUBJECT:** General - Societies - Marseilles - Prize essays.

1522 Recueil des pièces lues dan les séances publiques / Académie royale de Nîmes. - Nîmes: 1756-1777. **OTHER TITLE:** Mémoires, Académie de Nîmes. **SOURCES:** G446, OCLC **LOCATIONS:** BUC, ULS **SUBJECT:** General - Societies - Nîmes.

1523 Recueil des pièces qui ont concurru pour le prix de l'Académie royale de chirurgie, Paris. - v.1-5, 1733-1783. Paris: Delaguette, 1733-1783. **TRANSLATED AS:** Sammlung der Preisschriften der königlichen Parisischen Academie der Chirurgie. Publisher varies. **OTHER TITLE:** Mémoires sur les sujets proposés pour les prix, Académie royale de chirurgie, Paris. **SOURCES:** OCLC **LOCATIONS:** BUC, CCF, CCN, ULS **SUBJECT:** Surgery - Societies - Paris - Prize essays.

1524 Recueil des pièces qui ont remporté le prix / Académie royale des sciences, Paris. - v.1-9 (1720-1772) 1721-1777. Paris: 1721-1777. The volumes (except for v.6) are made up of memoirs separately printed, each with its own title page. Other editions are reported of some of the volumes (Ref.18). **CONTINUED BY:** Mémoires de mathématique et de physique presentés par divers savans. **MICROFORMS:** RM(3). **SOURCES:** G17g, OCLC, S1276j **LOCATIONS:** BUC, CCF, CCN, ULS, ZDB **SUBJECT:** Science - Societies - Paris - Prize essays.

1525 Recueil des procès-verbaux des séances / Chambre de commerce du Dunkerque. - (1700-1784?) **LOCATIONS:** CCF **SUBJECT:** Commerce - Societies - Dunkerque.

1526 Recueil d'observations de médecine des hopitaux militaires. - v.1-2, 1766-1772. Paris: Imprimerie royale, 1766-1772. **EDITOR:** François Richard Marie Claude de Hautesierck (Ref 108). **SOURCES:** GIII:23, OCLC **LOCATIONS:** BN, ULS **SUBJECT:** Military medicine - Periodicals - France.

1527 Recueil d'observations faites en plusieurs voyages pour perfectionner l'astronomie et la geographie / Académie royale des sciences, Paris. - v.1-9 (1679-1693) 1693. Paris: Imprimerie royale, 1693. Some of the works have individual title pages dated 1679-1683. **LOCATIONS:** BUC, NUC **SUBJECT:** Astronomy - Societies - Paris. **SUBJECT:** Geography - Societies - Paris.

1528 Recueil périodique de la Société de santé de Paris. - v.1-14, 1796-1802. Paris: Cruelbois, 1796-1802. **CONTINUATION OF:** Recueil périodique d'observations de médecine, de chirurgie et de pharmacie. **CONTINUATION OF:** Histoire de la Société royale de médecine, Paris. Publisher varies. **OTHER TITLE:** Recueil périodique de la Société de médecine de Paris. **CONTINUED BY:** Journal générale de médecine. Recueil périodique de littérature médicale étrangère. **INDEXES:** Table alphabétique et méthodique du matières in: Journal générale de médicine, v.1. **SOURCES:** G135a, OCLC **LOCATIONS:** BUC, CCF, CCN, ULS, ZDB **SUBJECT:** Medicine - Societies - Paris.

1529 Recueil périodique de littérature médicale étrangère. - v.1-2, 1799. Paris: Société de médecine, 1799. Issued as a supplement to Recueil périodique de la Société de médecine. **INDEXES:** Indexed in Journal générale de médecine v.1. **SOURCES:** OCLC **LOCATIONS:** BUC, CCF, ULS **SUBJECT:** Medicine - Societies - Paris - Reviews.

1530 Recueil philosophique et littéraire de la Société typographique de Bouillon. - v.1-10, 1769-1779. Bouillon: 1769-1779. **EDITORS:** Jean Louis Castilhon and Jean Baptise René Robinet (Ref.109). **LOCATIONS:** BUC, CCF, ULS, ZDB **SUBJECT:** General - Periodicals - France.

1531 Recueil pour les astronomes. - v.1-3, 1771-1776. Berlin: 1771-1776. A supplement was published in Berlin in 1779. **EDITOR:** Johann Bernouilli. **SOURCES:** B7654, S2376 **LOCATIONS:** BL, ULS **SUBJECT:** Astronomy - Periodicals - Germany.

1532 Redevoering uitgesproken ter opening van de jaarlijksche algemeene vergadering der Bataafsche maatschapij tot nut van't algemeen. - Amsterdam: 1788-1852. Subtitle varies. **OTHER TITLE:** Redevoering en aanspraaken uitgesproken ter opening van de jaarlijksche algemeene vergadering der Bataafsche maatschapij tot nut van't algemeen. **LOCATIONS:** BUC, CCF, CCN, ULS, ZDB **SUBJECT:** General - Societies - Amsterdam.

1533 Reflexionen und Erfahrungen für Bürger, Geistliche und junge Aerzte. - Düsseldorf: Gänzer, 1790. K5837 cites a source which indicates 1779 as the starting date. **SOURCES:** K5837 **SUBJECT:** Medicine - Periodicals - Germany - Popular. **SUBJECT:** General - Periodicals - Germany - Popular.

1534 Das Reich der Natur und Sitten, eine moralische Wochenshrift. - v.1-12, 1757-1762. Halle: 1757-1762. Weekly. **SOURCES:** B3898, S2735 **LOCATIONS:** BUC, CCF, ULS, ZDB **SUBJECT:** Natural history - Periodicals - Germany - Popular.

1535 Relationes de libris novis. - v.1-3 (4 issues each) 4(no.1) 1752-1755. Göttingen: Vandenhoek, 1752-1755. **OTHER TITLE:** Relationum de libris novis. **EDITOR:** Johann David Michaelis. **SOURCES:** G729, K185 **LOCATIONS:** BUC, CCF, CCN, ULS, ZDB **SUBJECT:** General - Periodicals - Germany - Reviews.

1536 Relations de l'Assemblée publique de l'Académie des sciences et belles lettres du jeudi. - v.1 (1723-1730) 1731. Beziers: Estinee Barbut, 1731. **LOCATIONS:** BL **SUBJECT:** General - Societies - Beziers.

1537 Repertorium chirurgischer und medicinischer Abhandlungen. - v.1-4, 1797-1801. Leipzig: Reinicke, 1797-1801. v.4 also as Neues Repertorium. Imprint varies. A new edition appeared in 1809. **OTHER TITLE:** Neues Repertorium chirurgischer und medicinischer Abhandlungen. **EDITOR:** Christian Friedrich Michaelis. **SOURCES:** G403, K3677 **LOCATIONS:** CCF, ULS **SUBJECT:** Medicine - Periodicals - Germany.

1538 Repertorium der medicinischen Litteratur. - v.1-5 (1789-1793) 1790-1795. Zürich: Ziegler, 1790-1795. **EDITOR:** Paul Usteri. **CONTINUED BY:** Medicinische Litteratur des Jahres. **SOURCES:** G311, K3666 **LOCATIONS:** BUC, CCF, CCN, ULS, ZDB **SUBJECT:** Medicine - Periodicals - Switzerland - Reviews.

1539 Repertorium der neuern wichtigsten Abhandlungen und Beobachtungen für Aerzte, Wundärzte und Apotheker. - v.1 (no.1-2) 1789. Guntersblum: F.C.L. Gegel, 1789. Monthly. **OTHER TITLE:** Repertorium für Aerzte. **SOURCES:** G312, K3655 **LOCATIONS:** CCF, ZDB **SUBJECT:** Medicine - Periodicals - Germany. **SUBJECT:** Pharmacy - Periodicals - Germany. **SUBJECT:** Surgery - Periodicals - Germany.

1540 Repertorium der neuesten Fortschritte der Physik. - No. 1, 1800. Breslau: 1800. **SOURCES:** S2455 **SUBJECT:** Science - Periodicals - Germany.

1541 Repertorium für Chemie, Pharmacie und Arzneimittelkunde. - v.1 (no.1-2) 1790. Leipzig: Hertel; Hildesheim: Georg Friedrich Schmidt, 1790. **CONTINUATION OF:** Magazin für Apotheker, Materialisten und Chemisten. New edition 1796 (Hannover, Hahn). **EDITOR:** Johann Caspar Philipp Elwert. **SOURCES:** B2763a, G214, K3705, S2805 **LOCATIONS:** BUC **SUBJECT:** Chemistry - Periodicals - Germany. **SUBJECT:** Pharmacy - Periodicals - Germany.

1542 Repertorium für die öffentliche und gerichtliche Arzneiwissenschaft. - v.1-3, 1789-1793. Berlin: Friedrich Viewig, 1789-1793. **CONTINUATION OF:** Magazin für die gerichtliche Arzneikunde und medicinische Polizei. **CONTINUATION OF:** Neues Magazin für die gerichtliche Arzneikunde und medicinische Polizei. **EDITOR:** Johann Theodor Pyl. **SOURCES:** G83, K3655 **LOCATIONS:** BUC, CCN, ULS, ZDB **SUBJECT:** Medical jurisprudence - Periodicals - Germany.

1543 Repertorium für Physiologie und Psychologie nach ihren Umfang und Verbindung. - v.1-2 (no.1) 1784-1786. Hof: Vierling, 1784-1786. **EDITOR:** Johann Hermann Pfingsten. **SOURCES:** B3924, G313, K3610, S2806 **SUBJECT:** Physiology - Periodicals - Germany. **SUBJECT:** Psychology - Periodicals - Germany.

1544 Repertory of arts and manufactures, consisting of original communications, specification of patent inventions and selections of useful practical papers from the transactions of the philosophical societies of all nations. - v.1-16, 1794-1802. London: G. and T.Wilkie, 1794-1802. Imprint varies.

CONTINUED BY: Repertory of arts, manufactures and agriculture. INDEXES: Index v.1-16 (1806). MICROFORMS: RP(1). SOURCES: B3934, OCLC LOCATIONS: BUC, CCF, CCN, ZDB SUBJECT: Technology - Periodicals - England. SUBJECT: Manufactures - Periodicals - England.

1545 Report / Literary and philosophical society of Newcastle-upon-Tyne. - v.1-83, 1793-1876. Newcastle: 1793-1876. Also published as Annual report and Year's report. SOURCES: S505b LOCATIONS: BUC, ULS SUBJECT: General - Societies - Newcastle-upon-Tyne.

1546 Reports of the Royal Humane Society, instituted for the recovery of persons apparently drowned. - No.1 (1779-1780) 1781, no.2(1781-1782) 1783, no.3(1783-1784) 1786, no.4(1785-1786) 1788, no.5(1787-1789) 1790. London: Rivington, 1781-1790. TRANSLATED AS: Avisos interessantes (Lisbon, 1788-1790). Issued also in 1775 (1774) as: Plans and reports. Publisher varies. OTHER TITLE: Plans and reports of the Royal Humane Society. CONTINUED BY: Transactions of the Royal Humane Society. SOURCES: G499 LOCATIONS: BUC, CF, ZDB SUBJECT: Resuscitation - Societies - London.

1547 Reports principally concerning the effects of nitrous acid in the venereal disease. - 3 v., 1797-1800. Bristol; London: 1797-1800. TRANSLATED AS: Neuesten Erfahrungen Britischer Aerzte. The reports were reprinted in 1801 (Ref.17:33). OTHER TITLE: Collection of testimonies respecting the treatment of veneral disease by nitrous acid (1799); Communications respecting the external and internal use of nitrous oxide (1800). EDITOR: Thomas Beddoes. SOURCES: G344, OCLC LOCATIONS: NUC SUBJECT: Medicine - Periodicals - England.

1548 Rheinisches Magazin zur Erweiterung der Naturkunde. - v.1, 1793-1794. Giessen: Georg Friedrich Heyer, 1793-1794. EDITOR: Moriz Balthazar Borkhauser. SOURCES: B4100, G730, K3351, S2643 LOCATIONS: BUC, CCF, ULS, ZDB SUBJECT: Natural history - Periodicals - Germany.

1549 Richerche critiche appartenenti all'Accademia del Pontaniano. - v.1, 1795. Naples: 1795. SOURCES: S1932a LOCATIONS: ULS SUBJECT: General - Societies - Naples.

1550 The Rise, minutes and proceedings / Medical society of New Jersey. - v.1 (1766-1800) 1875. Newark, N.J.: 1875. SOURCES: GIII:23a LOCATIONS: BUC, ULS SUBJECT: Medicine - Societies - New Jersey.

1551 Rit thess islenzka laerdóms lista félag. - v.1-15, 1781-1797. Copenhagen: 1781-1798. OTHER TITLE: Felagsritin gämlu. SOURCES: G731, S643 LOCATIONS: BUC, CCF, ULS SUBJECT: Science - Societies - Iceland.

1552 Rockenmedicin, eine Wochenschrift. - Wittenberg: 1788. Weekly. **EDITOR:** Johann Samuel Traugott Frenzel. **SOURCES:** K(1942)3442 (Ref.111). **SUBJECT:** Medicine - Periodicals - Germany.

1553 Rules and orders of the Society for promoting natural history. - v.1, 1790. London: 1790. **LOCATIONS:** BL **SUBJECT:** Natural history - Societies - London.

1554 Rules and orders of the Society instituted at Bath for the encouragement of agriculture, arts, manufactures and commerce. - Bath: R. Cuttwell, 1780. **LOCATIONS:** BL **SUBJECT:** Agriculture - Societies - Bath. **SUBJECT:** Economics - Societies - Bath.

1555 Russische Bibliothek zur Kenntniss des gegenwärtigen Zustandes der Literatur in Russland. - v.1-11 (6 issues each) 1772-1789. St. Petersburg: Hartknoch, 1772-1789. Imprint varies. **EDITOR:** Hartwig Ludwig Christian Bacmeister. **INDEXES:** Index in v.11, 1789. **SOURCES:** K292, OCLC, S3736 **LOCATIONS:** BUC, CCF, ZDB **SUBJECT:** General - Periodicals - Russia - Reviews.

1556 Saggi / Accademia degli unanimi, Turin. - v.1-2, 1793. Turin: 1793. **SOURCES:** S2054 **LOCATIONS:** CCF, ULS, ZDB **SUBJECT:** General - Societies - Turin.

1557 Saggi di dissertazione accademiche lette nella Accademia Etrusca di Cortona. - v.1-10 (9 pts.) 1735-1791. Florence and Rome: 1735-1791. New ed. of v.1 in 1742. **OTHER TITLE:** Dissertazioni, Accademia Etrusca di Cortona. **SOURCES:** S1816 **LOCATIONS:** BUC, CCF, CCN, ULS, ZDB **SUBJECT:** Antiquities - Societies - Cortona.

1558 Saggi di dissertazione dell' Accademia palermitana del buon gusto. - v.1-2 (1755-1791) 1755-1800. Palermo: 1755-1800. v.1-2 (1755-1780) as Atti. **OTHER TITLE:** Atti, Accademia palermitana. **INDEXES:** Cumulative index (1755-1889) in ser.2, v.1. **SOURCES:** G495, S1974 **LOCATIONS:** BUC, CCF, ULS, ZDB **SUBJECT:** General - Societies - Palermo.

1559 Saggi di naturali esperienze fatte nell'Accademia del cimento. v.1, 1667. Florence: Guiseppe Cocchini, 1667. **TRANSLATED AS:** Essayes of natural experiments, made in the Academia del Cimento. **TRANSLATED AS:** Tentamina experimentorum naturalium captorum in Academia del Cimento. There were 13 editions in the original language dating from 1691 to 1957, as well as 1 in English, 2 in Latin and 2 in French (Ref.45:347-54). Translated into French in Collection académique, v.1. **EDITOR:** Lorenzo Magalotti. **CONTINUED BY:** Atti e memorie inedite notizie, Academia del Cimento, Florence. **SOURCES:** GII:5, OCLC, S1821c **SUBJECT:** Science - Societies - Florence.

1560 Saggi ed osservazioni di medicina, della Società di Edinburgo. - v.1-7, 1751-1762. Venice: Storti, 1751-1762. **TRANSLATION OF:** Medical essays and

observations, Medical Society in Edinburgh. **LOCATIONS:** BUC, NUC
SUBJECT: Medicine - Societies - Edinburgh - Translations.

1561 Saggi scientifici e litterari / Accademia di scienze, lettere ed arte di Padova. -
v.1-3 (1779-1794) 1786-1794. Padua: 1786-1794. v.3 is in two separatly
paged parts. **CONTINUED BY:** Memorie della Accademia di scienze, lettere
ed arti di Padova. **INDEXES:** R. Accademia di scienze, lettere ed arte di
Padova. Indice generale per ordine alfabetico di autori e di materie dell'anno
1779 a tutto l'anno accademico 1899-1900. Padua, Randi, 1901.
MICROFORMS: RM(3). **SOURCES:** G494, OCLC, S1964f **LOCATIONS:**
BUC, CCF, CCN, ULS, ZDB **SUBJECT:** General - Societies - Padua.

1562 Saggio delle transazioni filosofiche, compendiate, tradotte dall' inglese. - v.1-5,
1729-1734. Naples: Moscheni, 1729-1734. A translation of the abridgment by
John Lowthorp covering the first 50 v. of the Transactions. **TRANSLATION
OF:** Philosophical transactions. **SOURCES:** S1960 **LOCATIONS:** BUC
SUBJECT: Science - Societies - London - Translations.

1563 Samling af rön och afhandlingar vörande landtbruket / Kungliga
Vetenskapsakademien, Stockholm. 1775-1783. A collection of agricultural
experiments and transactions (Ref.48,p.312). **LOCATIONS:** ZDB
SUBJECT: Science - Societies - Stockholm. **SUBJECT:** Agriculture -
Societies - Stockholm.

1564 Samling af rön och uptaker i senare ider uti physik. - Göteborg: 1781.
SOURCES: S607 **SUBJECT:** Science - Periodicals - Sweden.

1565 Der Sammler, eine gemeinnützige Wochenschrift für Bündten / Gesellschaft
landwirthschaftlicher Freunde in Bündten. - Chur: Bernhard Otto, 1779-1784.
EDITOR: Johann Georg Amstein. **SOURCES:** K2926, S2146
LOCATIONS: BUC, ULS **SUBJECT:** Agriculture - Societies - Chur.

1566 Der Sammler, eine Wochenschrift zur Kenntniss der Naturgeschichte, zur
Verbesserung der Land-und Stadtwirthschaft, der Polizey und des Finanzwesens
etc. - v.1-2, 1773-1774. Königsberg: Zeisens und Hartung, 1773-1774.
Weekly. Subtitle varies. **OTHER TITLE:** Preussische Sammler. **EDITOR:**
Friedrich Samuel Bock. **SOURCES:** B3759, K298, S2871 **SUBJECT:**
Natural history - Periodicals - Germany. **SUBJECT:** General - Periodicals -
Germany.

1567 Sammlung academischer Streitschriften die Geschichte und Heilung der
Krankheiten betreffend. In einem vollständigen Auszug gebracht und mit
Anmerckungen versehn von Lorenz von Crell. - Bd.1-Bd.3, 1779-1780.
Helmstädt: Johann Heinrich Kühnlein, 1779-1780. Imprint varies. **EDITOR:**
Albrecht von Haller. **TRANSLATION OF:** Disputationes ad morborum
historiam et curationem facientes. **SOURCES:** G353a **SUBJECT:** Medicine
- Periodicals - Switzerland - Dissertations.

1568 Sammlung astronomischer Abhandlungen, Beobachtungen und Nachrichten / Königliche Akademie der Wissenschaften, Berlin. - Bd.1-4, 1793-1808. Berlin: 1793-1808. Supplements the academy's Astronomisches Jahrbuch. **INDEXES:** Namen-und Sach-register der Berliner astronomischer Jahrbucher von 1776 bis 1829. Berlin, Dummler, 1829 (Ref.126). **SOURCES:** S2370k **LOCATIONS:** CCF, CCN, ZDB **SUBJECT:** Astronomy - Societies - Berlin.

1569 Sammlung auserlesene Landgesetze welche das Polizei und Cameralwissen zum Gegenstände haben. - No. 1-10, 1783-1793. **EDITOR:** Johann Beckmann. **SOURCES:** (Ref.112). **SUBJECT:** Agriculture - Periodicals - Germany.

1570 Sammlung auserlesener Abhandlungen über die interessantesten Gegenstände der Chemie, aus dem lateinischen mit einigen Anmerkungen gegleitet. - Leipzig; Böhme: Cnobloch, 1793. **EDITOR:** Carl Friedrich August Hochheimer. **SOURCES:** B4182, G732, S3009 **SUBJECT:** Chemistry - Periodicals - Germany - Dissertations.

1571 Sammlung auserlesener Abhandlungen zum Gebrauch practischer Aerzte. - Ser.1, v.1-24, 1774, 1774-1807, ser.2, v.25-41, 1815-1835. Leipzig: Dyck, 1774-1835. v.25-41 also as Neue Sammlung (v.1-17). There was a 2nd edition of v.1-12 in 6 v.in 1781-1785, a 3rd ed. in 1800 in 4 v.and an abbreviated version of v.1-21 by C.M. Koch in 7v.in 1800. **EDITORS:** Christian Erhart Kapp (v.1-24) and Carl Gottlob Kuhn (v.25-41) (1835). **CONTINUED BY:** Neue Sammlung auserlesener Abhandlungen zum Gebrauch practischer Aerzte. **INDEXES:** There were 3 indexes: 1) Allgemeines Register über die erste zwolf Bande (Leipzig, Koch, 1789; 2) to v.13-24 in 1805, and 3) to v.25-36 in v.12 of the new series (1829). **SOURCES:** G406, K3567 **LOCATIONS:** BUC, CCF, CCN, ULS, ZDB **SUBJECT:** Medicine - Periodicals - Germany.

1572 Sammlung auserlesener Schriften von Staats-und Landwirthschaftlichem Inhalte, Oeconomische Gesellschaft, Bern. - v.1-3, 1762-1775. Berne: Neue Buchhandlung, 1762-1775. **SOURCES:** K2875 **LOCATIONS:** ZDB **SUBJECT:** Agriculutre - Societies - Berne.

1573 Sammlung auserlesener Wahrnehmungen aus der Arzney- Wissenschaft, der Wund-Arzney- und der Apotheker-Kunst, aus dem Französischen übersetzt. - v.1-9, 1757-1765. Frankfurt; Leipzig: Johann Gottfried Bauer, 1757-1765. **TRANSLATION OF:** Journal de médecine, chirurgie et de pharmacie. **CONTINUED BY:** Neue Sammlung auserlesener Wahrnehmungen aus allen Theilen der Arzneywissenschaft. **SOURCES:** GIII:12a, K3539 **LOCATIONS:** BUC, CCF, ZDB **SUBJECT:** Medicine - Periodicals - France - Translations.

1574 Sammlung brauchbarer Abhandlungen aus Rozier's Beobachtungen über die Natur und Kunst. - v.1-2, 1775-1776. Leipzig: 1775-1776. An unauthorized translations of Rozier's Journal by Wünsch (Ref.p.103,p.435). **TRANSLATION OF:** Introduction aux observations sur la physique. **SOURCES:** B4183, S3010 **LOCATIONS:** ZDB **SUBJECT:** Science - Periodicals - France - Translations.

1575 Sammlung chemisch-pharmacologischer Aufsätze und kleiner Schriften. - v.1, 1786. Frankfurt; Leipzig: Kühnlin, 1786. **SOURCES:** G213 **SUBJECT:** Pharmacy - Periodicals - Germany. **SUBJECT:** Chemistry - Periodicals - Germany.

1576 Sammlung chemischer Experimente (Berlin). - v.1-6, 1759-1761. Berlin: 1759-1761. **LOCATIONS:** ZDB **SUBJECT:** Chemistry - Periodicals - Germany.

1577 Sammlung chemischer Experimente (Leipzig) - v.1-2, 1793. Leipzig: 1793. **LOCATIONS:** ZDB **SUBJECT:** Chemistry - Periodicals - Germany.

1578 Sammlung chirurgisches Bemerkungen. - v.1-5, 1758-1778. Altenburg: 1758-1778. **OTHER TITLE:** Sammlung chirurgisches Wahrnehmungen, aus verschiedenen Sprachen übersetzt. **LOCATIONS:** BUC **SUBJECT:** Surgery - Periodicals - Germany.

1579 Sammlung der auserlesensten und neuesten Abhandlungen für Wundärzte. St. 1-St.7, 1778-1783. Leipzig: Weygand, 1778-1783. There was an edition in Mannheim of this title and its successors in 1802 in 7 parts (Ref.21:25:628). **EDITOR:** Johann Christoph Sommer. **CONTINUED BY:** Neue Sammlung der auserlesensten und neuesten Abhandlungen für Wundärzte. **CONTINUED BY:** Neueste Sammlung der auserlesensten und neuesten Abhandlungen für Wundärzte. **CONTINUED BY:** Ausgesuchte Beiträge für die Entbindungskunst. **SOURCES:** G407, K3579 **LOCATIONS:** BUC, CCF, ULS, ZDB **SUBJECT:** Surgery - Periodicals - Germany.

1580 Sammlung der besten und neuesten Reisebeschreibungen, in einem ausführlichen Auszüge worinnen eine genau Nachricht von der Religion, Regierungsverfgassung, Handlung, Sitten, Naturgeschichte und anderen merkwürdigen Dingen verschiedener Länden und Völker gegeben wird. - v.1-35, 1763-1802. Berlin: Mylius, 1763-1802. v.25-35 also as Neue Sammlung. **OTHER TITLE:** Neue Sammlung der besten und neuesten Reisebeschreibungen. **EDITOR:** Johann Friedrich Zückert (v.1-7). **TRANSLATION OF:** World displayed, or, a curious collection of voyages. **SOURCES:** B4193, OCLC **LOCATIONS:** ZDB **SUBJECT:** Geography - Periodicals - England - Translations. **SUBJECT:** Travel - Periodicals - England - Translations.

1581 Sammlung der deutschen Abhandlungen, welche in der königliche Akademie der Wissenschaften vorgelesen worden. - Bd.1(1788-89)1793, Bd.2(1790-91)1796), Bd.3(1792-97)1799, Bd.4(1798- 1800)1803, Bd.5(1801-02)1805, Bd.6(1803)1806. Berlin: Georg Decker, 1793-1806. **CONTINUATION OF:** Miscellanea Berolinensis. **CONTINUATION OF:** Histoire avec les memoires Preussische-königliche Akademie der Wissenschaften. **CONTINUATION OF:** Nouveaux mémoires de l'Académie royale des sciences et belle-lettres, Berlin. Publisher varies. **OTHER TITLE:** Abhandlungen der königlichen Akademie der Wissenschaften. **CONTINUED BY:** Abhandlungen der königlichen Akademie der Wissenschaften. **INDEXES:** Kohnke, Otto, ed. Gesamtregister über die in den Schriften der Akademie von 1700-1899 erscheinen wissenschaftlichen Abhandlungen und Festreden. Berlin, 1900. **MICROFORMS:** SMS(1), IDC(1). **SOURCES:** GXIII:487f, K449, OCLC, S23701 **LOCATIONS:** BUC, CCF, CCN, ULS, ZDB **SUBJECT:** General - Societies - Berlin.

1582 Sammlung der Gemeinnützigsten practischen Aufsätze und Beobachtungen aus den Schriften der kgl. medicinischen Gesellschaft zu Paris. - v.1, 1784. Halle: Johann Jacob Gebauer, 1784. **TRANSLATION OF:** Histoire de la Société royale de médecine, Paris. **SOURCES:** G41a, K3611 **SUBJECT:** Medicine - Societies - Paris - Translations.

1583 Sammlung der Natur und des Geistes. - Vienna: Gahelen, 1783. **SOURCES:** K5636 **SUBJECT:** Natural history - Periodicals - Austria - Popular.

1584 Sammlung der neueren Schriften über die Vieharzneikunst, in vollständigen Auszügen und Uebersetzungen. - St.1-2, 1783-1785. Stendal: Franzen und Grosse, 1783-1785. "Ist eine Sammlung, keine Zeitschrift." (Ref.111:195a). **EDITOR:** Johann Conrad Hennemann. **SOURCES:** G224 **SUBJECT:** Veterinary medicine - Periodicals - Germany.

1585 Sammlung der neuesten Beobachtungen englisher Aerzte und Wündärzte. - v.1-4, 1790-1794. Frankfurt: Andrea, 1790-1794. **TRANSLATION OF:** London medical journal. **SOURCES:** G46b **SUBJECT:** Medicine - Periodicals - England - Translations.

1586 Sammlung der Preisschriften der königlichen Parisischen Academie der Chirurgie. - Bd.1, 1756. Altenberg: Richter, 1756. A translation by Johann Ernset Zeyher of selected prize essays published by the Académie royale de chirurgie, Paris. **SOURCES:** G9d **SUBJECT:** Surgery - Societies - Paris - Translations.

1587 Sammlung der vorzüglichsten Schriften aus Heilarznei. - v.1, 1785-1786. Prague: Driesbach, 1785-1786. **EDITOR:** Joseph Knobloch. **SOURCES:** K3624 **SUBJECT:** Medicine - Periodicals - Czechoslovakia.

1588 Sammlung der vorzüglichsten Schriften der Thierarznei. - v.1-2, 1785-1786. Prague: Joseph Emmanuel Driesbach, 1785-1786. Imprint varies. **EDITOR:** Joseph Knobloch. **INDEXES:** Indexed in Wimmel (Ref.7). **SOURCES:** K3625 **LOCATIONS:** BL **SUBJECT:** Veterinary medicine - Periodicals - Czechoslovakia.

1589 Sammlung einiger Abhandlungen aus der Oekonomie, Cameralwissenschaft, Arzneikunde und Scheidekunst. - Leipzig: 1777. **EDITOR:** Christian Friedrich Reuss. **SOURCES:** K327 **SUBJECT:** Economics - Periodicals - Germany. **SUBJECT:** Pharmacy - Periodicals - Germany. **SUBJECT:** Medicine - Periodicals - Germany.

1590 Sammlung einiger ausgesuchten Stücke / Gesellschaft der freien Künste, Leipzig. - v.1-3, 1754-1756. Leipzig: Breitkopf, 1754-1756. **EDITOR:** Johann

Christoph Gottsched. **SOURCES:** G467, K201 **LOCATIONS:** BUC, ULS, ZDB **SUBJECT:** General - Societies - Leipzig.

1591 Sammlung interesanter Aufsätze und Beobachtungen für practische Aerzte und Wundärzte, nebst einigen Briefen über das Brownisch-Weikardische System. - v.1, 1797. Elberfeld: Comptoir für Litteratur, 1797. **EDITOR:** Gerhard Wilhelm von Eicken. **SOURCES:** G255 **SUBJECT:** Medicine - Periodicals - Germany. **SUBJECT:** Brunonianism - Periodicals - Germany.

1592 Sammlung kleiner academischer Schriften über Gegenstände der gerichtlichen Arzneigelahrtheit. - v.1-2 (no. 1-8) 1793-1797. Altenburg: Richter, 1793-1797. Imprint varies. Friedrich August Waitz (Weiz). **CONTINUED BY:** Neue Sammlung kleiner academischer Schriften über Gegenstände der gerichtlichen Arzneigelahrtheit. **CONTINUED BY:** Neueste Sammlung kleiner academischer Schriften über Gegenstände der gerichtlichen Arzneigelahrtheit. **SOURCES:** G410 **LOCATIONS:** CCF, CCN **SUBJECT:** Medical jurisprudence - Periodicals - Germany - Dissertation

1593 Sammlung medicinisch-practischer Beobachtungen und Abhandlungen, zum Nutzen und Unterricht. - v.1, 1798. Ulm: Stettinsche Buchhandlung, 1798. **EDITOR:** Melchoir Adam Weickard. **SOURCES:** G257, OCLC **LOCATIONS:** NUC **SUBJECT:** Medicine - Periodicals - Germany. **SUBJECT:** Brunonianism - Periodicals - Germany.

1594 Sammlung medicinischer-chirurgischer Originalabhandlungen aus dem Hannoverische Magazin. - v.1-3 (1750-1786) 1786-1787. Hannover: Helwig, 1786-1787. **CONTINUATION OF:** Hannoverische Magazin. **EDITOR:** Ludolf von Guckenberger. **SOURCES:** G659b, K3629 **LOCATIONS:** BUC, ZDB **SUBJECT:** Medicine - Periodicals - Germany.

1595 Sammlung medicinischer und chirurgischer Anmerkungen. - v.1-8, 1747-1763. Berlin: Haude und Spener, 1747-1763. **EDITOR:** Johann Friedrich Henkel. **CONTINUED BY:** Neue medicinische und chirurgische Anmerkungen. **SOURCES:** K3527 **SUBJECT:** Medicine - Periodicals - Germany. **SUBJECT:** Surgery - Periodicals - Germany.

1596 Sammlung naturhistorischer und physicalischer Aufsätze. - v.1, 1796. Nürnberg: Rasesch, 1796. **EDITOR:** Franz von Paul Schrank. **SOURCES:** B4186, G734, OCLC, S3171 **LOCATIONS:** NUC **SUBJECT:** Science - Periodicals - Germany.

1597 Sammlung neuer und merkwürdiger Reisen zu Wasser und zu Lande. - No. 1-11, 1750-1764. Göttingen: 1750-1764. **SOURCES:** B4187 **SUBJECT:** Travel - Periodicals - Germany. **SUBJECT:** Geography - Periodicals - Germany.

1598 Sammlung nützlicher Aufsätze und Nachrichten die Baukunst betreffend für angehende Baumeister und Freunde der Architectur, herausgegeben von mehreren Mitglieder des Kgl. Preussischen Ober-Bau Departements. - v.1-6 (6

St. each) 1797-1806. Berlin: Maurer, 1797-1806. Several editions appeared. Imprint and title vary slightly. **EDITOR:** David Gilly (v.6). **SOURCES:** K4101 **LOCATIONS:** BUC, CCF, CCN, ULS, ZDB **SUBJECT:** Building - Periodicals - Germany. **SUBJECT:** Architecture - Periodicals - Germany.

1599 Sammlung nützlicher und bewährter Beiträge zum Besten der Stadt-, Haus-, und Landwirthschaft, Gärtnerei, des Jagd-und Forstwesens und des Technologie. - v.1-4, 1799-1803. Augsburg: Platzer, 1799-1803. Imprint varies. **OTHER TITLE:** Sammlung nützlicher und bewährter Beiträge für die Oeconomie. **SOURCES:** K3027 **LOCATIONS:** BUC, ZDB **SUBJECT:** Technology - Periodicals - Germany. **SUBJECT:** Agriculture - Periodicals - Germany. **SUBJECT:** Economics - Periodicals - Germany.

1600 Sammlung nützlicher Unterrichte / Gesellschaft des Ackerbaus und nützlicher Künste im Herzogthums Krain. - v.1-4, 1770-1779. Laibach: Johann Friedrich Eger, 1770-1779. **SOURCES:** K2903 **LOCATIONS:** ZDB **SUBJECT:** Agriculture - Societies - Krain. **SUBJECT:** Technology - Societies - Krain.

1601 Sammlung öconomisches Schriften / Gesellschaft des Ackerbaues und der nützlichen Künste in dem Herzogthums Steyermark. - v.1, 1764. Graz: 1764? **SOURCES:** (Ref.6,p.289). **SUBJECT:** Agriculture - Societies - Steyermark. **SUBJECT:** Technology - Societies - Steyermark.

1602 Sammlung physicalisch-ökonomischer Aufsätze zur Aufnähme der Naturkunde und der damit verwandten Wissenschaften in Böhmen. - v.1, 1795. Prague: Calve, 1795. **EDITOR:** Franz Wilibald Schmidt. **SOURCES:** B4190, G736, K3363, S3509 **SUBJECT:** Science - Periodicals - Czechoslovakia. **SUBJECT:** Economics - Periodicals - Czechoslovakia.

1603 Sammlung physicalischer Aufsätze, besonders die Böhmische Naturgeschichte betreffend. - v.1-5, 1791-1798. Dresden: Walther, 1791-1798. **EDITORS:** Johann Mayer and Franz Alexander Reuss. **SOURCES:** G735, OCLC **LOCATIONS:** BUC, CFF, CCN, ZDB **SUBJECT:** Natural history - Periodicals - Germany.

1604 Sammlung vermischter Abhandlungen jezt lebender Scheidekünstler. - v.1, 1782. Hamburg: Hoffmann, 1782. **SOURCES:** G738 **SUBJECT:** Pharmacy - Periodicals - Germany. **SUBJECT:** Chemistry - Periodicals - Germany.

1605 Sammlung verschiedener medicinischer Responsorum und Sektionsberichte. - v.1-2, 1772. Halle: 1772. There was a 2nd edition, Halle, 1774. **EDITOR:** Philipp Conrad Fabricius. **SOURCES:** (Ref.96:283). **SUBJECT:** Medical jurisprudence - Periodicals - Germany.

1606 Sammlung von Abhandlungen für Thierärzte und Oeconomen, aus dem Dänischen. - v.1-6, 1795-1807. Copenhagen: 1795-1807. Publisher varies. **EDITOR:** Erik Nissen Viborg. **INDEXES:** Indexed in Wimmel (Ref.7). **SOURCES:** G231, K3697 **SUBJECT:** Veterinary medicine - Periodicals - Denmark - Translations.

1607 Sammlung von anatomischen Aufsätze und Bemerkungen zur Aufklärung der Fischkunde. - v.1, 1795. Leipzig: Schäfer, 1795. EDITOR: Johann Gottlob Schneider. SOURCES: B4195, G573, S3008   SUBJECT: Zoology - Periodicals - Germany.

1608 Sammlung von Beobachtungen aus der Arneigelahrtheit und Naturkunde. - v.1-5, 1769-1776. Nördlingen: 1769-1776. EDITOR: Johann August Philipp Gesner. SOURCES: B4196, G408, OCLC, S3144 LOCATIONS: BUC, CCN, ULS SUBJECT: Medicine - Periodicals - Germany. SUBJECT: Natural history - Periodicals - Germany.

1609 Sammlung von Natur- und Medicin- wie auch hierzu gehörigen Kunst- und Literatur- Geschichten. - v.1-9, 1717-1726. Breslau; Leipzig; Budissin: M. Hubert, 1717-1726. Imprint varies. Associated with Deutsche Akademie der Naturforscher. Reprinted in 1736 as Schatz-Kammer der Natur und Kunst. OTHER TITLE: Breslauer Sammlungen; Sammlung der Natur und Medizin; Annales physico-medica, oder Geschichte der Natur und Kunst; Acta Wratislaviensis; Schatz-Kammer der Natur und Kunst. EDITORS: Johann Kanold, Johann Georg Brunschwitz and Johann Christian Kundmann. CONTINUED BY: Miscellanea physico-medica-mathematica. 4 supplements were issued under the title: Supplementum curieuser und nützlicher Anmerkung von Natur und Kunstgeschicghte. INDEXES: Universal Register, 1736. MICROFORMS: RM(3). SOURCES: B4197, GII:7l, K3182, OCLC, S3011 LOCATIONS: BUC, CCF, CCN, ULS, ZDB   SUBJECT: Medicine - Periodicals - Germany. SUBJECT: Natural history - Periodicals - Germany.

1610 Sammlungen aus der Naturgeschichte, Oecomomie-, Polizei-, Cameral- und Finanzwissenschaft. - v.1, 1774. Dresden: Hilsch, 1774. EDITOR: Emmanuel Carl Heinrich Börner. SOURCES: B4198, S2526 LOCATIONS: BL SUBJECT: Natural history - Periodicals - Germany. SUBJECT: Economics - Periodicals - Germany.

1611 Sammlungen der medicinischen Societät in Budissin. - Altenburg: Neisse und Hennings, 1757. SOURCES: (Ref.113,p.492). SUBJECT: Medicine - Societies - Budissin.

1612 Sammlungen nützlicher und angenehmer Gegenstände aus allen Theilen der Natur-Geschichte, Arzneiwissenschaft und Haushaltungskunst. - v.1, 1773. Vienna: E.F. Bader, 1773. EDITOR: Franz Xaver August Wasserberg. SOURCES: B4138, K3251, S3012 SUBJECT: Natural history - Periodicals - Austria. SUBJECT: Medicine - Periodicals - Austria.

1613 Sammlungen von landwirthschaftlichen Dingen der Schweitzerischen Gesellschaft in Bern. - Bd.1-2, 1760-1761. Zürich: Heydegger, 1760-1761. Published simultaneously in French and German. OTHER TITLE: Recueil de mémoires concernant l'économie rurale; Schweitzerischen Gesellschaft in Bern Sammlungen von landwirthschaftlichen Dingen. EDITOR: Elie Bertrand (Ref.29,p.152). CONTINUED BY: Abhandlungen und Beobachtungen durch die ökonomische Gesellschaft zu Bern gesammelt. CONTINUED BY: Neue

Sammlung physisch-ökonomischer Schriften herausgegeben von der ökonomischen Gesellschaft des Kantons Bern. **CONTINUED BY:** Neueste Sammlung von Abhandlungen und Beobachtungen von der ökonomischen Gesellschaft in Bern. **SOURCES:** K2873 **LOCATIONS:** BUC, ULS, ZDB **SUBJECT:** Agriculture - Societies - Bern. **SUBJECT:** Economics - Societies - Bern.

1614 Sammlungen zur Physik und Naturgeschichte von einigen Liebhabern dieser Wissenschaften. - v.1-4 (6 issues each) 1778-1792. Leipzig: Dyck, 1778-1792. **OTHER TITLE:** Abhandlungen zur Physik und Naturgeschichte. **EDITORS:** Johann Samuel Traugott Gehler and Johann Carl Gehler. **MICROFORMS:** RM(3). **SOURCES:** B4199, G737, K3268, OCLC, S3013 **LOCATIONS:** BUC, CCF, ULS, ZDB **SUBJECT:** Science - Periodicals - Germany.

1615 Sanktpeterburgskie vrachnie vaidomosti. - No. 1-52, Nov.2, 1792-July 4, 1794. St. Petersburg: 1792-1794. **EDITOR:** Conrad Friedrich Uden (Ref.32). **SOURCES:** G107 **SUBJECT:** Medicine - Periodicals - Russia.

1616 Scelta d'opuscoli interessanti. - v.1-3, 1775-1777. Milan: Guiseppe Marelli, 1775-1777. There was a 2nd ed. of v.1-2 in 1775, and a new ed. of the 3rd v.1781-1784. **EDITORS:** Carlo Amoretti, Francesco Soave and others. **CONTINUED BY:** Opuscoli sulle scienze e sulle arti. **CONTINUED BY:** Nuova scelta d'opuscoli interessanti. **SOURCES:** B4211, S1912 **LOCATIONS:** BUC, CCF, CCN, ULS **SUBJECT:** Science - Periodicals - Italy.

1617 Schau-platz vieler ungereimter Meinungen und Erzahlungen, worauf die unter dem Titul der Magiae Naturalis so hoch gepriesene Wissenschaften und Künste vorgestellet, geprüfet und entdecket werden. - v.1-3 (St. 1-24) 1735-1742. Berlin; Leipzig: Haude, 1735-1742. **EDITOR:** Tharsander (i.e. Georg Wilhelm Wegner). **SOURCES:** (Ref.111:387). **LOCATIONS:** BUC **SUBJECT:** Alchemy - Periodicals - Germany.

1618 Schauplatz der Künste und Handwerke, oder, vollständige Beschreibung derselben/ Academie der Wissenschaften zu Paris. - Bd.1-20, 1762-1795. Berlin: Johann Heinrich Rüdiger, 1762-1795.Bd.1-4, translated by Johann Heinrich Gottlob von Justi, 5-13 by D.G. Schreiber, 14-15 by Johann Conrad Harrepeter, 16-18 by Johann Samuel Halle, 19 by C.G.D. Muller, and 20 by B.C. Rosenthal. Imprint varies. **TRANSLATION OF:** Description des arts et des metiers, Académie des sciences, Paris. **SOURCES:** (Ref.143,p.6). **SUBJECT:** Technology - Societies - Paris - Translations.

1619 Schauplatz der Natur und Künste, in vier Sprachen, Deutsch, Lateinisch, Französisch, und Italienisch. - v.1-10, 1774-1783. Vienna: 1774-1783. **TRANSLATED AS:** Zryelishche prirodu i khudozhestv. **LOCATIONS:** BUC, CCN, ULS, ZDB **SUBJECT:** Natural history - Periodicals - Austria.

1620 Scheikundige bibliotheek, door een Gezelschap van beminnaaren dezer weetenschap. - v.1-21, 1790-1798. Delft: 1790-1798. Subtitle varies.

CONTINUED BY: Nieuwe scheikundige bibliotheek. **SOURCES:** B4213, S801, G739 **SUBJECT:** Chemistry - Periodicals - Netherlands - Reviews.

1621   Schertz- und ernsthaffte, vernüfftige und einfältige Gedancken über allerhand lustige und nützliche Bücher und Fragen. - v.1-3 (no. 1-28), Jan. 1688-Apr. 1690. Frankfurt; Leipzig: Georg Weidmann, 1688-1690. Imprint varies. **OTHER TITLE:** Freimüthige lustige und ernsthaffte jedoch vernunfft- und gesetzmässige Gedanken; Monats-gespräche über allerhand, führnehmlich aber neue Bücher. **EDITOR:** Christian Thomasius. **SOURCES:** K4 **LOCATIONS:** ULS, ZDB **SUBJECT:** General - Periodicals - Germany - Reviews.

1622   Schlesische ökonomische Sammlungen. - v.1-4 (St. 1-24) 1754-1763. Breslau: 1754-1763. **SOURCES:** K2866 **LOCATIONS:** CCF **SUBJECT:** Economics - Periodicals - Poland.

1623   Schlesische provinzialblätter. - v.1-130, 1785-1849. Breslau: E. Trewendt, 1785-1867. **OTHER TITLE:** Streit's Schlesische Provincialblätter. **EDITOR:** Christian Garve (Ref.6, p.296). **CONTINUED BY:** Rübezahl. Literarische Beilage published 1795-1834. **SOURCES:** OCLC, S2457 **LOCATIONS:** BUC, CCF, ULS, ZDB **SUBJECT:** General - Periodicals - Poland.

1624   Schleswig-Glücksburgische Beiträge zur Aufnahme öconomischer Wissenschaften. - St. 1, 1758. Flensburg: Korte, 1758. **EDITOR:** Philipp Ernst Lüders. **SOURCES:** K2871 **SUBJECT:** Economics - Periodicals - Germany.

1625   Schriften der Berlinischen Gesellschaft Naturforschender Freunde. - Bd.1-Bd.11, 1780-1794. Berlin: F. Maurer, 17801794. **CONTINUATION OF:** Beschäftigungen der Berlinischen Gesellschaft naturforschender Freunde. Imprint varies. There is a modern index (Ref.53, v.3). **OTHER TITLE:** Beobachtungen und Entdeckungen aus der Naturkunde (1786- 1794). **CONTINUED BY:** Neue Schriften der Gesellschaft naturforschender Freunde zu Berlin. **INDEXES:** Index for vols 1-11, 1780-1794 in v.11. **SOURCES:** B909a, G469a, K3272, OCLC, S2292b **LOCATIONS:** BUC, CCF, CCN, ULS, ZDB **SUBJECT:** Natural history - Societies - Berlin.

1626   Schriften der Curfürstlichen deutschen Gesellschaft in Mannheim. - v.1-11, 1787-1794. Mannheim: Schwann und Löffler, 1787-1794. Imprint varies. **SOURCES:** K412 **LOCATIONS:** CCF, ULS **SUBJECT:** General - Societies - Mannheim.

1627   Schriften der Drontheimer Gesellschaft. - v.1-4, 1765-1770. Copenhagen: 1765-1770. **TRANSLATED AS:** Skrifter det Trondhjemske Selskab. **SOURCES:** G476a, K257, S642 **LOCATIONS:** BUC, CCN, ZDB **SUBJECT:** Science - Societies - Trondhjem - Translations.

1628   Schriften der Naturforschenden Gesellschaft zu Copenhagen. - v.1-2, 1793. Copenhagen: C.G. Prost, 1793. **TRANSLATION OF:** Skrivter, Naturhistorie-selskab, Copenhagen. **SOURCES:** S629a **LOCATIONS:** BUC,

CCN, ULS   SUBJECT: Natural history - Societies - Copenhagen - Translations.

1629 Schriften der ökonomische Societät zu Leipizg. - T.1-8, 1771-1790. Dresden: Walther, 1771-1790. CONTINUATION OF: Anzeigen von der Leipziger ökonomischen Societät. CONTINUED BY: Neue Schriften der ökonomische Societät zu Leipizg. SOURCES: K2906 LOCATIONS: BUC, CCF, ULS, ZDB SUBJECT: Economics - Societies - Leipzig.

1630 Schriften der physikalischen Classe / Kongelige Danske Videnskabers Selskab. - Bd.1-3, 1800-1805. Copenhagen: Schubothe, 1800-1805. Translated by Carl Gottlob Rafn. TRANSLATION OF: Selskabs skrifter, Koneglige Danske Videnskabers Selskab. SOURCES: S613c LOCATIONS: BUC, CCF, ZDB SUBJECT: Science - Societies - Copenhagen - Translations.

1631 Schriften der Regensburger botanischen Gesellschaft. - v.1, 1792. Regensburg: 1792. OTHER TITLE: Geschichte der Regensburger botanischen Gesellschaft. SOURCES: G567, K3348 LOCATIONS: BUC, CCF, ULS SUBJECT: Botany - Societies - Regensburg.

1632 Schwäbische Nachrichten von Oeconomie-, Cameral-, Polizei-, Handlungs-, Manufactur-, Mechanischen-, und Bergwerkssachen. - St.1-10, 1756-1757. Stuttgart: 1756-1757. EDITOR: Johann Jacob Moser. INDEXES: Index in St.10. SOURCES: K2715 LOCATIONS: CCF, ZDB SUBJECT: Economics - Periodicals - Germany.

1633 Schwäbisches Magazin von gelehrten Sachen. - v.1-6 (12 issues each) 1775-1780. Stuttgart: Erhard; Mannheim: Löffler, 1774-1780. CONTINUATION OF: Gelehrte Ergötzlichkeiten und Nachrichten. EDITOR: Balthasar Haug. CONTINUED BY: Zustand der Wissenschaften und Künste in Schwaben. SOURCES: B4230, G740, K312, S3262 LOCATIONS: CCF, CCN, ULS, ZDB SUBJECT: General - Periodicals - Germany - Reviews.

1634 Schwedische Annalen der Medicin und Naturgeschichte. - Heft 1-2, 1799-1800. Berlin; Stralsund: Gottlieb August, 1799-1800. Imprint varies. EDITOR: Carl Asmund Rudolphi. SOURCES: B4231, G159, K3733, S2384 LOCATIONS: BUC, CCF, ULS, ZDB SUBJECT: Medicine - Periodicals - Germany. SUBJECT: Natural history - Periodicals - Germany.

1635 Schwedisches Magazin, oder, gesammelte Schriften der gröszen Gelehrten in Schweden für die Liebhaber der Arzneiwissenschaft, der Naturgeschichte, Chemie und Oeconomie. - v.1-2, 1768-1770. Copenhagen: 1768-1770. EDITOR: Johann Carl Weber. CONTINUED BY: Neues Schwedisches Magazin kleiner Abhandlungen. SOURCES: B4232, K3551, S650 LOCATIONS: CCF, ZDB SUBJECT: Science - Periodicals - Denmark.

1636 Schwedisches Magazin, oder, Schriften aus der Naturforschung, Stadt- und Landwirthschaft. - v.1-2, 1788-1790. Copenhagen: Schubothe, 1788-1790.

SOURCES: B4233, G741, K2961, S651   SUBJECT: Natural history - Periodicals - Denmark.   SUBJECT: Agriculture - Periodicals - Denmark.

1637   Schwedisches ökonomisches Wochenblatt. - St. 1-4, 1765-1766. Griefswald: Knobloch, 1765-1766. Weekly. EDITOR: Johann Carl Dahnert. SOURCES: K2883 SUBJECT: Economics - Periodicals - Germany.

1638   Scientific receptable, containing problems, solutions, enigmas, rebuses, charades, anagrams & etc. - v.1-3 (no. 1-26) 1796-1819. London: T.N. Longman, 1791-1819. Imprint varies. EDITOR: Thomas Whiting. SOURCES: B4296 LOCATIONS: BUC, ULS SUBJECT: Mathematics - Periodicals - England - Popular.

1639   Scots farmer, or, select essays on agriculture, adapted to the soil and climate of Scotland. - v.1-2, 1772-1774. Edinburgh: W. Auld, 1772-1774. Reissued as Northern farmer in 2 v.London, 1778-1780. OTHER TITLE: Northern farmer. MICROFORMS: RP(1) (Goldsmiths' Kress Library). SOURCES: OCLC LOCATIONS: BUC SUBJECT: Agriculture - Periodicals - Scotland.

1640   Scriptores nevrologici minores selecti. - v.1-4, 1791-1795. Leipzig: Junius und Feind, 1791-1795. EDITOR: Christian Friedrich Ludwig. SOURCES: G412 LOCATIONS: DSG   SUBJECT: Neurology - Periodicals - Germany. SUBJECT: Medicine - Periodicals - Germany.

Scriptorum a Societas Havniensis bonis artibus promovendis, see no. 1668.

1641   Séance publique tenue par la Faculté de médecine en l'Université de Paris. - Paris: Quillau, 1778-1779. LOCATIONS: CCF SUBJECT: Medicine - Societies - Paris.

1642   Séances / Académie royale des sciences, arts et belles-lettres de Caen. - Caen: 1762. SOURCES: S1095f SUBJECT: Science - Societies - Caen.

1643   Séances des écoles normales, recueillies par des sténographes et revues par les professeurs. - No.1-10, 1800-1801. Paris: 1800-1801. There were several editions of some of the volumes. SOURCES: S1549 LOCATIONS: CCF SUBJECT: General - Societies - Paris.

1644   Séances publique / Académie des sciences, arts et belles lettres de Dijon. - Dijon: Frantin, 1773-1805. An unnumbered series published irregularly which was continued into the 19th century. 1804-1805 as Analyse des travaux. OTHER TITLE: Notice de la séances publique, Société libre de sciences, arts et agriculture, Dijon. INDEXES: Index for 1769-1869 in the society's Mémoires, ser.2,v.16. SOURCES: (Ref.114). SUBJECT: General - Societies - Dijon.

1645   Séances publique de l'Académie royale de chirurgie, Discours prononcé / Paris. - Paris: 1752-1793. OTHER TITLE: Discours prononcés à la séances publique de l'Académie royale de chirurgie, Paris; Discours prononcés aux écoles de chirurgie. LOCATIONS: CCF SUBJECT: Surgery - Societies - Paris.

1646 Séances publique de l'Académie royale de chirurgie / Paris. - Paris: 1773-1785. LOCATIONS: BL, CCF SUBJECT: Surgery - Societies - Paris.

1647 Séances publiques / Académie des sciences, belles-lettres et arts de Besançon. - Besançon: 1754-1858. OTHER TITLE: Procès-verbaux des séances publiques, Académie des sciences, belles-lettres et arts de Besançon. CONTINUED BY: Mémoires de l'Académie des sciences, belles-lettres et arts de Besançon. SOURCES: G450 LOCATIONS: BUC, CCF, ULS, ZDB SUBJECT: Science - Societies - Besançon.

1648 Select essays on commerce, agriculture, mines, fisheries and other useful subjects. - v.1, 1754. London: D. Wilson and T.Durham, 1754. TRANSLATION OF: Journal économique. LOCATIONS: BL SUBJECT: Science - Periodicals - France - Translations. SUBJECT: Technology - Periodicals - France - Translations. SUBJECT: Agriculture - Periodicals - France - Translations. SUBJECT: Medicine - Periodicals - France - Translations.

1649 Select papers on the different branches of medicine, by a society instituted for the improvement of physical knowledge, to be continued occasionally. - v.1, 1767. London: W. Griffin, 1767. SOURCES: OCLC, (Ref.17:11). LOCATIONS: DSG SUBJECT: Medicine - Societies - London.

1650 Select transaction of the honourable, the Society of improvers in the knowledge of agriculture in Scotland. - v.1, 1743. Edinburgh: Sands and others, 1743. Imprint and subtitle vary. EDITOR: Robert Maxwell. MICROFORMS: RP(1) (Goldsmiths' Kress Library). SOURCES: OCLC LOCATIONS: BUC SUBJECT: Agriculture - Societies - Scotland.

1651 Selecta medica francofurtensia, anatomen, imprimis practicam, chirurgiam, materiam medicam ipsamque universam medicinam tam clinicam quam forensem. - v.1-5 (each 6 issues) 1736-1747. Frankfurt: Johann Gottfried Conrad, 1736-1747. TRANSLATED AS: Anatomische merkwürdige Nachrichten. SOURCES: G417, K3524 LOCATIONS: BUC, ULS SUBJECT: Medicine - Periodicals - Germany. SUBJECT: Medical jurisprudence - Periodicals - Germany.

1652 Selecta physcio-oeconomica, oder, Sammlung von allerhand zur Naturforschung und Haushaltungskunst gehörigen Begebenheiten. - v.1-3 (v.1-17) 1749-1756. Stuttgart: Johann Christian Erhard, 1749-1756. EDITORS: Johann Albrecht Gesner and Johann Christian Erhard. SOURCES: B4317, K3206, S3263 LOCATIONS: CCF, ULS, ZDB SUBJECT: Natural history - Periodicals - Germany.

1653 Selskabs skrifter / Koneglige Danske Videnskabers Selskab. - v.1 (1800)1801-v.6 (1809-1812)1818. Copenhagen: 1801-1818. TRANSLATED AS: Schriften der physikalischen Classe, Koneglige Danske Videnskagbers Selskab. CONTINUATION OF: Skrifter, som udi det kjöbenhavnske selskab af laerdoms og videnskabs elstere ere fremlagte og oplaeste. CONTINUATION OF: Nye samling af det Kongelige Danske videnskabers skrifter. INDEXES:

Fortegnelse over de Koneglige danske Videnkabernes selskabs publicationer, 1742-1930. Ed. Asger Lomholt. Copenhagen, 1930. **MICROFORMS:** RM(3). **SOURCES:** G475e, OCLC, S615d **LOCATIONS:** BUC, CCF, ULS, ZDB **SUBJECT:** Science - Societies - Copenhagen.

1654 Seltenheiten der Natur und Oeconomie. - v.1-3, 1753. Leipzig: Friedrich Lankisch, 1753. Imprint varies. **EDITORS:** Johann Daniel Titius and Michael Christoph Hanow. **SOURCES:** K3217 **LOCATIONS:** BL **SUBJECT:** Natural history - Periodicals - Germany. **SUBJECT:** Economics - Periodicals - Germany.

1655 Seltsamer Naturgeschichten des Schweizerlandes Wochentliche Erzehlung. - Feb. 11, 1705-Dec. 28, 1707. Zürich: Schaufelbergaer und Hardmeier, 1705-1707. Weekly. **EDITOR:** Johann Jacob Scheuchzer. **SOURCES:** (Ref.141). **SUBJECT:** Natural history - Periodicals - Switzerland.

1656 Semanario de agricultura y artes dirigido a los parrocos. - v.1-23, 1797-1808. Seville; Madrid: 1797-1808. **LOCATIONS:** CCF, CCN **SUBJECT:** Agriculture - Periodicals - Spain.

1657 Semanario económico. - v.1-3 (no. 1-41), Apr. 11, 1765-Nov.12, 1767, ser. 2, v.1-4,Oct. 16, 1777-June 25, 1778. Madrid: 1765-1778. **LOCATIONS:** BUC, ULS **SUBJECT:** Economics - Periodicals - Spain.

1658 Semestre medico-clinico. - Barcelona: 1760-1765. **EDITOR:** Vicente Mitjavial y Finosell. **SOURCES:** (Ref.62,p.209). **SUBJECT:** Medicine - Periodicals - Spain.

1659 Sermones in solennibus conventibus / Academia scientiarum imperialis petropolitana, St. Petersburg. - v.1 -4, (1725-1751) 1726-1751. St Petersburg: 1726-1751. Title varies. **INDEXES:** Tableau générale méthodique et alphabétique de matières contenues dans les publications de l'académie depuis sa fondation, 1872. **SOURCES:** S3706f **SUBJECT:** Science - Societies - St. Petersburg.

1660 Siebenbürgische Quartalschrift. - v.1-7 (4 issues each) 1790-1801. Hermannstadt: Hochmeister, 1790-1801. **EDITORS:** Johann Filtsch, Joseph K. Elder and Johann Binder. **SOURCES:** G743, K1243 **LOCATIONS:** BUC, CCF, ULS, ZDB **SUBJECT:** Geography - Periodicals - Germany.

1661 Skrifter / Danske Landhusholdingselskab, Copenhagen. - v.1-7, 1776-1804. Copenhagen: 1776-1804. **LOCATIONS:** BUC, CCF, ULS **SUBJECT:** Agriculture - Societies - Copenhagen.

1662 Skrifter det Kongelige Norske videnskabers Selskab. - v.4-5, 1768-1774. Copenhagen: 1768-1774. **TRANSLATED AS:** Schriften der Drontheimer Gesellschaft. **CONTINUATION OF:** Skrifter det Trondhjemske Selskab. **CONTINUED BY:** Nye Samling det Kongelige Norske Videnkabers Selskab. **CONTINUED BY:** Nyeste Samling det Kongelige Norske Videnskabers

Selskab. **INDEXES:** Schmidt-Nielsen, Brynjulf. Fortegnelse over selskabets Skrifter det Kongelige Norske Videnkabers selskab. Trondhjem, 1912. **SOURCES:** G476b, OCLC, S663a **LOCATIONS:** BUC, CCF, CCN, ULS, ZDB **SUBJECT:** Science - Societies - Trondhjem.

1663 Skrifter det Trondhjemske Selskab. - v.1-3, 1761-1765. Copenhagen: 1761-1765. **TRANSLATED AS:** Schriften der Drontheimer Gesellschaft. **CONTINUED BY:** Skrifter det Kongelige Norske videnskabers Selskab. **CONTINUED BY:** Nye Samling det Kongelige Norske Videnkabers Selskab. **CONTINUED BY:** Nyeste Samling det Kongelige Norske Videnskabers Selskab. **INDEXES:** Schmidt-Nielsen, Brynjulf. Fortegnelse over selskabets Skrifter det Kongelige Norske Videnkabers selskab. Trondhjem, 1912. **SOURCES:** G476, OCLC, S663 **LOCATIONS:** BUC, CCF, ULS **SUBJECT:** Science - Societies - Trondhjem.

1664 Skrifter, som udi det Kjöbenhavnske selskab af laerdoms og videnskabs elstere ere fremlagte og oplaeste. - v.1-12 (1743-1779) 1745-1779. Copenhagen: Gottman Friderich Kisel, 1745-1779. **TRANSLATED AS:** Abhandlungen die von der königliche Dänischen Gesellschaft den Preis erhalten haben. **TRANSLATED AS:** Sriptorum a Societats Havniensis bonis artibus promovendis dedita danice editorum. Publisher varies. **CONTINUED BY:** Nye samling af det Kongelige Danske videnskabers skrifter. **CONTINUED BY:** Kongelige Danske videnskabers selskabs skrifter. **INDEXES:** Fortegnelse over det Kongelige danske Videnkabernes selskabs publicationer, 1742-1930. Ed. Asger Lomholt. Copenhagen, 1930. **MICROFORMS:** RM(3). **SOURCES:** G475, OCLC, S622 **LOCATIONS:** BUC, CCF, CCN, ULS **SUBJECT:** Science - Societies - Copenhagen.

1665 Skrivter af det Naturhistorie-selskab, Copenhagen. - v.1-6 (11 St.) 1790-1810. Copenhagen: N. Moller, 1790-1810. **TRANSLATED AS:** Schriften der Naturforschenden Gesellschaft zu Copenhagen. **OTHER TITLE:** Mémoires de la Société d'histoire naturelle de Copenhagen. **INDEXES:** Index in v.6, p. 135-175. **MICROFORMS:** IDC(8). **SOURCES:** G546, OCLC, S630 **LOCATIONS:** BUC, CCN, ULS **SUBJECT:** Natural history - Societies - Copenhagen.

1666 Soderzhanïe uchenykh razsuzhdenïi / Akademïi nauk, St. Petersburg. - v .1-4, 1748-1754. St. Petersburg: 1748-1754. **CONTINUATION OF:** Kratkoe opisaïe kommentariev Academia scientiarum imperiales Petropolitana. **CONTINUED BY:** Ezhemiesiachnyia sochinenïia uchenykh dielaks, Academia scientiarum imperiales Petropolitana. **LOCATIONS:** ULS **SUBJECT:** General - Societies - St. Petersburg.

1667 Le Spectacle de la nature, ou, entretiens sur les particularités de l'histoire naturelle. - v.1-9, 1732-1749. Paris: Estienne, 1732-1749. **TRANSLATED AS:** Spettacolo della natura. **TRANSLATED AS:** Schauplatz der Natur und Kunste in vier Sprachen. **TRANSLATED AS:** Neuer Schauplatz der Natur. **TRANSLATED AS:** Spectacle de la nature, or Nature displayed. An encyclopedic work on science which went through several editions and translations into English, German and Italian. **EDITOR:** Noel Antoine Pluche. **SOURCES:** G550, OCLC **LOCATIONS:** BL **SUBJECT:** Science - Periodicals - France - Popular.

1668 Scriptorum a Societats Havniensis bonis artibus promovendis dedita danice editorum. - Pars 1-3, 1745-1747. Copenhagen: 1745-1747. **TRANSLATION OF:** Skrifter, som udi det Kjöbenhavnske selskab af laerdoms og videnskabs elstere ere fremlagte og oplaeste. **SOURCES:** G475a, S652 **LOCATIONS:** BUC **SUBJECT:** Science - Societies - Copenhagen - Translations.

1669 Statistische-geographische Monatsschrift der Cosmographischen Gesellschaft in Wien. - Vienna: 1797. Monthly. **EDITOR:** Joseph Max von Liechtenstern. **SOURCES:** K1305, S3595a **SUBJECT:** Astronomy - Periodicals - Austria.

1670 Stockholmisches Magazin darinnen kleine schwedische Schriften, welche die Geschichte, Staatsklugheit und Naturforschung betreffen. - v.1-3, 1754-1756. Stockholm: Gottfried Kiesewetter, 1754-1756. Text in German and Swedish. **EDITOR:** Carl Ernst Klein. **MICROFORMS:** RM(3). **SOURCES:** B4383, K198, OCLC, S688 **LOCATIONS:** ULS **SUBJECT:** General - Periodicals - Sweden. **SUBJECT:** Natural history - Periodicals - Germany - Popular

1671 Stralsundisches Magazin, oder, Sammlungen auserlesener Neuigkeiten, zur Aufnahme der Naturlehre, Arzneiwissenschaft und Haushaltungskunst. - v.1-2 (six issues each) 1767-1776. Berlin; Stralsund: Gottlieb August Lange, 1767-1776. Subtitle varies. **SOURCES:** B4384, K3243, OCLC, S2389 **LOCATIONS:** BUC, CCF, ULS, ZDB **SUBJECT:** Natural history - Periodicals - Germany. **SUBJECT:** Medicine - Periodicals - Germany.

1672 Stuttgarder allgemeine Magazin. - Stuttgart: Johann Christian Erhard, 1767-1768. **CONTINUATION OF:** Etwas für Alle. **EDITOR:** Johann Christian Erhard. **SOURCES:** K5318 **SUBJECT:** Economics - Periodicals - Germany.

1673 Suite des mémoires de l'Académie royale des sciences. - Paris: Imprimerie royale, 1718-1768. v.1-19, 1702-1821. This series of separate monographs on various subjects appeared at irregular intervals, some of them in multiple editions. **MICROFORMS:** RM(3). **SOURCES:** OCLC, S1276k **LOCATIONS:** CCF, CCN, ULS **SUBJECT:** Science - Societies - Paris.

1674 Sundhed og underholdning. - Nos.1-184, 1787. Copenhagen: Horrebow, 1787. **CONTINUATION OF:** Sundstidende. **CONTINUATION OF:** Nye sundstidende. **EDITOR:** Johann Clemens Tode. **CONTINUED BY:** Hygaea og muserne. **CONTINUED BY:** Museum for sunheds og kundskabs elskere. **CONTINUED BY:** Hertha. **CONTINUED BY:** Sundhedsbog. **CONTINUED BY:** Medicinalblade. **CONTINUED BY:** Sundhedsjournal. **CONTINUED BY:** Sundhedsraad. **CONTINUED BY:** Sundhedsjournal et maanedskrift. **CONTINUED BY:** Nyeste sundhedsblade. **SOURCES:** G184 **LOCATIONS:** ULS **SUBJECT:** Medicine - Periodicals - Denmark.

1675 Sundhedsblade. - v.1-2, 1785-1786. Copenhagen: 1785-1786. **CONTINUATION OF:** Sundstidende. **CONTINUATION OF:** Nye sundstidende. **EDITOR:** Johann Clemens Tode. **CONTINUED BY:** Sundhed og underholdning. **CONTINUED BY:** Hygaea og muserne. **CONTINUED BY:** Museum for sunheds og kundskabs elskere. **CONTINUED BY:** Hertha. **CONTINUED BY:** Sundhedsbog. **CONTINUED BY:** Medicinalblade. **CONTINUED BY:** Sundhedsjournal. **CONTINUED BY:** Sundhedsraad. **CONTINUED BY:** Sundhedsjournal et maanedskrift. **CONTINUED BY:** Nyeste sundhedsblade. **SOURCES:** G182 **LOCATIONS:** ULS **SUBJECT:** Medicine - Periodicals - Denmark.

1676 Sundhedsbog. - v.1, 1789. Copenhagen: 1789. **CONTINUATION OF:** Sundstidende. **CONTINUATION OF:** Nye sundstidende. **CONTINUATION OF:** Sundhed og underholdning. **CONTINUATION OF:** Hygaea og muserne. **CONTINUATION OF:** Museum for sunheds og kundskabs elskere. **CONTINUATION OF:** Hertha. **EDITOR:** Johann Clemens Tode. **CONTINUED BY:** Medicinalblade. **CONTINUED BY:** Sundhedsjournal. **CONTINUED BY:** Sundhedsraad. **CONTINUED BY:** Sundhedsjournal et maanedskrift. **CONTINUED BY:** Nyeste sundhedsblade. **SOURCES:** G188 **SUBJECT:** Medicine - Periodicals - Denmark.

1677 Sundhedsjournal. - v.1-2, 1703-1795. Copenhagen: P. Poulsen, 1793-1795. **CONTINUATION OF:** Sundstidende. **CONTINUATION OF:** Nye sundstidende. **CONTINUATION OF:** Sundhed og underholdning. **CONTINUATION OF:** Hygaea og muserne. **CONTINUATION OF:** Museum for sunheds og kundskabs elskere. **CONTINUATION OF:** Hertha. **CONTINUATION OF:** Sundhedsbog. **CONTINUATION OF:** Medicinalblad. **EDITOR:** Johann Clemens Tode. **CONTINUED BY:** Sundhedsraad. **CONTINUED BY:** Sundhedsjournal et maanedskrift. **CONTINUED BY:** Nyeste sundhedsblade. **SOURCES:** G195 **LOCATIONS:** ULS **SUBJECT:** Medicine - Periodicals - Denmark.

1678 Sundhedsjournal et maanedskrift. - Nos.1-3, 1796. Copenhagen: 1796. **CONTINUATION OF:** Sundstidende. **CONTINUATION OF:** Nye sundstidende. **CONTINUATION OF:** Sundhed og underholdning. **CONTINUATION OF:** Hygaea og muserne. **CONTINUATION OF:** Museum for sunheds og kundskabs elskere. **CONTINUATION OF:** Hertha. **CONTINUATION OF:** Sundhedsbog. **CONTINUATION OF:** Medicinalblad. **CONTINUATION OF:** Sundhedsjournal. **CONTINUATION OF:** Sundhedsraad. **EDITOR:** Johann Clemens Tode. **CONTINUED BY:** Nyeste sundhedstidende. **SOURCES:** G199 **LOCATIONS:** ULS **SUBJECT:** Medicine - Periodicals - Denmark.

1679 Sundhedsraad, i anledning af ildebranden. - v.1, 1795-1796. Copenhagen: 1795-1796. **CONTINUATION OF:** Sundstidende. **CONTINUATION OF:** Nye sundstidende. **CONTINUATION OF:** Sundhed og underholdning. **CONTINUATION O:** Hygaea og muserne. **CONTINUATION OF:** Museum for sunheds og kundskabs elskere. **CONTINUATION OF:** Hertha. **CONTINUATION OF:** Sundhedsbog. **CONTINUATION OF:** Medicinalblad.

CONTINUATION OF: Sundhedsjournal. EDITOR: Johann Clemens Tode. CONTINUED BY: Sundhedsjournal et maanedskrift. CONTINUED BY: Nyeste sundhedsblade. SOURCES: G196 LOCATIONS: ULS SUBJECT: Medicine - Periodicals - Denmark.

1680 Sundstidende et medicinske ugeskrift af blandet inhoud. - v.1-3, 1778-1781. Copenhagen: 1778-1781. EDITOR: Johann Clemens Tode. CONTINUED BY: Nye sundhedstidende. CONTINUED BY: Sundhedsblade. CONTINUED BY: Sundhed og underholdning. CONTINUED BY: Hygaea og muserne. CONTINUED BY: Museum for sunheds og kundskabs elskere. CONTINUED BY: Hertha. CONTINUED BY: Sundhedsbog. CONTINUED BY: Medicinalblade. CONTINUED BY: Sundhedsjournal. CONTINUED BY: Sundhedsraad. CONTINUED BY: Sundhedsjournal et maanedskrift. CONTINUED BY: Nyeste sundhedsblade. SOURCES: G178 LOCATIONS: BUC, ULS SUBJECT: Medicine - Periodicals - Denmark.

1681 De Surinaamsche Arzt. - St. 1 (No. 1-13) Sept.18, 1786-Oct. 8, 1788. Paramoribo: W.H. Poppelmann, 1786-1788. A facsimile was issued in Utrecht, 1981. EDITOR: Jacob Voegen van Engelen. LOCATIONS: NUC, ZDB SUBJECT: Medicine - Periodicals - Paramoribo.

1682 Svenska vetenskapsakademien protokoll fur åren 1739, 1740 och 1741, med anmärkingar utfigna af E. W. Dahlgren. v.1-v.2, (1739-1741)1918. Stockholm: Almqvist & Wiksell, 1918. OTHER TITLE: Protokoll, Kongliga Svenska vetenskapsadademien, Stockholm. SOURCES: G477g LOCATIONS: ULS SUBJECT: Science - Societies - Stockholm.

1683 Synopsis observationum medicarum et physicarum quae decurias III ad centurieae X Ephemerides Academiae Caesareae Leopoldina Carolinae Naturae curiosorum ab anno 1670 usque ad annum 1722 publicatarum. - v.1, 1739. Nürnberg: Maurizz Endteri, 1739. EDITORS: Wilhelm Andreas Kellner and Andreas Elias Büchner. SOURCES: OCLC LOCATIONS: BL SUBJECT: Science - Societies - Halle - Abridgments.

1684 Syntagma commentationum societatis regiae scientiarum Gottingensis. - v.1, 1759, v.2, 1767. Göttingen, 1759-1767. EDITOR: Johann David Michaelis. SOURCES: G472b, S2694g SUBJECT: General - Societies - Göttingen.

1685 Table alphabétique des matières contenues dans l'histoire et les mémoires de l'Académie des sciences depuis 1666-1790. - T.1-10, 1729-1809. Paris: G. Martin, J.B. Coignard et H.L.Guerin, 1729-1809. Several editions of at least the first 6 volumes were published. Publisher varies. Edited by Godin, Pierre Demours and Louis Cotte. MICROFORMS: IDC(8). SOURCES: GII:17e, S12761 LOCATIONS: BUC, ULS, ZDB SUBJECT: Science - Societies - Paris - Indexes.

1686 Tagebuch für Liebhaber der Astronomie. - No. 1-2, 1793-1794. Nürnberg: 1793-1794. LOCATIONS: CCF SUBJECT: Astronomy - Periodicals - Germany.

1687 Tagebuch oder, monathliche Anweisung zum Gebrauch für
     Schmetterlingssammler. - No.1-3, 1800. Leipzig: 1800. Monthly.
     SOURCES: S3019 **SUBJECT:** Entomology - Periodicals - Germany.

1688 Tal hållne i...vid praesidiers afläggning / Kungliga svenska vetenskaps akademien.
     (1739-1792) 1747-1792. Stockholm: 1747-1792. **SOURCES:** S693t
     SUBJECT: Science - Societies - Stockholm

1689 Taschenbuch für der Gesundheit. - Berlin: Nicolai, 1784. **SOURCES:**
     (Ref.3:226). **SUBJECT:** Medicine - Periodicals - Germany - Popular.

1690 Taschenbuch für die deutschen Wundärzte. - v.1-5, 1783-1790. Altenburg:
     Richter, 1783-1790. Imprint varies. **EDITOR:** Friedrich Hoffmann, Johann
     Stephen Hausmann and Friedrich August Waitz (Weiz). **CONTINUED BY:**
     Medicinisch-chirurgische Aufsätze, Krankengeschichten und Nachrichten.
     SOURCES: G55, K3605 **LOCATIONS:** CCF, ULS **SUBJECT:** Surgery
     - Periodicals - Germany - Reviews.

1691 Taschenbuch für Freunde der Gebirgskunde. - No.1, 1798. Göttingen: 1798.
     SOURCES: B4417 **SUBJECT:** Geology - Periodicals - Germany.

1692 Taschenbuch für Freunde und Liebhaber der allgemeinen Weltkunde. - v.1-2,
     1797-1799. Quedlinburg: 1797-1799. **SOURCES:** B4419, S3189 **SUBJECT:**
     General - Periodicals - Germany.

1693 Taschenbuch für Gartenfreunde (Kiel). - v.1-6, 1782-1788. Kiel: Hirschfeld;
     Leipzig: Buchhandlung der Gelehrten, 1782-1788. Christian Cay Lorenz
     Hirschfeld. **CONTINUED BY:** Gartenkalender. **CONTINUED BY:** Kleine
     Gartenbibliothek. **SOURCES:** B4420, K2934 **LOCATIONS:** ZDB
     SUBJECT: Gardening - Periodicals - Germany.

1694 Taschenbuch für Gartenfreunde (Leipzig). - Leipzig: Vosz, 1795-1799.
     EDITOR: Wilhelm Gottlieb Becker. **SOURCES:** B4421 **LOCATIONS:**
     BUC, ZDB **SUBJECT:** Gardening - Periodicals - Germany.

1695 Taschenbuch für Natur- und Gartenfreunde. - v.1-12, 1795-1806. Tübingen: J.C.
     Cotta, 1795-1806. **OTHER TITLE:** Taschenkalender für Natur- und
     Gartenfreunde. **SOURCES:** B4422, G569, S3284 **LOCATIONS:** BUC,
     CCF, ULS, ZDB **SUBJECT:** Gardening - Periodicals - Germany.
     SUBJECT: Natural history - Periodicals - Germany - Popular.

1696 Taschenbuch für Pferdeliebhaber. - v.1-11, 1792-1802. Tübingen: Cotta,
     1792-1802. **OTHER TITLE:** Taschenkalender für Pferdeliebhaber. **EDITOR:**
     Franz Max Bouwinghausen von Wollnerode. **SOURCES:** K2983
     LOCATIONS: ZDB **SUBJECT:** Veterinary medicine - Periodicals -
     Germany.

1697 Taschenbuch zur Belehrung und Unterhaltung für Wald- und Jagdfreunde. -
     Gotha; Marburg: 1793-1812. Imprint varies. **OTHER TITLE:**

Neujahrsgeschenck für Jagd- und Forstliebhaber; Taschenbuch für Forst- und Jagdfreunde. **CONTINUED BY:** Sylvan. **LOCATIONS:** CCF, ULS, ZDB **SUBJECT:** Forestry management - Periodicals - Germany. **SUBJECT:** Hunting - Periodicals - Germany.

1698   Technologische Compendium. - v.1, 1787. Elbing: 1787. **LOCATIONS:** ZDB **SUBJECT:** Technology - Periodicals - Poland.

1699   Technologisches Taschenbuch für Künstler, Fabrikanten und Metallurgen. - Göttingen: Johann Christian Dieterich, 1786. **EDITORS:** Johann Dietrich Brandis and Johann Friedrich August Göttling. **SOURCES:** B4446, K3913 **LOCATIONS:** ZDB **SUBJECT:** Manufactures - Periodicals - Germany. **SUBJECT:** Technology - Periodicals - Germany.

1700   Technologisches Magazin. - v.1 (no. 1-3) 1790-1792. Memmingen: Andreas Seyler, 1790-1792. **EDITOR:** Christoph Wilhelm Jacob Gatterer. **CONTINUED BY:** Neues technologisches Magazin. **SOURCES:** B1827, G745, K2973, S3090 **LOCATIONS:** BUC, ULS, ZDB **SUBJECT:** Technology - Periodicals - Germany.

1701   Technologisches Museum zur Verteidigung des Künstler- und Gewerbstandes eröffnet in Wien. - Prague: Schönfeld, 1798. **SOURCES:** K3920 **SUBJECT:** Technology - Periodicals - Czechoslovakia.

1702   Tentamina experimentorum naturalium captorum in academia del Cimento. 2 v.in 1, 1731. Leyden: Joan. et Herm. Verbeek, 1731. There was an edition in Vienna in 1756 and another in Vienna, Prague and Trieste in 1761 (Ref.45:354). **EDITOR:** Petrus van Musschenbroek. **TRANSLATION OF:** Saggi di naturali esperienze fatta nell Accademia del Cimento. **MICROFORMS:RP(1). SOURCES:** OCLC **SUBJECT:** Science - Societies - Florence - Translations.

1703   Der Teutsche Obstgärtner, oder, gemeinnnütziges Magazin des Obstbaues in Teutschlands sämtlichen Kreisen. - v.1-22 (12 nos. each) 1794-1804. Weimar: Industrie Comptoir, 1794-1804. A selection was issued under the title: Abildungen aller Obstgarten. Weimar, n.d. (BL). **OTHER TITLE:** Deutsche Obstgärtner. **EDITOR:** Johann Volkmar Sickler. **CONTINUED BY:** Allgemeines teutsches Garten-Magazin. **INDEXES:** Index to v.1-22 (1794-1804) in v.22. **SOURCES:** B4476, K2999, OCLC **LOCATIONS:** BUC, CCF, CCN, ULS, ZDB **SUBJECT:** Fruit-culture - Periodicals - Germany. **SUBJECT:** Agriculture - Periodicals - Germany.

1704   Der Tirolische Arzt, eine Wochenschrift für seine Landsleute. - v.1-2, Mar. 1, 1791-May 22, 1792. Innsbruck: Trattner, 1791-1792. Weekly. **EDITORS:** Claudius Martin von Scherer and Franz Niedermaier. **SOURCES:** G99, K3670 **LOCATIONS:** ULS, ZDB **SUBJECT:** Medicine - Periodicals - Switzerland - Popular.

1705 The Topographer, containing a variety of original articles, illustrative of the local history and antiquities of England. - v.1-4 (1-27), April 1789-June 1791. London: Robson and Clarke, 1789-1791. Imprint varies. v.5 no.1, edited by Thomas Phillips, issued 1821. **EDITORS:** Samuel Egerton Brydges and Lawrence Stebbing Shaw. **CONTINUED BY:** Topographical miscellanies. **SOURCES:** OCLC   **LOCATIONS:** BUC, CCF, ULS   **SUBJECT:** Geography - Periodicals - England.

1706 Topographical miscellanies. - No.1-7, Aug. 1791-Feb. 1792. London: 1791-1792. **CONTINUATION OF:** Topographer. **EDITORS:** Samuel Egerton Brydges and Lawrence Stebbing Shaw. **LOCATIONS:** BUC, CCF, ULS **SUBJECT:** Geography - Periodicals - England.

1707 Topographisk journal for Norge / Topographiske selskabs for Norge. - v.1-10 (No.1-34) 1792-1808. Christiania: Jens Orbeg Berg, 1791-1808. **SOURCES:** B4572, OCLC, S595   **LOCATIONS:** BUC, CCF, ULS   **SUBJECT:** Geography - Societies - Christiania.

1708 Transactions / Medical Society of London. - v.1(part 1-2) 1789-1817. London: 1789-1817.   **SOURCES:** G78b   **LOCATIONS:** BUC, ULS, ZDB **SUBJECT:** Medicine - Societies - London.

1709 Transactions of a Society for the improvement of medical and chirurgical knowledge. - v.1,1793;2,1800;3,1812.   London: J. Johnson, 1793-1812. **TRANSLATED AS:** Abhandlungen der Londonschen Gesellschaft zur Vermehrung des medicinischen und chirurgischen Wissenschaften.Publisher varies.   **SOURCES:** G118 **LOCATIONS:** BUC, CCF, CCN, ULS, ZDB **SUBJECT:** Medicine - Societies - London.   **SUBJECT:** Surgery - Societies - London.

1710 Transactions of the American Philosophical Society / Philadelphia. - v.1-6 (1769-1804) 1771-1809.   Philadelphia: William and Thomas Bradford, 1771-1809.   Published in facsimile, Philadelphia, American Philosophical Society, 1969.   Kraus reprint.   **INDEXES:** General index to v.1-4 in v.4. **MICROFORMS:** UM(1). **SOURCES:** G457, S4240b **LOCATIONS:** BUC, CCF, CCN, ULS, ZDB   **SUBJECT:** General - Societies - Philadelphia.

1711 Transactions of the College of Physicians of Philadelphia. - v.1 (no.1-3), (1787-1793) 1795.   Philadelphia: T.Dobson, 1795.   **TRANSLATED AS:** Medicinische Verhandlungen, (College of Physicians of Philadelphia). Resumed publication in 1841. **SOURCES:** G111, OCLC **LOCATIONS:** BUC, CCF, ULS   **SUBJECT:** Medicine - Societies - Philadelphia.

1712 Transactions of the Dublin Society. - v.1-23, 1787-1907. Dublin: Graisbery and Campbell, 1787-1907.   Imprint varies.   **OTHER TITLE:** Transactions of the Royal Irish Academy, Dublin.   **INDEXES:** Index to the serial publications of the Royal Irish Academy, Dublin, Hodges, Figgis, 1912.   **SOURCES:** G501, OCLC, S93b **LOCATIONS:** BUC, CCF, CCN, ULS **SUBJECT:** Science - Societies - Dublin.

1713 Transactions of the Dublin society for promoting husbandry and other useful arts in Ireland. - v.1-6, 1799-1810. Dublin: Griasbery & Campbell, 1799-1810. The first volume is made up of short monographs which were to "be had separate at the publishers." SOURCES: G500a, OCLC, S85b LOCATIONS: BUC, ULS, ZDB SUBJECT: Agriculture - Societies - Dublin. SUBJECT: Science - Societies - Dublin.

1714 Transactions of the Linnean Society of London. - v.1-30, 1791-1875. London: Benjamin White, 1791-1875. INDEXES: Index v.1-25, 1791-1866. MICROFORMS: RM(3), IDC(8). SOURCES: G479, S333e LOCATIONS: BUC, CCF, CCN, ULS, ZDB SUBJECT: Natural history - Societies - London.

1715 Transactions of the Royal Humane Society from 1774 to 1784, with an appendix of miscellaneous observations on suspended animation to the year 1794. - v.1, 1795. London: Rivington, 1795. TRANSLATED AS: Abhandlungen der Londoner Gesellschaft zur Rettung Verunglückter und Scheintodter. CONTINUATION OF: Reports of the Royal Humane Society. SOURCES: G499a LOCATIONS: BUC, ULS SUBJECT: Resuscitation - Societies - London.

1716 Transactions of the Society for the encouragement of agriculture in Dumfries, Wigton and Kirk-Cudbright. - Nos.1-2, 1776. Dumfries: 1776. MICROFORMS: RP(1) (Goldsmiths' Kress Library). LOCATIONS: BUC SUBJECT: Agriculture - Societies - Dumfries.

1717 Transactions of the Society for the Promotion of agriculture, arts and manufactures, instituted in the state of New York. - v.1-4, 1792-1819. Albany: John Barber, 1792-1819. Imprint varies. A 2nd edition of v.1 was issued in 1801 in Albany, Charles and George Webster. CONTINUED BY: Transactions of the Albany Institute. SOURCES: G518, OCLC, S4219 LOCATIONS: BUC, ULS SUBJECT: Agriculture - Societies - Albany.

1718 Transactions of the Society for the promotion of agriculture / North Carolina. - Raleigh: 1792. LOCATIONS: BUC SUBJECT: Agriculture - Societies - North Carolina.

1719 Transactions of the Society instituted at London for the encouragment of the arts, manufactures and commerce. - v.1-v.55, 1783-1844. London: Phillips, 1783-1844. TRANSLATED AS: Auszug aus den Transaktionen der Societät zu London zur Aufmunterung der Künste, der Manufakturen und der Handlung. CONTINUATION OF: Museum rusticum et commerciale. CONTINUATION OF: Memoirs of agriculture and other economical arts. Title and printer vary. There were at least 3 editions of some of the earlier volumes. INDEXES: Index to v.1-25 (1783-1807) appended to v.26 of the Transactions, to v.26-40 (1808-1822) with v.40. (Ref.50). SOURCES: OCLC LOCATIONS: BUC, CCF, CCN, ULS, ZDB SUBJECT: Technology - Societies - London. SUBJECT: Economics - Societies - London. SUBJECT: Agriculture - Societies - London.

1720 Transactions / Philosophical Society of Oxford. - Oxford: (1683-1690) 1925. See v.4 of Early sciences in Oxford, edited by R.T.Gunther. **LOCATIONS:** NUC **SUBJECT:** Science - Societies - Oxford.

1721 Transactions philosophique de la Société royale de Londres. - v.1-16 (1731-1744) 1741-1760. Paris: 1734-1761. Translators: 1731-1736 François de Bremond, 1737-1744 Pierre Demours. **TRANSLATION OF:** Philosophical transactions. **SOURCES:** GII:4i, S1597d **LOCATIONS:** CCF, LC, ZDB **SUBJECT:** Science - Societies - London - Translations.

1722 Transactions / Royal Society of Edinburgh. - v.1- , 1788- . Edinburgh: 1788- . **TRANSLATED AS:** Philosophische und historische Abhandlungen, K. Gessellschaft der Wissenschaften in Edinburg. **INDEXES:** General index to the first thirty-four volumes (1788-1888), Edinburgh, Neill, 1890. **MICROFORMS:** PMC(1,5,6). **SOURCES:** G502, OCLC, S124b **LOCATIONS:** BUC, CCF, CCN, ULS, ZDB **SUBJECT:** Science - Societies - Edinburgh.

1723 Travaux de la société libre d'agriculture et d'économie interieure du department du Bas-Rhin séante à Strasbourg. - Strasbourg: 1800. **OTHER TITLE:** Verhandlungen der freien Landwirthschafts-Gessellschaft des Niederrheinischen Department. **LOCATIONS:** CCF **SUBJECT:** Agriculture - Societies - Strasbourg.

1724 Triumph der Heilkunst, oder, durch Thatsachen erläuterte practische Anweisung für Hülfe in den verzweiflungsvollsten Krankheitsfällen. - v.1-5, 1800-1804. Breslau: Hirschberg; Lissa: Johann Friedrich Korn, 1800-1804. **EDITOR:** Christian August Struve. **SOURCES:** G314 **LOCATIONS:** ULS, ZDB **SUBJECT:** Medicine - Periodicals - Poland.

1725 Trudy vol'nago ekonomicheskago obtchestva / St. Petersburg. - v.1-52, 1765-1798. St. Petersburg: 1765-1915. **TRANSLATED AS:** Abhandlungen der freyen ökonomischen Gesellschaft in St. Petersburg. **TRANSLATED AS:** Auswahl ökonomische Abhandlungen welche die freye ökonomische Gesellschaft in teutscher Sprache erhalten hat. **TRANSLATED AS:** Preisschriften und Abhandlungen der kaiserlichen freyen ökonomischen Gesellschaft in St. Petersburg. **OTHER TITLE:** Travaux de la Société économique libre, St. Petersburg. **MICROFORMS:** DA(1). **LOCATIONS:** BUC, CCF, ULS, ZDB **SUBJECT:** Economics - Societies - St. Petersburg.

1726 Twee-maandellijke uitreksels. - Rotterdam: Pieter Rabus, 1702-1798. **CONTINUATION OF:** Boekzaal van Europe. A rival publication was issued under the same title from 1702-1794 by Van der Staart. **EDITORS:** Pieter Rabus, Willem Sewel and Joan Van Gavern. **CONTINUED BY:** Boekzaal der geleerde wereld of tijdschrift voor leteerkundigen. **CONTINUED BY:** Maandelijke uitreksels of Boekzaal der geleerde wereld. **INDEXES:** Index to Deel 1-19 in 19. See also: Register der boeken, uytgetrokken in de Boekzaalen, of, Tweemandelyke uytreksels, vorrheenen gesticht in de jaaren 1692 tot 1708 van de heeren P. Rabus, W. Sewel, en J. van Gavern: en nu in ordre gebracht door J. Le Long. Amsterdam, 1716. **LOCATIONS:** BUC, CCN, ZDB **SUBJECT:** General - Periodicals - Netherlands - Reviews.

1727 Über den thierischen Magnetismus, in einem Brief an Herrn Geheimen Rathe in Mainz. - St.1-2, 1787. Tübingen: Jacob Friedrich Heerbrandt, 1787. **EDITOR:** Eberhard Gmelin. **CONTINUED BY:** Neue Untersuchungen über den thierischen Magnetismus. **CONTINUED BY:** Materialen für die Anthropologie. **SOURCES:** G237 **LOCATIONS:** BL **SUBJECT:** Animal magnetism - Periodicals - Germany.

1728 Übersicht der neuesten medicinischen Litteratur. - v.1 (St.1-3) 1795-1797. Chemnitz: Carl Gottlob Hofmann, 1795-1797. **CONTINUATION OF:** Medicinische Litteratur für practische Aerzte. **CONTINUATION OF:** Neue medicinische Litteratur. **EDITOR:** Johann Christian Traugott Schlegel. **SOURCES:** G301b, K3698 **LOCATIONS:** BUC **SUBJECT:** Medicine - Periodicals - Germany - Reviews.

1729 Ueber den Nutzen der Eideschen in Krebschäden der Lustseuche und verschiedenen Hautkrankheiten. - v.1, 1788. Leipzig: Friedrich Gotthold Jacobäer, 1788. Translated and augmented by Johann Jacob Römer. **TRANSLATION OF:** Raccolta di vari opuscoli publicati sin'ora interno. **SOURCES:** G399a **LOCATIONS:** BL **SUBJECT:** Medicine - Periodicals - Italy - Translations.

1730 Ueber die neueren Gegenstände in der Chemie. - No. 1-11, 1791-1802. Breslau: Hirschberg und Lissa, 1791-1802. **EDITOR:** Jeremias Benjamin Richter. **SOURCES:** B4597 **LOCATIONS:** ULS **SUBJECT:** Chemistry - Periodicals - Poland.

1731 Uebersetzungen und deutsche Abhandlungen welche bei der Churfürstlich-Mainzichen Akademie der Wissenschaft nach und nach übergeben worden. - v.1-2, 1762-1763. Erfurt: Johann Friedrich Hartung, 1762-1763. Imprint varies. **EDITOR:** Siegesmund Lebrecht Hadelilch. **MICROFORMS:** RM(3). **SOURCES:** K234, OCLC **LOCATIONS:** BUC **SUBJECT:** Science - Societies - Erfurt.

1732 Uebersicht der neuesten pomologischen Literatur. - St.1 (1797) 1799. Frankfurt: Galhaumann, 1799. **SOURCES:** K3015 **LOCATIONS:** ZDB **SUBJECT:** Fruit culture - Periodicals - Germany **SUBJECT:** Agriculture - Periodicals - Germany.

1733 Uitgeleeze naturkundige verhandelingen. - v.1-3, 1734-1764. Amsterdam: 1734-1764. **LOCATIONS:** CCN, ULS **SUBJECT:** Natural history - Periodicals - Netherlands.

1734 Uitgezochte genees- en heelkundige verhandelingen. - No.1-2, 1773-1774. Dordrecht: Pieter van Braem; Rotterdam: Abraham Bothall, 1773-1774. **EDITOR:** Albertus Lentfrinck. **SOURCES:** G315 **LOCATIONS:** ULS **SUBJECT:** Medicine - Periodicals - Netherlands.

1735 Uitgezogte verhandelingen uit de nieuwste werken van de societeiten der wetenschapen in Europa en van andere geleerde mannen. - v.1-10, 1757-1765. Amsterdam: F. Houttuyn, 1757-1765. INDEXES: General index to v.1-10 in 10. MICROFORMS: RM(3). SOURCES: G746, OCLC LOCATIONS: CCF, CCN, ULS SUBJECT: Science - Periodicals - Netherlands - Reviews.

1736 Unächter Acacien-Baum, zur Ermunterung d. allgemein, Anbaues dieser in ihrer Art einzigen Holzart. - St.1-6, 1794-1803. Leipzig: Gräff, 1794-1803. EDITOR: Friedrich Casimir Medicus. SOURCES: K2994 LOCATIONS: ZDB SUBJECT: Forestry management - Periodicals - Germany.

1737 Unbekannte, wie auch zu wenig bekannte Wahrheiten der Mathematik, physik und Philosophie. - St.1-9, Nov.1787-Jul. 1788. Stettin: S.F. Struck, 1787-1788. Monthly. EDITOR: Johann Jacob Meyen. SOURCES: K3235 LOCATIONS: ZDB SUBJECT: Mathematics - Periodicals - Germany. SUBJECT: Science - Periodicals - Germany.

1738 Ungarische Agricultur-Zeitung. - St.1-3, 1789. Pressburg: 1789. SOURCES: (Ref.116,p.189). SUBJECT: Agtriculture - Periodicals - Hungary.

1739 Ungarischer Land-und Hauswirthschafts Kalender. - Pest: Weinginad, 1791. EDITOR: J.P. von Plessing. SOURCES: (Ref.117,p.983). SUBJECT: Agriculture - Periodicals - Hungary.

1740 Ungrisches Magazın, oder, Beiträge zur ungrischen Geschichte, Geographie, Naturwissenschaft und dahin einschlagender Litteratur. - v.1-4, 1781-1787. Pressburg: Löwe, 1781-1787. EDITOR: Carl Gottlieb von Windisch. CONTINUED BY: Neues ungrisches Magazin. SOURCES: B4607, G747, K1117, OCLC, S3518 LOCATIONS: BUC, CCF, ULS, ZDB SUBJECT: Science - Periodicals - Austria - Popular.

1741 Universal magazine of knowledge and pleasure. - v.1, June 1747- 113, Dec. 1803. London: John Hinton, 1747-1803. CONTINUED BY: New universal magazine. MICROFORMS: RP(1) (Goldsmiths' Kress Library). SOURCES: Ref.118,p.54. LOCATIONS: BUC, CCF, CCN, ULS, ZDB SUBJECT: General - Periodicals - England.

1742 Der Unterhaltende Arzt über Gesundheitspflege, Schönheit, Medicinalwesen, Religion und Sitten. - v.1-4, 1785-1789. Copenhagen; Leipzig: Faber und Nitschke, 1785-1789. EDITOR: Johann Clemens Tode. SOURCES: G183, 3615 LOCATIONS: BL SUBJECT: Medicine - Periodicals - Denmark - Popular.

1743 Unterhaltungen aus der Naturgeschichte. - v.1-27, 1792-1824. Augsburg: Martin Engelbrecht, 1792-1824. A subscription book in parts, each on a different subject. EDITORS: Gottieb Tobias Wilhelm and Georg Adam Neuhofer. SOURCES: B4627, G551, OCLC, S2252 LOCATIONS: ULS SUBJECT: Natural history - Periodicals - Germany - Popular.

1744 Unterhaltungen für Conchylienfreunde und für Sammler der Mineralien. - No.1, 1789. Erlangen: Wolfgang Walther, 1789. EDITOR: Johann Samuel Schröter. SOURCES: K3332, OCLC LOCATIONS: ZDB SUBJECT: Mollusks - Periodicals - Germany. SUBJECT: Mineralogy - Periodicals - Germany.

1745 Unterhaltungen über verschiedene Gegenstände aus den physikalischen Wissenschaften zur geminnützigen Kenntniss der Mathematik. - Leipzig: 1781. EDITOR: Johann Sebastian Horrer. SOURCES: K3239 SUBJECT: Natural history - Periodicals - Germany. SUBJECT: Mathematics - Periodicals - Germany.

1746 Utvalda allmänt och nyare merendels rön och samlingar i medicin, pharmacie, chemie, naturkunnighet, landhushállning, handel och slögder, jämte utdrag af nöjsamare ämmen i naturhistorie, verlds-och resebeskrifiningar. - v.1-2, 1797-1801. Stockholm: 1797-1801. EDITOR: Anders Sparrman. SOURCES: B4646, G748, OCLC, S700 SUBJECT: Science - Periodicals - Sweden.

1747 Vaderlandsch kabinet van koophandel, zeevart, landbouw, fabrijken, enz. behelzende, eene verzameling van verhandelingen de uitbreiding en verbetering der gemelde vaderlandsche welvaart bronne ten deel hebende. - No.1-3, 1786-1790. Amsterdam: Arende Fokke Simonz, 1786-1790. LOCATIONS: CCF,CCN SUBJECT: Commerce - Periodicals - Netherlands. SUBJECT: Economics - Periodicals - Netherlands.

1748 Vaderlandsche bibliotheek van wetenschap, kunst en smaak. - Amsterdam: Martius de Bruijn, 1773-1796. CONTINUED BY: Nieuwe vaderlandsche bibliothek van wetenschap, kunst en smaak. SOURCES: B4648, S782 LOCATIONS: CCF, CCN SUBJECT: General - Periodicals - Netherlands - Reviews.

1749 Vaderlandsche letter-oefeningen of Tijdschrift. - v.1-115, 1761-1876. Amsterdam: 1761-1876. As Vaderlandsche letter-oefeningen 1761-1767, and after 1814; as Nieuwe vaderlandsche letter-oefeningen from 1768-1771; as Hedendaagsche vaderlandsche letter-oefenngen 1772-1778; as Algemeene vaderlandsche letter-oefeningen from 1779-1785 and 1791-1811; as Nieuwe allgemeene vaderlandsch letter-oefeningen from 1786-1790, and as Tijdschjrift van kunsten en wetenschappen 1812-1813. OTHER TITLE: Nieuwe vaderlandsche letter-oefeningen; Hedendaagsche vaderlandsche letter-oefeningen; Algemeen vaderlandsche letter-oefeningen; Tijdschrift van kunsten en wetenschappen. SOURCES: S722, OCLC LOCATIONS: BUC, CCF, CCN, ULS, ZDB SUBJECT: General - Periodicals - Netherlands.

1750 Veckoskrift för läkare och naturforskare. - v.1-7, 1781-1786. Stockholm: Strangnäs, 1781-1786. EDITORS: Andreas Johan Hagström, Johan Kraak, Johann Lorens Odhelius and Gabriel Lund. CONTINUED BY: Läkaren och naturforskaren. SOURCES: B4657, G47, S701 LOCATIONS: BUC, ULS SUBJECT: Medicine - Periodicals - Sweden. SUBJECT: Natural history - Periodicals - Sweden.

1751 Verhandelingen uitgegeven door de Hollandsche maatschappij der wetenschappen te Haarlem. - Deel 1-30, 30v.in 40. Haarlem: J. Bosch, 1754-1793. **TRANSLATED AS:** Abhandlungen der Holländischen Gesellschaft der Wissenschaften zu Haarlem. **TRANSLATED AS:** Abhandlungen aus der Naturgeschichte, praktischer Arzneikunst und Chirurgie, aus den Schriften der Haarlemer und anderer holländischen Gesellschaften. v.28-30 have the imprint Haarlem en Amsterdam; v.6 and v.7 were each issued in 2 parts paged continuously; v.7-8, v.10-11,13,16-17,20 were each in 2 parts paged separately; and v.9 and 19 were each in 3 parts paged separately. **CONTINUED BY:** Naturkundige verhandelingen van de Koninglijke Maatschappij der Wetenschappen te Haarlem. **INDEXES:** Register of te hoofdzaaklyke inhoud der Verhandelingen enz. in de twaalf eerste deelen. Haarlem, J. Bosch, 1772. Beredeneerd register of te hoofzaaklyke inhoud der Verhandelingen die inde xxviii deelen van de Hollandsche Maatschapij der Wetenschappen voorkomen. Door Johannes Florentius Martinet. Haarlem en Amsterdam, J. Allart, 1793. **SOURCES:** G474, OCLC, S849b **LOCATIONS:** BUC, CCF, CCN, ULS, ZDB **SUBJECT:** Science - Societies - Haarlem.

1752 Verhandelingen uitgegeven door de maatschappy ter bevordering van den landbouw te Amsterdam. - v.1-19, 1778-1832. Amsterdam: Jan Christiaan Seep, 1778-1806. An atlas was published 1823 to accompany v.18. **SOURCES:** OCLC **LOCATIONS:** BUC, CCN, ULS **SUBJECT:** Agriculture - Societies - Amsterdam.

1753 Verhandelingen uitgegeven door het zeeuwsch genootschap der wetenschappen te Vlissingen. - Deel 1-15, 1769-1792. Middleburg: Pieter Gelissen, 1769-1792. **TRANSLATED AS:** Abhandlungen der Seeländlischen Gesellschaft der Wissenschaften zu Vlissingen. Place and publisher vary. **CONTINUED BY:** Nieuwe verhandelingen, Zeeuwsch genootschap der wetenschappen te Vlissingen. **INDEX:** Beredeneerd register van alle Verhandelingen in Stukken. In Deel 12, St.2, 1788. **MICROFORMS:** RM(3). **SOURCES:** G522, S890c **LOCATIONS:** BUC, CCF, CCN, ULS ZDB **SUBJECT:** Science - Societies - Vlissingen.

1754 Verhandelingen uitgegeven door Teyler's tweede genootschap. - St.1-28, 1781-1857. Haarlem: Joh. Ensched en J. Van Wair, 1781-1857. St. 26 published 1866. Publisher varies. **MICROFORMS:** RM(3). **SOURCES:** G521, OCLC, S860 **LOCATIONS:** BUC, CCF, CCN, ULS, ZDB **SUBJECT:** Science - Societies - Haarlem.

1755 Verhandelingen van de koniglijke academie der heilkonst. - Hague: 1745. **TRANSLATION OF:** Mémoires de l'Académie royale de chirurgie de Paris. **LOCATIONS:** CCF **SUBJECT:** Surgery - Societies - Paris - Translations.

1756 Verhandelingen van de natuur-en geneeskundige correspondentie-societeid in der vereenigte Nederlanden. - v.1-4 (no.1-2) 1780-1795. Hague: Hoogstraten, 1780-1795. Imprint varies. **OTHER TITLE:** Verhandelingen omtrend de weergesteldheid en ziekten der vereenigte Nederlanden. **SOURCES:** G57,

OCLC, S832a   LOCATIONS: BUC, CCF, ULS   SUBJECT: Medicine -
Societies - Hague.   SUBJECT: Natural history - Societies - Hague.

1757 Verhandelingen van het Bataafsch Genootschap der Proefondervindelyke
wysbegeerte te Rotterdam. - Deel 1-12, 1774-1798.   Rotterdam: Reiner
Arrenberg, 1774-1798. Publisher varies. A 2nd ed. of v.1-4 was published in
Rotterdam, and another edition of v.1-3 in Batavia.   CONTINUED BY:
Nieuwe verhandelingen van het Bataafsch genootschap der Proefondervindelijke
wijsbegerte te Rotterdam.   INDEXES: Algemeene bladwyzer over de eerste
deelen (1774-1781) der Verhandelingen.   Rotterdam, Arrenberg, 1784.
MICROFORMS: RM(3).   SOURCES: G461, OCLC, S898c   LOCATIONS:
BUC, CCF, CCN, ULS, ZDB   SUBJECT: Science - Societies - Rotterdam.

1758 Verhandelingen van het Bataviaasch genootschap der konsten en vetenschapen. -
v.1-12, 1779-1830. Amsterdam: Jan Allart, 1779-1830. TRANSLATED AS:
Abhandlungen der Gesellschaft der Künste und Wissenschaften in Batavia.
v.1-4 reprinted in Amsterdam and Rotterdam, 1781-1786, and a 3rd ed. of
v.1-2 in 1825-1826. INDEXES: Alphabetisch register op de 41 eerste deelen
der Verhandelingen. Batavia, W. Bruining, 1881. MICROFORMS: IDC(8).
    SOURCES: G460, S725   LOCATIONS: BUC, CCF, CCN, ULS
SUBJECT: Science - Societies - Batavia.

1759 Verhandelingen van het genootschap ter bevordering der geneeskunde te
Amsterdam. - v.1-8, 1791-1813. Amsterdam: 1791-1805. CONTINUED BY:
Nieuwe verhandelingen van het genootschap ter bevordering der geneeskunde
te Amsterdam.   SOURCES: G103   LOCATIONS: BUC, CCF, CCN, ULS
SUBJECT: Medicine - Societies - Amsterdam.

1760 Verhandelingen van het Genootschap ter bevordering van genees-en heelkunde,
opgeregt tot Antwerpen, onder de Zinspreuk: Occidit, qui non servat. - v.1-3,
1798-1801. Antwerp: J. Schröter, 1798-1801.   CONTINUED BY: Nieuwe
verhandelingen van het genootschap ter bevordering van genees-en heelkunde.
SOURCES: G153, S937a   LOCATIONS: BUC, CCF, ULS   SUBJECT:
Medicine - Societies - Antwerp.

1761 Verhandelingen van het provinciall Utrechtsch genootschap van kunsten en
Wetenschappen / Haarlem. - v.1-11, 1781-1821. Utrecht: S. de Waal,
1781-1821. v.10 is a collection of treatises separately published at various dates.
Publisher varies.   CONTINUED BY: Nieuwe verhandelingen van het
provinciall Utrechtsch genootschap van kunsten en wetenschappen, Haarlem.
MICROFORMS: RM(3). SOURCES: G488, OCLC, S923c   LOCATIONS:
BUC, CCF, CCN, ULS, ZDB   SUBJECT: Science - Societies - Utrecht.

1762 Verhandelingen, werktuig-en scheikunde / Nederlandsche maatschappij ter
bevordering van nijverheid. - No.1-2, 1780-1822. Haarlem: 1780-1822.
SOURCES: S854a   LOCATIONS: ULS   SUBJECT: Chemistry - Societies
- Haarlem.

1763 Verhandelingen, zeevaart en visscherijen / Nederlandsche maatschappij ter bevordering van nijverheid. - No.1-7, 1780-1787. Haarlem: 1780-1787. **SOURCES:** S854b **SUBJECT:** Naval art and science - Societies - Haarlem.

1764 Verhandlungen der Gesellschaft landwirthschaftlicher Freunde in Bündten. - St.1-5, 1780-1782. Chur: 1780-1782. **SOURCES:** K2931 **LOCATIONS:** ZDB **SUBJECT:** Agriculture - Societies - Chur.

1765 Verhandlungen der Gesellschaft von Freunden der Entbindungskunst zu Göttingen. - No.1-2, Apr. 1796-Apr. 1798. Göttingen: J.G. Rosenbusch, 1796-1798. **EDITOR:** Friedrich Benjamin Osiander. **SOURCES:** G149 **SUBJECT:** Obstetrics - Societies - Göttingen.

1766 Verhandlungen der Helvetische Gesellschaft. - Zurich: 1764-1858. Imprint varies. Some of the series published in Schinznach, Olten, and Basle (Ref.95, p.17); 1850-1857 not published. **SOURCES:** K1023, S2114 **LOCATIONS:** ZDB **SUBJECT:** General - Societies - Zurich.

1767 Verhandlungen der Helvetische Gesellschaft in Olten. - Basle: Wilhelm Haas, 1786-1791. **SOURCES:** K1181 **LOCATIONS:** ZDB **SUBJECT:** General - Societies - Olten.

1768 Verhandlungen der Helvetische Gesellschaft in Zofingen. - Aarau: Sauerländer, 1800-1812. **SOURCES:** K464 **SUBJECT:** General - Societies - Zofingen.

1769 Verhandlungen der Helvetische Gesellschaft in Schinznach. - Zürich: 1771-1773. **LOCATIONS:** CCF **SUBJECT:** General - Societies - Schinznach.

1770 Verhandlungen der Liefländischen gemeinnützigen und ökonomischen Societät. - Leipzig: Hartknoch, 1795. **SOURCES:** K3006 **SUBJECT:** Economics - Societies - Liefländ.

1771 Verhandlungen und Geschichte der Helvetisch-Militarischen Gesellschaft (Ref.95, p.2). - Basle: 1779-1797. **SOURCES:** K3936 **LOCATIONS:** ZDB **SUBJECT:** Military art and science - Societies - Basle.

1772 Verhandlungen und Schriften der Hamburgischen Gesellschaft zur Beförderung der Künste und nützlichen Gewerbe. - v.1-7, 1792-1807. Hamburg: Carl Ernst Bohn, 1792-1807. **CONTINUATION OF:** Nachrichten der Hamburgischen Gesellschaft zur Beförderung der Künste und nützlichen Gewerbe. Continued as ns. v.1-3, 1844-1846. **EDITOR:** Johann Arnold Günther. **INDEXES:** Index v.1-3 in 3 and 4-6 in 6. **SOURCES:** K3915, OCLC, S2753a **LOCATIONS:** BUC, CCF, CCN, ULS, ZDB **SUBJECT:** Economics - Societies - Hamburg.

1773 Der Verkündiger, oder, Zeitschrift für die Fortschritte und neuesten Beobachtungen, Entdeckungen und Erfindungen in den Künste und Wissenschaften. - v.1-16, 1797-1812. Nürnberg: Expedition, 1797-1812. Subtitle varies. **EDITOR:** Johann Michael Leuchs. **SOURCES:** B4665,

K5933, G749, 3172   **LOCATIONS: CCF, ZDB   SUBJECT:** General -
Periodicals - Germany.

1774 Vermischte Schriften aus der Naturwissenschaft, Chemie und Arzneygelahrtheit.
- St.1-6, 1756-1758. Frankfurt: Paul Siegmund Gäbler, 1756-1758. **EDITOR:**
Friedrich August Cartheuser. **SOURCES:** B4669, K3224, S2611 **SUBJECT:**
Natural history - Periodicals - Germany. **SUBJECT:** Chemistry - Periodicals
- Germany. **SUBJECT:** Medicine - Periodicals - Germany.

1775 Vermischte Abhandlungen der Physisch-chemischen Warschauer Gesellschaft zur
Beförderung der praktischen Kenntnisse in der Naturkunde, Oekonomie,
Manufakturen und Fabriken, besonder in Absicht auf Polen. - No.1, 1768.
Warsaw; Dresden: Michael Grölle, 1768.   **SOURCES:** K2890, S3757
SUBJECT: Chemistry - Societies - Warsaw. **SUBJECT:** Science - Societies
- Warsaw. **SUBJECT:** Economics - Societies - Warsaw.

1776 Vermischte Abhandlungen der Westfällischen ökonomischen Societät zu Hamm.
- St.1-2, 1793-1794.   Halle: Händel, 1793-1794.   **OTHER TITLE:**
Abhandlungen zu Beforderung er Oekonomie, der Fabriken und Manufakturen,
der Handlung, der Künste und Gewerbe. **SOURCES:** K2989 **LOCATIONS:**
BUC, ZDB **SUBJECT:** Economics - Societies - Hamm.

1777 Vermischte Aufsätze aus der Oekonomie, Naturgeschichte und Chemie. - No.1,
1794.   Marburg: Krieger, 1794.   **OTHER TITLE:** Neues Journal für
Oekonomie, Naturgeschichte und Chemie.   **EDITOR:** Conrad Mönch.
SOURCES: B4667, K2996, S3084 **SUBJECT:** Economics - Periodicals -
Germany. **SUBJECT:** Natural history - Periodicals - Germany. **SUBJECT:**
Chemistry - Periodicals - Germany.

1778 Vermischte Aufsätze von einer Gesellschaft Gelehrter in Halle. - v.1-5,
1785-1786. Halle: Haug; Leipzig: Barth, 1785-1786. Weekly.   **OTHER
TITLE:** Hallische Wochenblatt zum besten der Armen ausgegeben.
SOURCES: B4668, G468, K5677, S3026 **SUBJECT:** General - Societies -
Halle.

1779 Vermischte Beiträge zur physicalischen Erdbeschreibung. - v.1-6 (St. 1-16)
1773-1787. Brandenburg: Halle, 1773-1787.   **EDITOR:** Johann Friedrich
Wilhelm Otto. **SOURCES:** K1068 **LOCATIONS:** CCF, ZDB **SUBJECT:**
Geography - Periodicals - Germany. **SUBJECT:** Geology - Periodicals -
Germany.

1780 Vermischte Bibliothek, oder, Auszüge verschiedenen zur Arzneigelahrtheit,
Chemie, Naturkunde, Oeconomie, zu Manufakturen und Künsten gehörigen
academischen Streitschriften und anderen Abhandlungen. - v.1-2, 1758-1760.
Braunschweig: Waisenhaus, 1758-1760. **EDITOR:** Carl Ludewig Neuenhahn.
SOURCES: K3229, OCLC   **LOCATIONS:** BL   **SUBJECT:** Science -
Periodicals - Germany - Dissertations.   **SUBJECT:** Economics - Periodicals
- Germany - Dissertations.

1781 Vermischte chirurgische Aufsätze. - No.1, 1794. Leipzig: Schwickert, 1794. **SOURCES:** G433, K3688 **SUBJECT:** Surgery - Periodicals - Germany.

1782 Vermischte chirurgische Schriften. - v.1-3, 1776-1782. Berlin: 1776-1782. **TRANSLATED AS:** Heelkundige mengelschriften. Imprint varies. There is a 2nd edition, Berlin und Stettin, Friedrich Nicolais, 1785-1797. **EDITOR:** Johann Leberecht Schmucker. **SOURCES:** G434 **LOCATIONS:** CCF, CCN, ULS, ZDB **SUBJECT:** Surgery - Periodicals - Germany.

1783 Vermischte medicinisch-chirurgische Cautelen für Wundärzte. - v.1-4, 1788-1791, v.5, 1797. Frankfurt: 1788-1797. v.5 as Sammlung. **OTHER TITLE:** Sammlung chirurgisch-practischer Vorfälle. **EDITOR:** Johann Christoph Jäger. **SOURCES:** G435 **SUBJECT:** Surgery - Periodicals - Germany.

1784 Vermischte medicinische und chirurgische Bemerkungen über verschiedene Krankheiten der Brust und des Unterleibes. - v.1-3, 1784-1786. Leipzig: Weygand, 1784-1786. **SOURCES:** K3605a **SUBJECT:** Medicine - Periodicals - Germany.

1785 Vermischte ökonomische Sammlung. - No.1-2, 1750. Leipzig: 1750. Abridgement of the Sammlung 1718-1736. **OTHER TITLE:** Sammlung von Natur und Medicin. **SOURCES:** GII:7m **SUBJECT:** Natural history - Periodicals - Germany - Abridgements.

1786 Vermischte Schriften, als eine Fortsetzung des medicinischen Journals. - No.1-5, 1772-1778. Chemnitz: 1772-1778. **CONTINUATION OF:** Medicinisches Journal über allerhand in die Arzneiwissenschaft. **EDITOR:** Gottwald Schuster. **SOURCES:** K3560 **SUBJECT:** Medicine - Periodicals - Germany.

Vermischte Schriften aus der Naturkunde, see no. 1774.

1787 Vermischte Schriften des Ackerbaugesellschaft in Tirol. - v.1-3, 1769? Innsbruck: 1769? **SOURCES:** (Ref.6,p.290). **SUBJECT:** Agriculture - Societies - Innsbruck.

1788 Vermischte Verbesserungs-Vorschläge und freie Gedanken über verschiedene den Nahrungszustand, die Bevölkerung und Staatswirthschaft der Deutschen. - v.1-2 (6 St. each) 1777-1778. Frankfurt: Esslinger, 1777-1778. **EDITOR:** Johann Friedrich von Pfeiffer. **SOURCES:** K2919 **SUBJECT:** Economics - Periodicals - Germany.

1789 Versuch einer Beschreibung historischer und natürlicher Merkwürdigkeiten der Landschaft Basel. - St.1-23 (1748-1763) 1763-1771. Basle: Emanuel Thusrneisen, 1763-1771. **EDITOR:** Daniel Brückner. **SOURCES:** B4672, S2129 **LOCATIONS:** BL **SUBJECT:** Geography - Periodicals - Switzerland.

1790 Versuch einer gründlichen Erläuterung der merkwürdigsten Begebenheiten in der Natur. - St.1-4, 1723. Halle: Johann Adam Spörl. Subtitle varies. **EDITOR:**

Ludwig Philipp Thümmig. SOURCES: K3183, OCLC LOCATIONS: ZDB SUBJECT: Natural history - Periodicals - Germany.

1791 Versuch einer landwirthschaftlicher Geographie, oder, Nachrichten von der Landwirthschaft einzelner Lder und Landgutten. - v.1, 1795. Leipzig: 1795. LOCATIONS: BUC SUBJECT: Geography - Periodicals - Germany.

1792 Versuch eines Magazin für die Arithmetik. - St.1-2, 1785-1787. Celle; Leipzig: W. Nauck, 1785-1787. EDITOR: Georg Friedrich Petersen. SOURCES: K3308 LOCATIONS: ZDB SUBJECT: Mathematics - Periodicals - Germany.

1793 Versuche und Abhandlungen der naturforschenden Gesellschaft zu Danzig. - v.1-3, 1747-1756. Danzig: Schreiber, 1747-1756. Imprint varies. CONTINUED BY: Neue Sammlung von Versuchen und Abhandlungen der naturforschenden Gesellschaft in Danzig. MICROFORMS: RM(3). SOURCES: G482, K3199, S2485d LOCATIONS: BUC, CCF, CCN, ULS, ZDB SUBJECT: Natural history - Societies - Danzig.

1794 Vervolg der prijsvragen / Nederlandsche maatschapij voor nijverheid en handel, Haarlem. - v.1-47, 1778-1885. Haarlem: 1778-1885. OTHER TITLE: Prijsvragen, Nederlandsche maatschapij voor nijverheid en handel, Haarlem. LOCATIONS: BUC, ULS SUBJECT: Economics - Societies - Haarlem - Prize essays.

1795 Verzameling van berichten over eenige onderwerpen der navigatie, meest uit rapporten en journaalen van Nederlands de zielieder getrokken. - v.1-2, 1788-1822. Amsterdam: Van Keulen, 1788-1822. CONTINUED BY: Berigten en verhandelingen over eeinge onderwerpen der zeevaart. LOCATIONS: BUC, CCN, ULS, ZDB SUBJECT: Naval art and science - Periodicals - Netherlands.

1796 Verzameling van genes- heel- en artzneneykundige aanmerkungen en waarnemingen over opmerkelyke en zeldzame zaaken en gevallen betreffend deese weetenschappen in't Fransch te Parys uiltgegeven. - v.1-2, 1755-1756. Delft: Reiner Bortet, 1755-1756. Compiled by Jean François Simon. SOURCES: G316 SUBJECT: Medicine - Periodicals - France - Translations.

1797 Verzeichniss Oberlausitzischer Urkunden / Oberlausitzische Gesellschaft der Wissenschaften, Görlitz. - v.1-2(nos.1-20)1799-1824. Görlitz: 1799-1824. LOCATIONS: ULS SUBJECT: General - Societies - Görlitz.

1798 Vetenskaps-handlingar för läkare och fältskärer. - v.1-7, 1793-1805. Stockholm: Johan C. Holberg, 1793-1805. EDITOR: Sven Anders Hedin. CONTINUED BY: Vetenskaps-journal for läkare och fältskärer. SOURCES: G117 LOCATIONS: BUC, ULS SUBJECT: Medicine - Periodicals - Sweden.

1799 Le Vigneron champenois. - No.1-42, 1774-1915. Chalons-sur Marne: 1774-1915. LOCATIONS: ULS SUBJECT: Agriculture - Periodicals - France.

1800 Der Volksfreund, eine Zeitschrift für den Handwerks und Landmann. - No.1-36, Jan 1- May 3, 1794. Gera: Heinsius, 1794. EDITOR: Johann Michael Armbruster. SOURCES: K3000 LOCATIONS: BUC, ULS SUBJECT: Agriculture - Periodicals - Germany. SUBJECT: Economics - Periodicals - Germany.

1801 Der Volkslehrer, Unterhaltungen aus der Religion, Natur, Kunst, Menschenkunde für Bürger und Landmann. - No.1-6, 1798-1800. Zittau: Schöps, 1798-1800. SOURCES: K5953 SUBJECT: Science - Periodicals - Germany - Popular.

1802 Vollständige Auszüge aus den besten chirurgischen Dispüten aller Akademien, nebst Anzeigen der neuesten chirurgischen Bücher. - v.1-6, 1769-1776. Leipzig; Budissin: Jacob Deinzen, 1769-1776. EDITOR: Friedrich August Weiz (Waitz). SOURCES: G317, K3555, OCLC LOCATIONS: BL, CCF, NUC SUBJECT: Surgery - Periodicals - Germany - Dissertations. SUBJECT: Surgery - Periodicals - Germany - Reviews.

1803 Vollständige Auszüge aus neueren Dissertationen, physikalisch-medizinischen Inhalts. - v.1-2, 1775-1776. Bremen: 1775-1776. EDITOR: Georg Heinrich Weiz (Waitz). SOURCES: K3572, OCLC SUBJECT: Medicine - Periodicals - Germany - Dissertations. SUBJECT: Science - Periodicals - Germany - Dissertations.

1804 Vollständige Einleitung in die Monaths-Schriften der Deutschen. - v.1-2, 1747-1753. Erlangen: Johann Caspar Müller, 1747-1753. SOURCES: K154 SUBJECT: General - Periodicals - Germany - Reviews.

1805 Vollständiger astronomischer Kalendar nach die verbesserten Stylo / Akademie der Wissenschaften, Berlin. - Berlin: 1747. LOCATIONS: ZDB SUBJECT: Astronomy - Periodicals - Germany.

1806 Vorlesungen der Churfpälzischen physicalisch- ökonomischen Gesellschaft. - Bd.1-Bd.5, (1784-1790) 1786-1791. Mannheim: Schwan und Götz, 1786-1791. CONTINUATION OF: Bemerkungen der Churpfälzischen physikalisch-ökonomischen Gesellschaft, Mannheim. SOURCES: K2952, OCLC, S3075b LOCATIONS: BUC, CCF, CCN, ULS SUBJECT: Science - Societies - Mannheim. SUBJECT: Bee culture - Societies - Mannheim.

1807 Die Vorzüglichten Vorlesungen welche der könglichen schwedischen Akademie der Wissenschaften zu Stockholm gehalten worden sind. - v.1-2, 1794-1795. Leipzig: 1794-1795. EDITOR: Gröning. TRANSLATION OF: Handlingar, Kungliga vetenskapsakadaemien, Stockholm. SOURCES: B4709, S3014 SUBJECT: Science - Societies - Stockholm - Translations.

1808 Vytgeleezene naturkundige verhandelingen. - Amsterdam: 1735. SOURCES: B4716, S787 SUBJECT: Natural history - Periodicals - Netherlands.

1809 Weekeljks discours over de pest en alle pestilentiale ziekten. - No. 1-30, Nov.17, 1721- June 9, 1722. Amsterdam: J. Oosterwijk, 1721-1722. Weekly.

OTHER TITLE: Weekeljks discours over de grasserende, mitagaders de byzondere land-kivaalen van elke natie enz.   SOURCES: GIII:4 LOCATIONS: ULS   SUBJECT: Medicine - Periodicals - Netherlands. SUBJECT: Smallpox - Periodicals - Netherlands.

1810 Weekly memorials for the ingenious (Chiswell ed.). - No.1-29, Mar. 20-Sept.20, 1682. London: R. Chiswell, Thomas Basset, W. Crook and Sam Crouch, 1682. Weekly.   LOCATIONS: BUC, ULS   SUBJECT: General - Periodicals - England.   SUBJECT: Science - Periodicals - England.

1811 Weekly memorials for the ingenious (Faithorne ed.). - No.1-50, Jan. 16, 1682-Jan. 16, 1683.   London: Henry Faithorne and John Kersey, 1682-1683. Weekly.   MICROFORMS: UM(1), RP(1).   LOCATIONS: BUC, CCF, CCN, ULS   SUBJECT: Science - Periodicals - England.   SUBJECT: Medicine - Periodicals - England.

1812 Weekly miscellany for the improvement of husbandry, trade, arts and sciences. - No.1-22, Jul. 4-Nov.28, 1727. London: Samuel Richardson, 1727. Weekly. In part a reprint of Collection of letters for the improvement of agriculture (Ref.121,p.82).   EDITOR: Richard Bradley (Ref.120, p.34).   LOCATIONS: BUC, ULS   SUBJECT: Agriculture - Periodicals - England.

1813 Weekly observations / Dublin Society. - No.1-52, Jan. 4-June 28, Oct. 11, 1737-Apr. 4, 1738. Dublin: R. Reilly, 1737-1738. Weekly. July-Sept. 1737 not published. Originally printed in the Dublin newsletter (Ref.122). Reprinted in 1739 and in other editions in 1740, 1753, 1753 and 1756, as well as under the title: Essays and observations (London, 1739) and in a French ed. (Paris, 1740) (Ref.106) Glasgow edition in 1756 (Ref.6, p.286).   OTHER TITLE: Dublin society's weekly observations.   SOURCES: G500, OCLC   LOCATIONS: BUC, ULS   SUBJECT: Agriculture - Societies - Dublin.

1814 Die Welt in Bildern, forzüglich zum Vergnügen und Unterricht der Jugend. - v.1-8, 1787-1793. Vienna: Baumeister, 1787-1793. A book issued in weekly parts to make 8 vols.   EDITOR: Joseph Anton Ignaz Baumeister.   SOURCES: K3321, OCLC   SUBJECT: Science - Periodicals - Austria - Popular.

1815 Werken / Genootschap der beschouwende en werkdadige wiskunde, onder de spreuk mathesis scientiarum genitrix. - v.1, 1794. Leyden: 1794.   SOURCES: S871a   SUBJECT: Mathematics - Societies - Leyden.

1816 Werken / Maatschappij der wis-, bouw-, natuur-, en teekenkunde, onder de spreuk: de wiskunde is de moeder der wetenschappwen. - v.1, 1796. Leyden: 1796.   SOURCES: S873a   SUBJECT: Technology - Societies - Leyden. SUBJECT: Science - Societies - Leyden.

1817 Western Apiarian Society transactions. - Pt.1-7, 1800-1809. Exeter: Trewman, 1800-1809.   LOCATIONS: BUC   SUBJECT: Bee culture - Societies - Exeter.

1818 Westphälisches Magazin zur Geographie, Historie und Statistik. - v.1-4 (no.1-16) 1784-1788. Lemgo; Leipzig: Meyer, 1784-1788. Imprint varies. **EDITOR:** Peter Florenz Weddigen. **CONTINUED BY:** Neues Westphälisches Magazin zur Geographie, Historie und Statistik. **SOURCES:** K1153 **LOCATIONS:** BUC, CCF, ZDB **SUBJECT:** Geography - Periodicals - Germany.

1819 Wiener medicinische Monatsschrift. - v.1-4 (St. 1-12) 1789. Vienna: Joseph Stahael, 1789. Monthly. **EDITOR:** Georg Ernst Kletten. **SOURCES:** G82, K3654 **LOCATIONS:** ULS, ZDB **SUBJECT:** Medicine - Periodicals - Austria.

1820 Wiener ökonomische Zeitung, oder, allmonatliche Anzeigen heutiger Erfahrungen und Erfindungen. - v.1-6, 1785-1790. Vienna: Gottfried Herrmann, 1785-1790. Monthly. Imprint varies. **OTHER TITLE:** Gelehrte Nachrichten über Arbeiten deutscher Gelehrter. **EDITOR:** Johann Joseph Kausch. Supplement published 1786-1787 under the title: Gelehrte Nachrichten über die Arbeiten deutscher Gelehrter zur ökonomie, Physik und Medicien. **SOURCES:** B4772, K2940, K3313 **LOCATIONS:** ULS **SUBJECT:** Economics - Periodicals - Austria. **SUBJECT:** Medicine - Periodicals - Austria. **SUBJECT:** Science - Periodicals - Austria.

1821 Wienerische Beiträge zur praktischen Arzneikunde, Wundarneikunst und Geburtshülfe. - v.1-2, 1781-1783. Vienna: Rudolph Gräfer, 1781-1783. **EDITOR:** Josef Jacob Mohrenheim. **SOURCES:** G45, K3589 **LOCATIONS:** ULS, ZDB **SUBJECT:** Medicine - Periodicals - Austria. **SUBJECT:** Surgery - Periodicals - Austria. **SUBJECT:** Obstetrics - Periodicals - Austria.

1822 Wienerische Nachrichten und Abhandlungen aus dem Oeknomie und Commerzwesen. - v.1-3, 1767-1769. Vienna: Trattner, 1767-1769. **LOCATIONS:** ZDB **SUBJECT:** Commerce - Periodicals - Austria.

1823 Wienerische Realzeitung, oder, Beiträge und Anzeigen von gelehrten und Kunstsachen. - v.1-15, 1771-1785. Vienna: Heubner, 1771-1785. **EDITOR:** Hohann Reutenstauch and others. **SOURCES:** (Ref.21,25:456). **SUBJECT:** General - Periodicals - Austria - Reviews.

1824 Der Wirth und die Wirthin, eine ökonomische und moralische Wochenschrift. - v.1 (St.1-12) 1757. Braunschweig: Schröder, 1757. Weekly. **OTHER TITLE:** Braunschweigische Sammlung von ökonomischen Dingen. **SOURCES:** K4982 **SUBJECT:** Economics - Periodicals - Germany - Popular.

1825 Wiskunstig mengelwerk in eene aaneenschakeling van uitgeleezene vorstellen / Genootschap der mathematische wetenschappen,onder despreuk: een onvermoeide arbeid komt alles te boven. - v.1-2, 1796-1802. Amsterdam: 1796-1802. **CONTINUATION OF:** Wiskuntsige verlustiging in een aaneenschakeling van uitgeleezene voorstellen. **OTHER TITLE:** Mengelwerk van uitgeleezene en andere wis-en naturkundige verhandelingen. **SOURCES:**

OCLC, S735c **LOCATIONS:** BUC, CCF, CCN **SUBJECT:** Mathematics - Societies - Amsterdam.

1826 Wiskuntsige verlustiging in een aaneenschakeling van uitgeleezene voorstellen / Genootschap der mathematische wetenschappen, onder de spreuk: een onvermoeide arbeid komt alles te boven. - v.1-2, 1793-1795. Amsterdam: 1793-1795. **CONTINUED BY:** Wiskunstig mengelwerk in eene aaneenschakeling van uitgeleezene vorstellen. **MICROFORMS:** RM(3). **SOURCES:** OCLC, S735d **LOCATIONS:** BUC, CCF, CCN **SUBJECT:** Mathematics - Societies - Amsterdam.

1827 Wissenschaftliches Magazin für Aufklärung. - v.1-3, 1785-1788. Leipzig: Jacobäer, 1785-1788. Imprint varies. **EDITOR:** Ernst Ludwig Posselt. **SOURCES:** K390 **LOCATIONS:** ULS, ZDB **SUBJECT:** Science - Periodicals - Germany - Popular.

1828 Wittenbergisches Magazin für die Liebhaber der philosophischen und schönen Wissenschaften. - v.1-4, 1781-1784. Berlin: Christian Ludwig Stahlbaun; Wittenberg: Carl Christian Dürr, 1781-1784. **EDITOR:** Johann Jacob Ebert. **SOURCES:** B4782, K4519, S2398 **LOCATIONS:** CCN, ULS, ZDB **SUBJECT:** Science - Periodicals - Germany - Popular.

1829 Wittenbergisches Wochenblatt zum Aufnahmen der Naturkunde und des ökonomischen Gewerbes. - v.1-25, 1768-1792. Wittenberg: Dürr, 1768-1792. **EDITORS:** Johann Daniel Titius and Carl Heinrich Ludwig Pölitz. **CONTINUED BY:** Neues Wittenbergisches Wochenblatt. **SOURCES:** B4783, K2895, S3324 **LOCATIONS:** CCF, ULS, ZDB **SUBJECT:** Natural history - Periodicals - Germany. **SUBJECT:** Economics - Periodicals - Germany.

1830 Wochenblatt des aufrichtigen Volksarztes. - v.1-6 (12 issues each) 1796-1798. Eisenberg; Leipzig: Grieshammer, 1796-1798. **EDITOR:** Daniel Collenbusch. **CONTINUED BY:** Rathgeber für alle Stände in Angelegenheiten die Gesundheit. **SOURCES:** G202, K3706 **LOCATIONS:** CCF **SUBJECT:** Medicine - Periodicals - Germany - Popular.

1831 Wochenblatt für die Liebhaber der Geschichte der Naturkunde, der Weltweisheit und der schönen Wissenschaften. - Pressburg (?): 1774-1787. Weekly. **SOURCES:** (Ref.116,p.215). **SUBJECT:** Science - Periodicals - Hungary - Popular.

1832 Wochenschrift für die Liebhaber der Geschichte, der Erdbeschreibung, der Naturkunde, der Weltweissheit und der schönen Wissenschaften. - v.1, 1779. Cenburg: J.J. Siesz, 1779. Weekly. **SOURCES:** K5537 **SUBJECT:** Science - Periodicals - Germany - Popular.

1833 Wochenschrift für Kaufleute. - No.1,1795. Berlin: Langhoff, 1795. Weekly. **EDITOR:** J.M.F. Schulze. **SOURCES:** K2774 **SUBJECT:** Commerce - Periodicals - Germany.

1834 Wöchentliche Anzeigen zum Vortheil der Liebhaber der Wissenschaften und Künste. - v.1-3, 1764-1766. Zürich: Heidegger, 1764-1766. **CONTINUATION OF:** Freimüthige Nachrichten von neuen Büchern und anderen zur Gelahrtheit gehörigen Sachen. **EDITOR:** Johann Conrad Füssli. **SOURCES:** G752, K241 **LOCATIONS:** BUC, ZDB **SUBJECT:** General - Periodicals - Switzerland - Reviews.

1835 Wöchentliche Hallische Anzeigen. Halle: 1729-1810. Weekly. A selection of its medical articles was published in 2 v.in Halle, 1788- 1789 under the title: Auszüge verscheidener arzneiwissenschaftliche Abhandlungen aus den wöchentlichen Hallischen Anzeigen (Ref.49). **EDITOR:** Johann Peter von Ludewig. **SOURCES:** B4795, S2739 **LOCATIONS:** CCF, ZDB **SUBJECT:** General - Periodicals - Germany.

1836 Wöchentliche Nachrichten für die deutschprechenden Einwohner Frankreichts, besonders für die Handwerker und Bauern. - Strasbourg: 1790-1791. Weekly. **LOCATIONS:** CCF **SUBJECT:** Economics - Periodicals - France.

1837 Wöchentliche Nachrichten von neuen Landcharten, geographischen, statistischen und historischen Büchern und Sachen. - v.1-16, 1773-1788. Berlin: Haude und Spener, 1773-1788. Weekly. **OTHER TITLE:** Nachrichten von neuen Landcharten. **EDITOR:** Anton Friedrich Büsching. **SOURCES:** K1069 **LOCATIONS:** ULS, ZDB **SUBJECT:** Geography - Periodicals - Germany Reviews.

1838 Wöchentliche oberdeutsche Oekonomie-, Kunst-, Gewerbe- und Gesundheits-Universal-Zeitung. - Vienna: 1788. Weekly. **SOURCES:** (Ref.111:3430) **SUBJECT:** Economics - Periodicals - Austria. **SUBJECT:** Medicine - Periodicals - Austria - Popular.

1839 Wöchentliche Relationen der merkwürdigsten Sachen aus dem Reiche der Natur, der Staaten und der Wissenschaft. - v.1-2, 1794. Berlin: 1794. Weekly. **SOURCES:** B4796, S2399 **SUBJECT:** General - Periodicals - Germany - Reviews.

1840 Wöchentliche Unterhaltungen über die Charakteristik der Menschheit. - v.1-6, 1789-1791. Berlin: Mauer, 1789-1791. Weekly. **CONTINUATION OF:** Wöchentliche Unterhaltungen über die Erde und ihre Bewohner. **EDITOR:** Georg Wilhelm Bartholdy and Johann Friedrich Zöllner. **SOURCES:** B4798a, G244, K3333, S2400 **LOCATIONS:** BL, ZDB **SUBJECT:** Geography - Periodicals - Germany.

1841 Wöchentliche Unterhaltungen über die Erde und ihre Bewohner. - v.1-10, 1784-1788. Berlin: Mauer, 1784-1788. Weekly. **EDITORS:** Georg J.S. Lange and Johann Friedrich Zöllner. **CONTINUED BY:** Wöchentliche Unterhaltungen über die Charakteristik der Menschheit. **SOURCES:** B4798, G242, K3305, S2400 **LOCATIONS:** BUC, CCF, ZDB **SUBJECT:** Geography - Periodicals - Germany.

1842 The World displayed, or, a curious collection of voyages and travels, selected from the writers of all nations. - v.1-20, 1759-1761. London: J. Newberry, 1757-1761. Several editions are recorded. LOCATIONS: NUC SUBJECT: Geography - Periodicals - England. SUBJECT: Travel - Periodicals - England.

1843 Der Wundarzt, eine Wochenschrift zur Beförderung der Kenntniss des menschlichen Körper. - v.1-2, 1789-1793. Leipzig: Schweikert, 1789-1793. Weekly. Imprint varies. EDITOR: Christian Gottfried Roth. SOURCES: G80, K3647 SUBJECT: Surgery - Periodicals - Germany.

1844 Würzburger gelehrte Anzeigen. - v.1-11, 1786-1796. Würzburg: Rienner, 1786-1796. Weekly. CONTINUED BY: Würzburger wöchentliche Anzeigen von gelehrten und andern gemeinnützigen Gegenständen. CONTINUED BY: Neue Würzburger gelehrte Anzeigen. SOURCES: B4808, K397, S3333 LOCATIONS: ZDB SUBJECT: General - Periodicals - Germany - Reviews.

1845 Würzburger wöchentliche Anzeigen von gelehrten und andern gemeinnützigen Gegenständen. - v.1-2, 1797-1798. Würzburg: Köl, 1797-1798. Weekly. CONTINUATION OF: Würzburger gelehrte Anzeigen. CONTINUED BY: Neue Würzburger gelehrte Anzeigen. SOURCES: B4808a, K452, S333.2 LOCATIONS: ZDB SUBJECT: General - Periodicals - Germany - Reviews.

1846 Zapiski russkich doktorov.- SOURCES: (Ref.124,p.134). SUBJECT: Medicine - Periodicals - Russia.

1847 Zeitschrift für die Naturkunde. - v.1, 1792-1795. Leipzig: 1792-1795. EDITOR: Johann Christian Schedel. SOURCES: G754 SUBJECT: Natural history - Periodicals - Germany.

1848 Zeitschrift für die spekulative Physik. - v.1-2, 1800-1801.Jena; Leipzig: Christian Ernst Gabler, 1800-1801. CONTINUATION OF: Gnothi sauton. CONTINUATION OF: Psychologisches Magazin. EDITOR: Friedrich Wilhelm Joseph von Schelling. CONTINUED BY: Neue Zeitschrift für die Spekulative Physik. MICROFORMS: RM(3). SOURCES: B4884, G252a, K587, OCLC, S2837 LOCATIONS: BUC, CCN SUBJECT: Psychology - Periodicals - Germany.

1849 Zeitung für Fabriken und Manufakturen in dem Kaiserlichen Königlichen Staaten. - Weekly. SOURCES: K2753 SUBJECT: Manufactures - Periodicals - Germany.

1850 Zodiaco Lusitanico-delphico, anatomico, botanico, chirurgico, chymico, dendrologico, ictyologico, leithologico, medico, meteorologico, optico, ornithologico, phamaceutico e zoologico / Academia dos Esconidos, Porto. - No.1, 1749. Porto: 1749. SOURCES: (Ref.125). SUBJECT: Science - Periodicals - Portugal.

1851   Zodiacus medico-gallicus. - v.1-5 (1679-1783) 1680-1685.   Geneva: Leonard
Choét, 1680-1685.   Translated by Théophile Bonet.   **TRANSLATION OF:**
Nouvelles déscouvertes sur toutes les parties de la medécine. **MICROFORMS:**
RM(3).   **SOURCES:** B4933, GI:4a, S2164-65   **LOCATIONS:** BUC, CCF,
CCN, ULS   **SUBJECT:** Medicine - Periodicals - France - Translations.

1852   Der Zoologe, oder, compendieuse Bibliothek der Wissenswürdigsten aus der
Thiergeschichte und allgemeinen Naturkunde. - No.1-8, 1795-1797.   Eisenach;
Halle: Johann Jacob Gebauer, 1795-1797.   **EDITOR:** Christian Carl André.
SOURCES: B4936, G574, S2540   **SUBJECT:** Zoology - Periodicals -
Germany.

1853   Zoologische Annalen. - **CONTINUATION OF:** Magazin für die Thiergeschichte,
Thieranatomie und Thierarzneikunde.   **EDITOR:** Friedrich Albrecht Anton
Meyer.   **CONTINUED BY:** Zoologisches Archiv.   **INDEXES:** Indexed in
Wimmel (Ref.7).   **SOURCES:** B2791a, G575, K3353, OCLC, S3313
LOCATIONS: BUC, ULS, ZDB   **SUBJECT:** Veterinary medicine -
Periodicals - Germany.

1854   Zoologisches Archiv.- v.1-2, 1795-1796.   Leipzig: Dyck, 1795-1796.
CONTINUATION OF: Magazin für die Thiergeschichte, Thieranatomie und
Thierarzneikunde.   **CONTINUATION OF:** Zoologische Annalen.   **EDITOR:**
Friedrich Albrecht Anton Meyer.   **INDEXES:** Indexed in Wimmel (Ref.7).
SOURCES: B2791b, G576, K3365, S3048 **LOCATIONS:** BUC, ULS, ZDB
SUBJECT: Veterinary medicine - Periodicals - Germany.

1855   Zryelishche prirodui i khudozhestv.- v.1-10, 1784-1790.   St. Petersburg:
1784-1790.   **TRANSLATION OF:** Schauplatz der Natur und Kunst.
LOCATIONS: BUC   **SUBJECT:** Natural history - Periodicals - Germany -
Translations.

1856   An die Zürcherische Jugend / Naturforschende Gesellschaft in Zürich. - St.1-76,
1799-1874.   Zürich: 1799-1874.   Annual.   Educational pamphlets addressed to
children.   **OTHER TITLE:** Neujahrsblatt, Naturforschende Gesellschaft in
Zürich.   **SOURCES:** OCLC, S2206g   **LOCATIONS:** BUC, CCF, CCN,
ULS, ZDB   **SUBJECT:** Science - Societies - Zürich - Popular.

1857   Zustand der Wissenschaften und Künste in Schwaben. - St. 1-3, 1781-1782.
Augsburg: Stage; Mannheim: Löffler, 1781-1782.   **CONTINUATION OF:**
Gelehrte Ergötzlichkeiten und Nachrichten.   **CONTINUATION OF:**
Schwäbisches Magazin von gelehrten Sachen.   **EDITOR:** Balthasar Haug.
SOURCES: B4953, G755, K356, S2253   **LOCATIONS:** BL, CCF, ULS
SUBJECT: General - Periodicals - Germany - Reviews.

1858   Zuverlässigen Nachrichten der Litteratur in Europa, Veränderung und Wachstum
der Wissenschaften. - v.1-13, 1740-1757.   Leipzig: Gleditsch, 1740-1757.
CONTINUATION OF: Deutsche acta eruditorum oder, Geschichte der
Gelehrten.   Subtitle varies.   **EDITOR:** Christian Gottlieb Jöcher.   **SOURCES:**

B4954, K511, OCLC, S3051 **LOCATIONS:** CCF, ULS, ZDB **SUBJECT:** General - Periodicals - Germany - Reviews.

SUBJECT INDEX

Agriculture - Societies - Rennes 500
Agriculture - Societies - Rouen 511
Agriculture - Societies - St. Petersburg
  - Translations 7
Agriculture - Societies - Scotland
  1650
Agriculture - Societies - Seine-et-Oise
  1065
Agriculture - Societies - Seine
  inferieure 1467
Agriculture - Societies - Steyermark
  1601
Agriculture - Societies - Stockholm
  1563
Agriculture - Societies - Strasbourg
  1723
Agriculture - Societies - Tours 1510
Agriculture - Societies - Turin 175
  1123
Agriculture - Societies - Udine 1125
Agriculture - Societies - Venice 1483
Agriculture - Societies - Zürich 17
Agriculture - Societies - Zürich - Prize
  essays 151
Alchemy - Periodicals - England 244
  493
Alchemy - Periodicals - France 227
Alchemy - Periodicals - Germany
  219 459 524 732 733 1141
  1617
Alchemy - Periodicals - Germany -
  Reviews 1212 1247
Anatomy - Periodicals - Germany
  294
Anatomy - Societies - Amsterdam
  1359
Animal magnetism - Periodicals -
  Germany 210 332 961 1255
  1727
Animal magnetism - Societies -
  Strasbourg 177
Anthropology - Periodicals - England
  962
Anthropology - Periodicals - Germany
  330 941 967 1283 1436
Anthropology - Periodicals - Russia
  1239

Antiquities - Societies - Cortona 1557
Archeology - Societies - Edinburgh
  200
Archeology - Societies - London 199
  773
Archeology - Societies - Naples 192
Architecture - Periodicals - Germany
  1598
Architecture - Societies - Paris 1469
Asia - Periodicals - England 1391
  1392
Astrology - Periodicals - England
  244 423 430 493
Astrology - Periodicals - France 227
  1178
Astronomy - Periodicals - Austria
  556 704 1669
Astronomy - Periodicals - England
  423 1208
Astronomy - Periodicals - France 557
Astronomy - Periodicals - Germany
  487 1135 1165 1531 1686 1805
Astronomy - Periodicals - Italy 547
  679 1371
Astronomy - Periodicals - Netherlands
  1314
Astronomy - Societies - Berlin 246
  1568
Astronomy - Societies - Lisbon 561
Astronomy - Societies - London 245
Astronomy - Societies - Marseilles
  1083
Astronomy - Societies - Paris 179
  189 494 583 1513 1527
Astronomy - Societies - San Fernando
  134
Astronomy - Societies - Spain 1357
Balloon ascension - Periodicals -
  France 119
Balloon ascension - Periodicals - Italy
  678
Balneology - Periodicals - Germany
  642
Bee culture - Periodicals - Germany
  125 837 1411
Bee culture - Societies - Exeter 1817

PERSONAL NAME INDEX

INSTITUTIONAL NAME INDEX

TITLE INDEX

APPENDIX I: MICROFORM PUBLISHERS AND FORMATS

Formats (Ref. No. 150):

1. 35mm microfilm on reels
2. Micro-opaque cards, 75 x 125mm (3x5")
3. Micro-opaque cards, 6x9"
4. Micro-opaque cards, 5x8"
5. 16mm microfilm on reels
6. 16mm microfilm in cartridge or cassette'
7. Microfiche 75x125mm (3x5")
8. Microfiche, 9x12cm
9. Microfiche 105x148cm (4x6")
10. Micro-opaque cards, 105x148mm (4x6")
13. 35mm microfilm in cartridge or cassette

Publishers:

ACR   A.C.R.P.P.(Association pour la Conservation et la Reproduction Photographique de la Presse) 4, rue de Louvois, 75002 Paris, France. (North American Distributor: Clearwater Publishing Co. 1995 Broadway, New York, N.Y. 10023).

BHP   Brookhaven Press, P.O.Box 1653, La Crosse, Wisconsin, 54601

DA   Datamics, Inc. 114 Liberty Street, New York, N.Y. 10006

GO   Georg Olms Verlag AG, Olms Microform System, Hagentorwall 6-7, D-3200 Hildesheim, West Germany. (North American Distributor: Clearwater Publishing Co. Inc., 1995 Broadway, New York, N.Y. 10023).

IDC   Inter Documentation Company AG, Order Dept. Poststrasse 14, 6300 Zug, Switzerland.

LOC   Library of Congress, Photoduplication Service, Department C, 10 First Street, S.E. Washington, D.C. 20540

MHA   Microeditions Hachette, 5 Rue du Pont de Lodi, 75006 Paris, France.

MIM   Microforms International Marketing Corporation (Pergamon Press Inc.) Maxwell House, Fairview Park, Elmsford, N.Y. 10523

NYP   New York Public Library, Photograpahic Service, Room 316, Fifth Ave. and 42nd Street, New York, N.Y. 10016

PMC   Princeton Microfilm Corporation, P.O.Box 2073, Princeton, New Jersey, 08540

RM   Readex Microprint Corporation, 58 Pine Street, New Canaan, Connecticut, 06840

RP   Research Publications Inc. 12 Lunar Drive, Drawer AB, Woodbridge, Connecticut, 06525. (Research Publications Ltd., P.O.Box 45, Reading RG1, 8HF England).

SMS   Schnase Microfilm Systems, 120 Brown Road, P.O.Box 119, Scarsdale, New York. 10583.

UM   University Microfilms International, 300 North Zeeb Road, Ann Arbor, Michigan 48106 (30-32 Mortimer St. London W1N 7RA, England).

UWM   University of Wisconsin-Madison, Memorial Library, 728 State Street, Madison Wisconsin, 53706.

Y   Kraus Microform, A division of Kraus-Thomson Organization, One Water Street, White Plains, New York 10601.

APPENDIX II: ABBREVIATIONS

B   Bolton, Henry Carrington. A catalogue of scientific and technical periodicals (Ref.128)
BL   British Library, London
BN   Bibliothèque Nationale, Paris
BUC   *British union-catalogue of periodicals* (Ref.133)
CCF   *Catalogue collectif des periodiques*, Paris (Ref.134)
CCN   *Centrale catalogus van periodieken*, 'sGravenhage (Ref.144)
DSG   U.S. National Library of Medicine, Bethesda, Maryland
G   Garrison, Fielding H. "The medical and scientific periodicals of the 17th and 18th centuries." (Ref.129)
GB   Göttingen. Universitätsbibliothek
K   Kirchner, Joachim. *Bibliographie der Zeitschriften des deutschen Sprachgebietes.* (Ref.130)
LC   U.S. Library of Congress, Washington, D.C.
NUC   *The National Union Catalog; pre-1956 imprints.* London: Mansell, 1968-1981. 754v.
OCLC   Online Computer Library Center, Dublin, Ohio
S   Scudder, Samuel H. *Catalogue of scientific serials* (Ref.131)
ULS   *Union list of serials in the libraries of the United States and Canada.* (Ref.135)
ZDB   *Zeitschriften-Datenbank* (Ref.146)

REFERENCES

1. Vezényi, Pál, ed. *Bibliographia Academica Germaniae*. Münich-Pullach und Berlin: Verlag Dokumentation, 1971.
2. Hensing, Ulrich, "Acta eruditorum (1682-1782)". In: Fischer, Heinz-Dietrich, ed. "Deutsche Zeitschriften des 17 bis 20 Jahrhunderts." Pullach bei Münich: Verlag Documentation, 1973, p.29-47.
3. Beutler, J.H.C. und Gutsmuth, J.C.F.*Allgemeines Sachregister über die wichtigsten deutschen Zeit-und Wochenschriften*. Leipzig:1790.
4. Schimank, Hans. "Ludwig Wilhelm Gilbert und die Anfänge des "Annalen der Physik." *Sudhoffs Archiv*, 1963, 47:360-372.
5. Mémoires de l'Institut Nationale des Sciences et Arts. 1, 1797/8.
6. Müller, Hans Heinrich. *Akademie und Wirthschaft in 18. Jahrhunderts*. Berlin: Akademie Verlag, 1975.
7. Wimmel, Bernd and Armin Geuss, eds. *Die Tiermedizinischen Zeitschriften des 18. Jahrhunderts*. Stuttgart: Hiersemann, 1981.
8. Griffin, F.J. "The *Archiv der Insectengeschichte* of J.C. Fuessly," Heft 1-8, 1781-1796." *Journal of the Society for Bibliography of Natural History*. 1936, 1: 83-85.
9. Beckwick, Frank, *"The Bibliothèque Britannique."* *The Library*, 1932, 4th ser. 12:75-82.
10. Delaunay, Paul. *Le monde médicale Parisien aux dix-huitième siècle*. 2d ed. Paris: Rousset, 1906.
11. Saussure, R. de. "French psychiatry after the eighteenth century." *Ciba Symposium*, 1950, 11:1222-1252.
12. Archibald, Raymond C. "Notes on some minor English mathematical serials." *Mathematical Gazette*, 1929, 18:379-400.
13. Reesink, Hendrika J. *L'Angleterre et la litterature Anglais dans les trois plus anciens périodiques Française de Hollande de 1684 à 1709*. Paris: Libraire Ancienne, 1931.
14. Hannaway, Caroline C.F. *Medicine, public welfare and the state in eighteenth century France : The Société Royale de Médecine of Paris (1776-1793)*. Ph. D. Dissertation, Johns Hopkins University, 1974.
15. Gillispie, Charles C. *Science and polity in France at the end of the Old Regime*. Princeton: N.J. Princeton University Press, 1980.
16. Wilkinson, Thomas. "Mathematical periodicals." *Mechanicks Magazine*, a series of articles in; vs. 48-56, 1848-1852.
17. Lefanu, W.R. *British periodicals of medicine, 1640-1899*. Oxford: Wellcome Unit for the History of Medicine, 1984.
18. Jaeggli, Alvin E. "Recueil des pieces qui ont remporté le prix de l'Académie des Sciences depuis 1720 jusqu'en 1772." *Gesnerus*, 1975, 34:408-414.
19. Hahn Roger. *The anatomy of a scientific institution; the Paris Academy of Sciences, 1666-1803*. Berkeley, Calif.: University of California Press, 1971.
20. Barnes, Sherman B. *The beginnings of learned journalism, 1665-1730*. Ph. D. Dissertation, Graduate School, Ithaca, N.Y., Cornell University, 1934.

21. Callisen, Adolph Carl Peter. *Medicinisches Schriftsteller-lexicon der jetzt lebenden Aerzte, Wundaerzte, Geburtshelfer, Apotheker und Naturforscher aller gebildeten Völker.* Copenhagen: 1830-1845.

22. Gascoigne, Robert M. *A historical catalogue of scientific periodicals, 1665-1900.* New York: Garland, 1985.

23. Lustenberger, Fridolin. *Schweizerische Medizinisch-naturwissenschaftlilche Zeitschriften von 1751-1871, ihre Bedeutung und ihre Zeile.* Zürich: Leemann, 1927.

24. Strandell, Berger. ["An early overlooked Acta medica publication from Scandinavia; Acta medicorum Suecicorum."] In Danish. *Nordisk Medicinhist. Arsb.* 1969, Suppl.11 :30-36.

25. University of Oklahoma Library. *The catalogue of the history of science collections of the University of Oklahoma Libraries* [by] Duane H.D.Roller [and] Marcia M. Goodman. London : Mansell, 1976. 2v.

26. Kronick, David A. *A history of scientific and technical periodicals.* 2d ed. Metuchen, N.J.: The Scarecrow Press, 1976.

27. Andrade, E.N. da C. "The birth and early days of the Philosophical Transactions." *Notes and Records, Royal Society,* 1965, 20:9-27.

28. Rosen, Richard Leonard. *The Academy of Science of the Institute of Bologna, 1690-1804.* Ph. D. Dissertation, Cleveland, Ohio : Case-Western Reserve University, 1971.

29. Couperus, Marianne, ed. *L'étude des périodiques anciens. Colloque d'Utrecht.* Paris: Nizet, 1972,

30. Gore, J.H. "The Dutch Society for General Welfare." *Bulletin U.S. Department of Labor,* 1897, no.9: 130-148.

31. Franke, H. "Konrad Friedrich Uden und die erste medizinische Zeitschrift in russischer Sprache." *Zeitschrift für ärtzliche Fortbildung,* 1970, 64:736-738.

32. Müller-Dietz, Hans von. "Die erste medizinische Zeitschrift in russischer Sprache : Sankt-Peterburgskie Vedomosti (1792-1794)." *Sudhof's Archiv,* 1962, 46:229-250.

33. Harff, Horst. *Die Entwicklung der deutschen chemischen Fachzeitschrift bis zum Beginn des neunzehnten Jahrhunderts.* Dissertation, Friedrich Wilhelms-Universitäts zu Berlin, 1940.

34. Silva Pereira, A.X. da "Resenha bibliografica das publicaões mencionades." *Memorias da Academia das Ciencias de Lisboa, Classe Lettres,* 1941, 4:265-276.

35. Grant-Carteret, John. *Les almanachs francais; bibliographie-iconographie des almanachs, années, annuaires, calendriers, chansonniers, étrennes, états, heures, listes, livre d'addresse, tableaux, tablettes et autres publications annuelle edittes à Paris (1600-1895).* Paris: J. Alisie, 1896.

36. Guitard, Eugene. *Deux siècles de presse au service de la pharmacie et cinquante ans de "L'Union Pharmaceutique."* Paris: Pharmacie Central de France, 1913.

37. Kohring, Hans, ed. *Bibliographie der Almanache, Kalender und Taschenbücher für die Zeit von ca 1750-1860.* Hamburg: 1929.

38. Ballenger, Claude, et.al. eds. *Histoire gènèrale de la presse francaise.* Tome. I. *Des origines à 1814.* Paris: Presses Universitaires de France, 1967.

39. Rudio, F. *Festschrift der Naturforschenden Gesellschaft in Zürich, 1746-1896.* Zürich: Zurcher und Furrer, 1896.

40. Rubberdt, Rudolph. *Die ökonomische Sozietäten; ein Beitrag zur Wissenschaftsgeschichte des XVIII Jahrhunderts.* Würzburg: Trilitsch, 1934.

41. Biographische Lexikon der hervorragenden Aerzte aller Zeiten und Völker. 3d ed. Münich: Urban und Schwarzenberg, 1962,

42. Snorrason, E. "Danish physicians and the periodicals of the seventeenth and eighteenth centuries." *Danish Medical Bulletin*, 1958, 5:200-209.

43. Deniker, Joseph. *Bibliographie des travaux scientiques publiès par les sociétés savantes de la France depuis l'origine jusqu'en 1888.* Paris: Imprimerie Nationale, 1918-1922. 2v.

44. Guerra, Francisco. "El germen del periodismo Americano." *MD en Español*, 1978, June, p.13.

45. Middleton, W.E.Knowles. *The experimenters; a study of the Accademia del Cimento.* Baltimore: The Johns Hopkins Press, 1971.

46. [A critical review of medical journalism.] *Medicinisches Argos*, 1839, 1 :61ff.

47. Smeaton, W.A. "*L'Avantcoureur*; the journal in which some of Lavoisier's earliest research was published." *Annals of Science*, 1957, 13 :219-234.

48. Almhult, Artur, Arne Holmberg and Adolf Schüch, *Academies in Sweden.* Gdynia: Baltic Institute, 1937.

49. Sudhoff, Karl Friedrich Jacob. "Das medizinische Zeitschriftenwesens in Deutschland bis zur Mitte der 19 Jahrhunderts." *Münchener medizinischer Wochenschrift*, 1903, 50:455-463.

50. Haskell, Daniel C. *A checklist of cumulative indexes to individual periodicals in the New York Public Library.* New York: New York Public Library, 1942.

51. Engelhardt, Dietrich von. *Die chemischen Zeitschriften des Lorenz von Crell.* Stuttgart: Hiersemann, 1974.

52. Querner, Hans. "Naturgeschichte an der Kuhrpfälzischen Akademie der Wissenschaften." *Heidelberg Jahrbuch*, 1977,21:47-63.

53. Geuss, Armin, ed. *Indice naturwissenschaftlaich-medizinischer Periodica bis 1850.* Stuttgart: Hiersemann, 1971-

54. Lundstedt, Bernhard. *Sveriges periodiska Literatur, 1645-1899.* Stockholm: Bokförlegt Rediviva, 1969. 3v. (Reprint of 1895 ed.)

55. Hocks, Paul and Peter Schmidt. *Index zu deutschen Zeitschriften des Jahre 1773-1830.* Nendeln: KYO Press, 1979.

56. Belozubov, Leonid. *L'Europe savante (1718-1726).* Paris: A.G.Nizet, 1968.

57. Barblan, Marc A. "La santé publique vue par les rédacteurs de la *Bibliothèque Britannique* (1796-1815)." *Gesnerus*, 1975, 32: 129-146.

58. Bickerton, David M. "A scientific and literary periodical, the *Bibliothèque Britannique* 1796-1815; its foundation and early development." *Revue de literature comparée.* 1972, 4:527-547.

59. Hubrig, Hans. *Die patriotischen Gesellschaften des 18. Jahrhunderts.* Weinheim: Julius Beltz, 1957. (Göttingen Studien zu Pädogogik, Heft 36)

60. Fisch, Max H. "The Academy of Investigators." In: Underwood,E.A. ed. *Science, medicine and history.* London: Oxford University Press, 1953, v.1, p.521-63.

61. Berthelot, M. "Sur les publications de la Société Philomathique et sur ses origines." *Journal du savants*, 1888, 477-493.

62. Guinard, Paul J. *La presse Espagnole de 1737 à 1791, formation et signification d'un genre*. Paris: Centre de Recherche Hispanique, Institut d'Études Hispanique, 1973.

63. Blumer, George. "Some remarks on 'Cases and observations' by the Medical Society of New Haven County." *Bulletin of the Chicago Society for the history of medicine*, 1921, 2:205-18.

64. Solomon, Howard M. *The innocent inventions of Théophraste Renaudot; public welfare, science and propaganda*. Ph. D. Dissertation, Northwestern University, 1969. pp.53-93.

65. Heym, Gerard. "An alchemical journal of the eighteenth century." *Ambix*, 1937/8, 1:197-99.

66. Widdess, John David Henry. "An unrecognized medical periodical." *Irish Journal of Medical Science*, 1965, Ser.6, No.356:367-79.

67. Peumery, Jean Jacques. "Conversations médico-scientifique de l'Académie de l'Abbé Bourdelot (1610-1685)." *Histoire de sciences médicale*, 1978, 12:127-35.

68. Mendez Alvaro, Francisco. *Breves apuntes para historia del periodismo medico y farmaceutico en España*. Madrid: Enrique Teodoro, 1883.

69. American Philosophical Society, Philadelphia. *Proceedings*, 1884, v.22.

70. Dressler, Adolph. *Geschichte der italienische Presse*. 2d ed. Münich: R. Oldenbourg, 1933.

71. Traumüller, Friedrich. *Die Mannheimer meteorologische Gesellschaft (1780-1795)*. Leipzig: Dürr, 1884.

72. Couper, William James. *Edinburgh periodical press*. Stirling, Scotland: Mackay, 1908.

73. Fussell, G.E. "Old farming journals." *Economic Review*, 1932, 3:417-22.

74. Sprugge, Squire. "Medical journalism." *Glasgow Medical Journal*, 1928, 109:110-19.

75. Delprat, C.C. "De geschiedenis der Nederlandsche geneeskundige tidjschriften van 1689 tot 1857." *Neder. Tschr. Geneesk*. Eerste Helft A, 1927, 71:3-116.

76. *Catalgo dei periodici possedutti dell' Università dalla Biblioteca Laurenziana, dalle Academia e da oltre istitutzione de Firenze*. Firenze: Universià degli studi di Firenze, 1963.

77. Schrimpf, Hans Joachim. "Das *Magazin zur Erfahrungseelenkunde*." *Zeitschrift für Deutsche Philologie*, 1980, 99:161-87.

78. Todd, William B. "A bibliographic account of the *Gentleman's Magazine*, 1731-1754." *Studies in Bibliography*, 1965, 18:81-109.

79. Querner, Hans. "Naturgeschichte in der Kuhrpfälzischen Akademie der Wissenschaften." *Heidelberg Jahrbuch*, 1977, 21:47-63.

80. Bond, Richard P. ed. *Studies in the early English periodicals*. Chapel Hill: University of North Carolina Press, 1957.

81. Howe, M.A. *The Humane Society of the Commonwealth of Massachusetts*. Boston: 1918.

82. Cust, L. *History of the Society of Dilletanti*, London. London: Macmillan, 1898.
83. Court, Susan. "Fourcroy and the *Journal de la Société de Pharmaciens de Paris*." *Ambix*, 1979, 26:39-55.
84. Sgard, Jean. "Chronologie des Mémoires de Trévoux." *Dix-huitieme siècle*, 1976, 8:189-192.
85. Doneaud du Plan, "L'Académie de marine de 1752 à 1785." *Revue maritime et coloniale*, 1878, 58:476-509.
86. Charlier, Gustav and Roland Mortier. *Le Journal encyclopédique (1756-1793)*. Paris: Nizet, 1952.
87. Scheler, Lucien. *Lavoisier et la revolution Française*. Paris: Hermann, 1957.
88. Smeaton, W.A. "The Lycée and the Lycée des Arts." *Annals of Science*, 1955, 11:257-67, 1956, 12:309-19.
89. Morgan, Betty R. *Histoire du Journal des Sçavans depuis 1665 jusqu'en 1701*. Paris: Presse Universitaires de France, 1929.
90. Duveen, Denis I. and Herbert S. Klickstein. "*Le Journal polytype des sciences et des arts*." *Papers of the Bibliographical Society of America*, 1954, 48:402-13.
91. Hatin, Eugéne. *Bibliographie historique et critique de la presse périodique Francaise*. Paris: Firmin-Didot, 1866. (Reprinted 1965)
92. Levi-Malvino, Ettore. "Les éditions toscana de l'Éncyclopedie." *Revue de litterature comparée*, 1924, 3:215-56.
93. Delépine, Marcel. "Les Annales de chimie de leur fondation à la 173e année de leur parution." *Annales de chimie*, 1962, ser.13, 7:1-11.
94. Franke, H. "Die ersten deutsch-sprachigen medizinischen Periodica in Russland." *Zeitschrift für ärztliche Fortbildung*, 1970, 64:313-18.
95. Brandstetter, J. L. *Bibliographie des Gesellschaften, Zeitungen und Kalender in der Schweiz*. Bern: 1896.
96. Nemec, Jaroslav. *International bibliography of medicolegal serials, 1753-1967*. Bethesda, MD: National Library of Medicine, 1969.
97. Lopez Pinero, J.M. "The relation between the 'Alte Wiener Schule' and the Spanish medicine of the Enlightenment." *Clio Medica*, 1974, 9:109-23.
98. Levi-Valensi, J. and J. Tellier. "Nicolas Blegny, journaliste." *Aesculape*, 1934,24:170-3.
99. Kieslich, G. "Monatliche neueröffnete Anmerckungen über alle Theile der Arztney-Kunst." *Arztliche Mitteilungen*, 1961, 3:379-83.
100. Commander, John, Comp. Gilbert White's year; passages from the Garden Kalendar and the Naturalist's Journal, 1757-1793. London: The Scolar Press, 1979.
101. Schatzberg, Walter. *Scientific themes in the popular literature and poetry of the German Enlighenment*. Berne: Herbert Lange, 1973.
102. Fayet, Joseph. *La revolution française et la science, 1789-1795*. Paris: Riviere, 1960.
103. McClellan, James III. "The scientific press in transition; Rozier's Journal and the scientific societies in the 1770's." *Annals of Science*, 1979, 36:425-49.

104. Kronick, David A. "Notes on the printing history of the early *Philosophical Transactions*". *Libraries and Culture*, 1990, 25:243-68.
105. Drake, Stillman. "The Accademia dei Lincei." *Science*, 1966, 151:1194-1200.
106. Clarke, Desmond J. comp. *Publications of the Royal Dublin Society*. 2d ed. Dublin: 1953.
107. Kaiser, Wolfram Ernst "Jeremias Neifeld (1721-1773), Herausgeber der ersten medizinischen Fachzeitschrift in Polen." *Wissenschaft. Zeitschrift, Universität Halle*, 1971, 26:123-31.
108. Cilleuls, Jean des. "Un precursor de la presse médicale en France." *Histoire des sciences médicales*, 1972,6:153-9.
109. Bingham, Alfred J. "The *Recueil philosophique et litteraire*, 1769-1779." *Studies in Voltaire*, 1961, 18:113-128.
110. Tigerstedt, Robert. "Études sur la presse médicale en Suède." In: *Comptes rendus de la premier Congrès internationale de la presse médicale*. Paris: 901, pp.228-245.
111. Kirchner, Joachim. *Das deutsche Zeitschriftenwesens*. Leipzig: Harrasowitz, 1942.
112. Exner, Wilhelm Franz. *Johann Beckmann, Begründer der technologischen Wissenschaft*. Vienna: Gerold, 1878.
113. Engelmann, Wilhelm. *Bibliotheca medico-chirurgica et anatomico-physiologia*. Leipzig: 1848. (Reprinted, Hildesheim: George Olms), 1965.
114. Milsand, P. *Notes et documents pour servir à l'histoire de l'Académie des sciences, arts et belles-lettres de Dijon*. Dijon: J.E.Rabutot, 1874.
115. *Tableau générale methodique et alphabétique des matières contenue dans les publications de l'Académie imperiale des science de Ste. Petersbourg depuis sa fondation*. St. Peterburg: 1872.
116. Carroll, Dewey Eugene. *Newspaper and periodical publications in countries of Europe, 1600-1950*. Ph.D.Dissertation, University of Illinois, 1966.
117. Szinnyei, Jósef. *Bibliotheca Hungarica historiae naturalis et matheosis, 1472-1875*. Budapest: Az Athenaeum, 1878.
118. Puravs, Grace et.al. *Accessing English literary periodicas; a guide to the microfilm collection*. Ann Arbor, MI: University Microfilms International, 1981.
119. Johnston-Saint, P. "The first English medical journal". *Medical press and circular*, 1939, 201:117-8.
120. Harris, Michael. "London printers and newspaper production during the first half of the eighteenth century." *Printing History Journal*, 1977, 12:38-51.
121. Fussell, G.E. *The old English farming books, from Fitzherbert to Tull*. London: Lockwood, 1947.
122. Kirpatrick, T.D.C. "The periodical publications of science in Ireland." *Bibliographical Society of Ireland*, 1921, 2:33-58.
123. Lang, Carl Ludwig. *Die Zeitschriften der deutschen Schweiz bis zum Ausgang des 18. Jahrhunderts (1694-1795)*. Leipzig: Harrasowitz, 1939.
124. Stahnke, J. "Synopsis zur neueren Medizingeschichte Ruszlands." *Medizin - historisches Journal*, 1979, 15:124-53.

125. Silva Carvalho, A. da. "Histoire de la presse médicale au Portugal." *Medicina contemporanea*, 1938,54:299-304.

126. *Verzeichnis der Abhandlungen der Königlich-Preussischen Akademie der Wissenschaften (1710-1870)*. Berlin: 1871.

127. Prutz, Robert E. *Geschichte des deutschen Journalismus*. Hannover: C.F.Ruis, 1845.

128. Bolton, Henry Carrington. *A catalogue of scientific and technical periodicals (1665-1895), together with chronological tables and a library checklist*. 2d ed. Washington, D.C.: Smithsonian Institution, 1897.

129. Garrison, Fielding H. "The medical and scientific periodicals of the 17th and 18th centuries, with a revised catalogue and check-list." *Bulletin of the Johns Hopkins University Institute of the History of Medicine*, 1934, 2:285-343.

130. Kirchner, Joachim. *Bibliographie der Zeitschriften des deutschen Sprachgebietes bis 1900*. Stuttgart: Hiersemann, 1969-

131. Scudder, Samuel H. *Catalogue of scientific serials of all countries, including the transactions of learned societies in the natural, physical and mathematical science, 1663-1876*. Cambridge, Mass: Library of Harvard Universitry. 1879.

132. Kronick, David A. "Bibliographic dispersion of early periodicals." *Serials Librarian*, 1987, 12:55-59.

133. *British union catalogue of the periodicals of the world; from the seventeenth century to the present day in British libraries*. London: Butterworth, 1955-58. 4v.

134. *Catalogue collectif des périodiques du debut du XVIIe siècle à 1939 dans les bibliothèques de Paris et dans les bibliothèques universitaire des departements*. Paris: Bibliothèque Nationale, 1967-1977. 5v.

135. *Union list of serials in the libraries of the United States and Canada*. 3d ed. New York: Wilson, 1965. 5v.

136. *Allgemeine Literature-zeitung, Intelligenzblatt*, 1789, p.590.

137. *Bibliothek der praktischen Heilkunde*, 1800, 2:234ff.

138. *Italienische Bibliothek*, 1778, 1.

139. Société libre d'emulation, du commerce et de l'industrie de la Seine-Inferieure. *Table générale du Bulletin publié par la Société de 1797 à 1899*. Rouen: E.Cagniard, 1900.

140. *Journal encyclopédique*, 1771,4:215-24.

141. Salathé, René. *Die Anfänge der historischen Fachzeitschrift in der deutschen Schweiz, 1694-1813*. Basel: Helbing & Lichtenhaven, 1959. (Basler Beiträge zur Geschichtswissenschaft, Bd.76)

142. Kronick, David A. "The Fielding H. Garrison list of medical and scientific periodicals of the 17th and 18th centuries; addenda et corrigenda." *Bulletin of the History of Medicine*, 1958, 32:456-74.

143. Roller, Duane H. and Marcia M. Goodman, eds. *The catalogue of the history of science collections of the University of Oklahoma libraries*. London: Mansell, 1976. 2v.

144. *Centrale catalogus van periodieken en seriewerken in Nederlands bibliotheken*. 'sGravenhage: Koninklijke Bibliotheek, 1978. 7v.

145. *Catalogue des périodiques Russes des origins à 1970 conservés à la Bibliothèque Nationale*, Paris: Bibliothèque Nationale, 1978.
146. *Zeitschriften-Datenbank* (ZDB). Berlin: Deutsches Bibliotheksinstitut Staatsbibliothek Preussisches Kulturbesitz, October, 1988.
147. Armytage, W.H.G. *The rise of the technocrats; a social history*. London: Routledge and Kegan Paul, 1965.
148. Hahn, Roger. "The application of science to society; the societies of arts." *Studies in Voltaire*, 25:829-36.
149. Vierhaus, Rudolph, ed. *Deutsche patriotische und gemeinnützige Gesellschaften*. Münich: Kraus, 1980. (Wolfenbüttler Forschungen, Bd.8)
150. *Guide to microforms in print; incorporating international microforms in print*. Lynn Sabol, ed. Westport, Conn. : Meckler Publishing Company.
151. *Guide to reprints; an international bibliography of reprints*. Ed. Ann S. Davis. Kent, Conn. : Guide to reprints Inc., 1990.
152. Gelbart, Nina R. The French Revolution as medical event: the journalistic gaze. *History of European Ideas*, 1989, 10:417-427.